NATURAL DISASTER RESEARCH, PREDICTION AND MITIGATION

CURRENT INTERNATIONAL PERSPECTIVES ON WILDLAND FIRES, MANKIND AND THE ENVIRONMENT

Natural Disaster Research, Prediction and Mitigation

Additional books in this series can be found on Nova's website under the Series tab.

Additional e-books in this series can be found on Nova's website under the e-book tab.

NATURAL DISASTER RESEARCH, PREDICTION AND MITIGATION

CURRENT INTERNATIONAL PERSPECTIVES ON WILDLAND FIRES, MANKIND AND THE ENVIRONMENT

BRIGITTE LEBLON
AND
MARTIN E. ALEXANDER
EDITORS

nova publishers
New York

Additional color graphics may be available in the e-book version of this book.

Library of Congress Cataloging-in-Publication Data

Current international perspectives on wildland fires, mankind and the environment / editors, Brigitte Leblon, Faculty of Forestry and Environmental Management, University of New Brunswick, Canada, and Martin E. Alexander, Faculty of Agricultural, Life and Environmental Sciences, University of Alberta, Canada.
pages cm. -- (Natural disaster research, prediction, and mitigation)
Includes index.
ISBN 978-1-63463-682-7 (hardcover)
1. Wildfires--Prevention and control. 2. Fire ecology. I. Leblon, Brigitte, II. Martin E. Alexander.
SD421.C87 2015
628.9'2--dc23

2014046080

Published by Nova Science Publishers, Inc. † New York

CONTENTS

PREFACE

... there is one overriding challenge to fire management: that of maintaining full respect for the power of fire and effects of this power on both wildland environments and the people who live and work in these environments. Barrows (1974)[1]

Fire has and continues to impact both positively and negatively humans and our environment. In this respect, fire has been variously referred to as a "good servant but a bad master" and as a "two-edged sword." Being able to co-exist with fire on the landscape is becoming increasingly more important as man expands his use of the world's ecosystems. This is clearly evident by the major wildfire incidents with their devastating consequences that have occurred, for instance, in Chile, Arizona (USA), Colorado (USA), Alberta (Canada), Russia, and Israel during the past five years.

Non-fiction books serve to capture the science and culture associated with our world at a given moment in time. The first book on the subject of forest or vegetation fires entitled *Theory and Practice of Forest Fire Protection in the United States*, written by A.D. Folweiler, was published in 1937. Since then a good many other wildland fire-related works have appeared in print.

Within the past decade, Nova Science Publishers (https://www.novapublishers.com/catalog/) have produced four books dealing with various aspects of wildfires – *Wildfires: Issues and Consequences* (2004), *Wildfires, Fuels and Invasive Plants* (2010), *Wildfires and Wildfire Management* (2010), and *Wildfire Risk: Perceptions and Mitigation Options* (2012). *Current International Perspectives on Wildland Fires, Mankind and the Environment* represents the most recent addition to the series of books on wildland fires produced by Nova Science Publishers. The current book is in keeping with the company's focus on science and social science globally.

Our purpose for undertaking this book project was to make a contribution towards the scholarship surrounding the subject of wildland fires. We thank Nova Science Publishers for making this opportunity possible.

Potential chapter contributors were sought out based on the network of personal contacts we have established over time. This, the final product, is a reflection of those that had the inclination and time to undertake a contribution. No attempt was made to attain a comprehensive overview of wildland fire issues. In this regard, the book should be considered more of a "smorgasbord" of papers.

[1] Barrows JS (1974) The challenges of forest fire management. *Western Wildlands* 1(3), 3-5.

The present volume represents a collection of nine chapters covering fire science and technology topics that support fire management and on the management of wildfires and prescribed fires, written by specialists in their respective fields. Readers will find the chapters quite wide ranging and thoroughly engaging as a result of the international flavour of the contributions from both the northern and southern hemispheres.

In the opening (Chapter 1), Dale Wade, Steve Miller and Roger Underwood articulate why we need fire and the consequences of policies of attempted fire exclusion to try and manage the wildfire problem. They then offer a compelling success story of re-introducing fire in a wildland-urban interface ecosystem in the state of Florida (USA).

Next, Brigitte Leblon, Laura Bourgeau-Chavez, Jesús San-Miguel-Ayanz give an accounting of the current state-of-knowledge of the role of remote sensing technologies in managing wildfires (Chapter 2). This review centers on fuel moisture monitoring, fire detection, and burn scar mapping.

In Chapter 3, Maria Dentoni, Miriam Muñoz, and Dave Marek describe how they have gradually introduced the Canadian Forest Fire Danger Rating System (CFFDRS) into Argentina. Their chapter also includes a description of the role of human considerations in the success they have achieved to date.

Chapter 4, written by Marty Alexander and Rob Thorburn, outlines the development of a safety system for wildland firefighters, as termed by the acronym *LACES*, that comprises lookout(s), anchor point(s), communication(s), escape routes, and safety zone(s). This system evolved from *LCES*, as conceived by the late Paul Gleason, a veteran wildland firefighter in the USA and the first international "student of wildland fire."

With Chapter 5, co-authors Cassandra Hansen, Paul Dorhety, Samuel Lanier and Spencer Adkisson provide us with a new tool for wildfire mapping – cloud-based geographic information system (GIS). They do this in the form of a case study involving the 2013 Silver Fire in southern California.

We find in Chapter 6 Gavriil Xanthopoulos recounts the deaths that have occurred amongst rural firefighters and members of the public in Greece since 1977. The factors associated with an alarming trend in the annual number of wildland fire-related fatalities are in turn discussed.

The next two chapters deal with the history, ecology and management of fire in two very large regions of South America. Guillermo Defossé, María Godoy, Lucas Bianchi, Natalia Lederer, and Carlos Kunst give a good accounting of the Argentine Patagonia (Chapter 7) as does Carlos Kunst, Roxana Ledesma, Sandra Bravo, Ada Albanesi, Analía Anriquez, Juan Silberman, Guillermo Defossé, and Joaquín Casillo for the Argentine Chaco (Chapter 8).

In the concluding chapter (Chapter 9), Cornelis de Ronde documents an assessment undertaken of wildland firefighter fatalities in South Africa over the past twenty years or so. This includes an analysis of the degree of success/failure achieved in the use of counter or suppression firing in wildfire containment.

In closing, we would like to thank the following individuals who undertook peer-reviews of the chapters that collectively form this book: Al Beaver, Kelly Close, Cornelis de Ronde, Jason Greenlee, Christelle Hély, Pedro Jaureguiberry, Carlos Kunst, Brad Mayhew, George Mitri, Wesley Page, Steve Taylor, Dave Thomas, Daniel Thompson, Morgan Varner, and Dale Wade. Their contributions proved invaluable to ensuring the quality of the product you see before you.

Finally, we acknowledge the use of imagery for the front cover provided from the Fire Information for Resource Management System (FIRMS) operated by the NASA/GSFC/Earth Science Data and Information System (ESDIS, https://earthdata.nasa.gov) with funding provided by NASA/HQ.

Brigitte Leblon
Fredericton, New Brunswick, Canada

Martin E. Alexander
Edmonton, Alberta, Canada

November 2014

In: Current International Perspectives
Editors: Brigitte Leblon and Martin E. Alexander
ISBN: 978-1-63463-682-7

Chapter 1

THE NEED FOR APPROPRIATE FIRE REGIMES AND A PATHWAY FORWARD

Dale D. Wade[1*], Steven R. Miller[2] and Roger J. Underwood[3]

[1]USDA Forest Service (retired), Hayesville, North Carolina, US
[2]St. Johns River Water Management District, Palatka, Florida, US
[3]The Bushfire Front Inc. of Western Australia, Subiaco Western Australia

ABSTRACT

Fire has, and continues to play a pivotal role in shaping Earth's floral and faunal communities. The close epochal association between humans and fire is discussed, as is human use and misuse of this natural force. American and Australian fire management is highlighted because both countries unequivocally demonstrated large-scale intentional periodic fire would minimize catastrophic fire, but then abandoned the practice and are now struggling with mega-fires, an unintended but predictable consequence of extended fire exclusion. We provide suggestions to help reintroduce appropriate fire regimes at the landscape level including packaging it as the ecological necessity that it is. Regulatory agencies should facilitate rather than hinder fire use; they need to recognize the lopsided tradeoffs between ephemeral air quality degradation and minor human inconveniences on the one hand and ecosystem perpetuation, reduced greenhouse gas emissions, increased carbon sequestration and dramatically reduced potential for catastrophic fire on the other. Fewer catastrophic wildfires will in turn reduce suppression costs and damage while significantly reducing public exposure to this threat. We close with an example of a large-scale fire reintroduction success story at the most difficult location of all, the wildland urban interface that demonstrates its technical, social and political feasibility.

Keywords: Catastrophic fire, fire effects, fire exclusion, fire management, fire policy, hazard reduction; landscape-level, mega-fire, prescribed fire

* Corresponding Author address: P.O. Box 491, Hayesville, North Carolina 28904. Email: pyrowade@gmail.com

INTRODUCTION

This chapter has three objectives. The first is to show the necessity of fire to sustain structure and function in terrestrial ecosystems. The second is to explore the ramifications of human intervention, the results of which can vary from sustaining plant and animal diversity (pyrodiversity) to creating biologically impoverished ecosystems. The third is to discuss how the diverse objectives of stakeholders can be united to move forward as exemplified by a southern US success story which demonstrates an effective and practical solution that is embraced by surrounding residents.

We don't discuss the mechanics of executing prescribed burns, as several excellent 'how-to' publications are available including Weir (2009) and Waldrop and Goodrick (2012) written for the southern US, Silva et al. (2010) written for Europe, and Anonymous (2004) written for Western Australia. All four cover wildland fire fundamentals and outline firing techniques that pertain worldwide. We lean heavily on the southeastern US and Australia as examples for several reasons: both locations have survived the transition from feral to domestic fire; prescription fire is embraced by many (although not all) local, state and federal natural resource management agencies as well as by individual land owners and private organizations in both rural and urban settings; it is used in both countries to achieve a diverse suite of objectives from hazard reduction to timber, wildlife and range management to maintaining ecosystem health. Lastly these are the regions the coauthors know best, although all have extensive fire management experience outside their regions of domicile.

Fire is a natural process, but it is unlike other potentially catastrophic events. For example, volcanoes, earthquakes, hurricanes and floods can all be defined in strictly physical or mechanical terms and can take place without life. Fire cannot be so described; it requires life to produce the fuel that enables it to occur. Although we can mathematically describe combustion and how the heat energy released is transferred through its surroundings, no combination of these equations can describe fire behavior and fire effects without adding vegetative parameters.

There are two other very important distinctions: 1) humans can summon fire on command and 2) vegetation and fire interact with each other in a feedback loop. Both the decision to ignite, and the characteristics of the fuel complex to be burned result from human decisions. Culture is thus also an integral component of fire occurrence and behavior. We concur with Pyne (2007) that

> "The story of fire on Earth is increasingly the story of what people do and don't do, directly or indirectly, with regard to fire and its setting."

Roos et al. (2014) call this pyrogeography and state that it encompasses interconnections among organic evolution, the biosphere, and human economies, ecologies and ideologies, as well as the connections among the traditional academic disciplines that study them. Moritz et al. (2014) discuss these intertwined problems in the context of coupled socioecological systems.

Fire was the social and cultural hub of earlier generations and many of us still enjoy the camaraderie around an evening campfire. Fire was a group activity not only tied to ecological objectives such as fruit production and maintenance of grasslands, but also to ceremonies such as celebrating the advent of summer, or to fulfilling a religious ritual. Life in many less economically developed cultures still revolves around these age-old practices. The bottom line is that our perception of fire and our attitude towards it result from our personal experiences (whether actual or vicarious) with both structural and wildland fire. When our distant ancestors began using fire to manipulate vegetation to suit their needs, humans assumed responsibility for how this natural force sculpts plant and animal communities.

Fire in Deep Time

As researchers decipher the fossil evidence of fire through time, the body of literature that shows Earth is truly a fire planet continues to expand. Plants appeared roughly 420 million years ago and have had to cope with lightning and resultant fire ever since. Fire has, and continues to play a dominant role in the biotic evolution of Earth and its effects cascade to aquatic and marine sites not directly affected by flames. Fire forces living organisms, both floral and faunal, to adapt to its presence, confine themselves to habitats rarely visited, or be extirpated. Many ecologists believe plant and animal fire tolerance/survival traits developed through a long close association with wildland fire (e.g., Bond and Midgley 2000). He et al. (2012) links the evolutionary pathway of fire-adapted traits in the genus *Pinaceae* with paleoatmospheric conditions over the past 250 million years and concludes plant-fire response strategies and the evolution of seed plants in general, were strongly influenced by the fiery environments of the Cretaceous period.

Earth's carbon footprint is preserved in its geologic history, polar ice caps, soil profiles, lake sediments, long-lived trees, historical writings, and more recently in satellite data archives. Analyses of these records show that, up until the industrial revolution, fire was the major force interacting with climate to determine Earths floristic landscape. Robinson (1987) presented an in-depth look at the historical and future importance of global fire and stated it has played a major role in the evolution of our planet's atmosphere on epochal time scales. She developed a hypothesis that fire established, and continues to maintain atmospheric oxygen levels which allow humans to survive on Earth, but could not find enough data to test her hypothesis. Since then, additional fossil evidence, facilitated by advances in technology and computing power have proved her correct. A review of fire in deep time and the complex interaction between fire and global oxygen levels through time can be found in Scott et al. (2014).

Fire Regimes

We use the terms 'fire-dependent' and 'fire-adapted' interchangeably albeit there are nuances; these communities require fire for their perpetuation although the required dosage and frequency of application can vary widely between species and physiographic location. We also imply that there are 'good' and 'bad' fire regimes, but this is from a human perspective – ecosystem structure and composition continually change over geologic time as the climate changes. But when dramatic changes occur that differ from historically recorded conditions, including species extirpation as a response to human actions, such communities are viewed as degraded.

Fire regimes are an attempt to separate species with roughly the same natural range in fire return interval and general fire behavior traits into distinct groups. Interested readers can find information pertaining to the autecology and synecology of many North American plants and plant communities in the US Forest Service (USFS) online data base (http://www.feis-crs.org/beta/). The Joint Fire Science Program (http://www.firescience.gov/index.cfm) contains hundreds of fire and fuel related research summaries and sponsors a Fire Exchange Network (http://www.firescience.gov/JFSP_exchanges.cfm) which contains one-to three page discussions of various fire topics and 'how to' instructions for various fire related models etc. for every region of the US. Tall Timbers fire library and ecology proceedings contain a wealth of fire related information pertaining to plants and ecosystems worldwide (http://www.talltimbers.org/fire-ecology-conferences.html). Jason Greenlee's Fire Research Institute has over 22,000 fire-related papers in electronic format and is thus a good place to initiate a search for an article or topic (http://www.fireresearchinstitute.org). Miller (2000) provides a good review of fire related autecology and includes a table that categorizes fire sensitivity and sprouting ability of many North American species. Publications that discuss fire at the ecosystem level include Brown and Smith (2000), Keeley et al. (2009), Bond and Parr (2010), Silva et al. (2010), Jurskis and Underwood (2013), and Scott et al. (2014). It is evident from the material in these repositories that a myriad of terrestrial ecosystems and the species therein require fire at some point for their perpetuation. Information regarding species mentioned below can be found in one or more of these sources. In many cases we list additional references.

The Nature Conservancy (TNC) grouped Earth's ecosystems into three broad fire regimes (Hardesty et al. 2005) which we modify slightly in the paragraphs that follow. Figure 1 shows their global distribution.

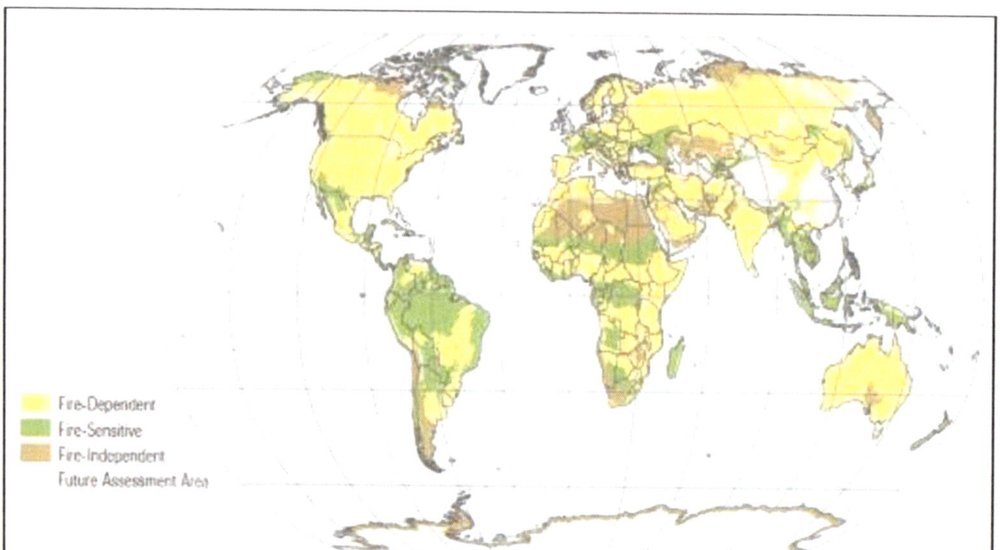

Figure 1. Global distribution of fire regime types. Fire independent ecosystems in tropical and mid latitudes are primarily deserts which do not support adequate vegetation for fire spread (from Shlisky et al. 2007).

Fire-dependent: Ecosystems where species evolved in the presence of recurrent fire and where fire is an essential process for conserving pyrodiversity (Martin and Sapsis 1992), whether it be grassland, savanna, woodland or forest. Some plant species developed physical characteristics such as thick, porous bark and large protected buds (Byram 1958; Hare 1961; Reifsnyder et al. 1976; Scott and Jones 1994) which allow individuals to survive low to moderate intensity fires (typically less than ~300 Btu/sec-ft (1038 kW/m)). An example is longleaf pine (*Pinus palustris*), a long-lived overstory tree endemic to the southern US. Longleaf is unusual in that it has numerous fire adaptations rather than just relying on one or two (Wahlenberg 1946; Wade 1996; Glitzenstein et al. 2003) that help it survive fire from seed fall through maturity. A more typical scenario is for a species to be fire sensitive as a seedling and increase its tolerance to fire over time as individuals mature; examples that fit this mold include gymnosperms such as western Larch (*Larix occidentalis*), Douglas-fir (*Pseudotsuga menziesii*) and giant sequoia (*Sequoiadendron giganteum*); angiosperms include many oaks (*Quercus spp.)*, eucalypts (*Eucalyptus* spp.), and wattles (*Acacia* spp.).

Other species developed individual and/or community characteristics to facilitate fire spread. Individual plants are typically geometrically configured and grouped within a community to allow fire to rapidly move through them, reducing residence time and thus the duration of meristematic and soil heating (think roots and seed bank); plant communities dominated by these species tend to have packing ratios (a measure of fuelbed compactness defined as the fraction of the fuel array volume occupied by fuel) that promote fire spread. Herbaceous examples include wiregrass *(Aristida stricta)* and cordgrass *(Spartina alterniflora)* of the south and eastern US. Examples of woody plant communities are epitomized by Mediterranean scrub ecosystems: chaparral in Southern California, maquis along both sides of the Mediterranean seacoast, mattorral in central Chile, fynbos in the South African Cape region and mallee in southwest and southern Australia.

Many hardwoods and some conifers have the ability to resprout from the base and/or from subterranean organs after being top-killed, particularly when young (Christensen et al. 1981). Bond and Midgley (2001) discuss the importance of this mechanism in maintaining ecosystem diversity and the tradeoffs between this strategy and that of regeneration from seed. Once sprouters are old enough to flower and set viable seed, their ability to produce basal or below-ground suckers typically decreases. When regenerating from seed, many, but not all woody plants put most of their energy into rapid height growth; exceptions include some Eastern US oaks which initially develop a strong root system instead, and many eucalypts such as jarrah (*Eucalyptus marginata*) which develop a highly fire-resistant lignotuber. The benefit of this strategy in an ecosystem characterized by frequent fire is that if the next fire occurs within a few years, this well-developed root system facilitates rapid sustained sprout growth which gives them a competitive edge (see Hutchinson et al. 2012; Brose et al. 2014). A few conifers such as longleaf and South Florida slash pine (*Pinus elliotti* var. *densa*) employ a variation of this tactic going through a grass stage whereby they develop a deep taproot before bolting up 3-5 feet (1-1.5 m) in a single year in an effort to get the terminal bud above the flames of the next fire.

A substantial number of plant species world-wide produce copious seed crops almost every year. Adults and sprouts are often killed by recurrent fire over time and may or may not continue to resprout. The fire and/or increased sunlight stimulate seed germination and establishment, thereby perpetuating the species. Genera with species that rely on this strategy include *Acacia*, *Acer*, *Betula*, *Eucalyptus*, *Fraxinus*, *Liriodendron*, *Tilia* and Ulmus to name a few. Other species produce fruits that wildlife favor, the seeds of which pass through the gut before being deposited.

Some species developed strategies to take advantage of higher intensity fires (which often involve live crowns), or high severity fires which consume the forest floor down to mineral soil thereby thermally killing stems and destroying the seed bank. Plants with survival strategies based on these types of fire either produce large seed crops that mature annually and/or that retain mature seed in serotinous cones (Midgley 2000; Ne'eman et al. 2004; Buma et al. 2014). When serotinous cones are subjected to high heat, the thermal seal holding the cone scales closed melts, allowing the cones to open over the next few days releasing their seed. Most species of *Eucalyptus* will only regenerate after fire which simultaneously opens the seed capsules, prepares a seed bed covered with ash (fertilizer), and temporarily removes competing understory and seed-eating insects (White and Underwood 1974).

One woody plant, *(Melaleuca quinquenervia)* which has naturalized in sub-tropical US, stores its seed in capsules that tend to remain closed as long as the limb they are attached to is alive. When a branch dies (regardless of the reason), the capsules open releasing literally millions of minute seed, most of which rapidly germinate resulting in a dense carpet of seedlings (see Wade 1981). The stems of this highly invasive species also develop thick papery bark (several inches is not uncommon) with very poor heat conductivity and diffusivity. Thermal defoliation stimulates advantageous buds located throughout the above-ground cambium to produce epicormic sprouts. Another tree that employs the epicormic sprout tactic is pond pine (*Pinus serotina),* a species endemic to southeastern US wetlands (Collingwood 1940; Wenger 1958). This tree also has serotinous cones as a backup strategy, the combination giving it a strong competitive advantage over other plants after thermal defoliation. See Meier et al. (2012) for an in-depth discussion of epicormic buds. A few species developed more than one strategy depending on location. Examples include slash pine (*Pinus elliottii* var. *elliottii* [stems elongate normally] and *P. elliottii* var. *densa* [pseudo-grass stage]) (Little & Dorman 1954), sand pine (*P. clausa* var. *clausa* [serotinous cones], and *P clausa* var. *immuginata* [non-serotinous cones]) (Little & Dorman 1952), and lodgepole pine (*Pinus contorta)* which can be divided into varieties that differ in serotiny (Lotan 1975).

Fire-sensitive: Ecosystems where most species are killed by fire. Some species such as mangrove (*Rhizophora mangle, Avicennia germinans and, Languncularia racemosa)* occupy sites that are inundated year-round except during severe drought and produce a closed canopy that precludes flammable surface and understory species. They have no known strategies to cope with fire. During extended droughts, even very low intensity fires will result in 100% mortality. Fire is also the mechanism that prevents them from colonizing adjacent inland ecosystems such as coastal marshes and wet prairies; when fire is withheld, such species readily invade these sites (Wade et al. 1980). Upland fire sensitive examples include eastern red cedar (*Juniperus virginiana*).

Fire Independent: Ecosystems where fire is such a rare event it can be ignored. They include cloud forests of Costa Rica; tundra ecosystems on the coast of Antarctica; temperate rain forests of the Olympic peninsula, northwestern US; and tropical rainforests of Australia. Prior to the 21st century, tropical rain forests in Africa, Southeast Asia and South America fit this category, but with uncontrolled agricultural fire on recently converted sites which spreads unchecked into adjacent forest, or as a precursor to desired conversion by arsonists, many tropical rain forests could now be placed in the fire-sensitive category (Cochrane 2002).

The three general ecosystem types mentioned above are all arguably fire-maintained – one by its presence and two by its absence. Every fire-adapted plant community has an array of fire frequencies, fire behaviors (e.g., flame lengths. fireline intensities, etc.) and severities that characterize it and determine the assemblage of fauna that occupy and use it. Fireline intensity can be defined as a measure of heat released in the flame front, while fire severity refers to the impact(s) the released heat energy has on the site (Keeley 2009).

Historical Overview of Fire Frequency

Lightning has been (Belcher et al. 2013) and is (Figure 2) a major ignition source throughout much of the world ever since vegetation appeared; the drier the fuel it strikes, the more likely it is to result in combustion, and the more dry fine surface fuels present, the more likely fire is to spread. Climate change can alter the distribution and effectiveness of this fire source, enhancing it in some locations and reducing its role in others as precipitation and lightning patterns change. For example, Westerling et al. (2006) showed that the 1°F temperature increase during the growing season between 1970 and 2003 in the western US resulted in earlier snowmelt and longer summers which in turn contributed to a 650% increase in fires larger than ~1,000 acres (400 ha) on federally managed coniferous forests. We don't doubt that increased temperature exacerbates the situation, but the underlying primary cause is Americas failed fire exclusion policy. At some point in the distant past, our ancestors observed that fire could enhance their standard of living and they learned to use it to mold the landscape to their liking. Brain and Sillent (1988) and Berna et al. (2012) present evidence that shows hominins had domestic fire 1 to 1.5 million years BP. Our ability to start and use fire is unique, separating us from all other animals (although there are recent Australian observations of birds of prey carrying lighted sticks and dropping them into unburnt fuel).

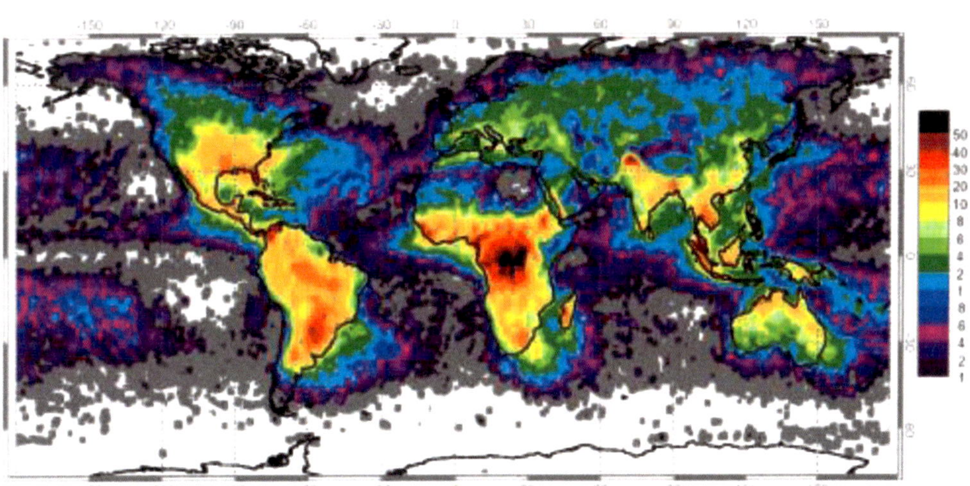

Figure 2. World-wide lightning activity. Data from space-based optical sensors reveal the uneven distribution of worldwide lightning. The actual number of strikes is lower because these sensors cannot distinguish between cloud-to-cloud and cloud-to-ground flashes. Units are flashes/km^2/yr. (Figure from http://science.nasa.gov/headlines/y2001/ast05dec_1.htm. Image credit: NSSTC Lightning Team).

By 300-400 thousand years ago, fire was part of the hominin tool kit worldwide. Successive generations right down to the current one continue to expand the tasks that can be efficiently accomplished with fire. In some locations, including the southeastern US, the practice has persisted in spite of statutes banning its use (Schiff 1962; Pyne 1982, 2010). But even here, a substantial amount of this storehouse of accumulated fire knowledge is being lost as people born before WWII pass on without having handed their fire knowledge on to their children. Reasons include reduced rural area due to unbridled urban sprawl and personal choices regarding employment and relocation. Local, state and federal government agencies are having trouble finding applicants with more than minimal field credentials when experienced fire management employees retire. Moreover, agency budgets have decreased substantially over the past several years which has resulted in many positions being abolished or remaining vacant; in some instances experienced personnel have been transferred to non-fire positions. The skillful application of fire is as much art as science and thus cannot be learned from a book. To make matters worse, the fire expertise many current academic people profess to have is primarily derived from computer models and visual media products. Video clips can help bridge the gap between field and classroom, but for training to be truly effective, the instructor must be able to explain the 'why' of what is being viewed. Hands-on experience is the only way to become a proficient practitioner able to burn with finesse.

Although this chapter focuses on the ecological necessity of fire, not the economic incentives for its use, one reason it was a near universal practice across the rural southern United States at the beginning of the 20th century (see Eldredge 1911; Mattoon 1915) was because it was the only practical tool available to protect ones property and turpentine orchards, and to provide early spring browse for free-ranging 'piney woods' cattle. With the advent of plantation forestry in the South following WWII, pulp and paper companies throughout the region embraced its use to reduce wildfire damage to their pine plantations. Southern managers that used fire achieved a cost to benefit ratio of better than 2 to 1, which means a company managing any of the southern yellow pines with fire would more than double their return on the expenditures involved with well planned and executed fires (Wade and Moss 1999). Southern forest managers have used fire extensively for centuries to reduce subsequent wildfire damage, the benefits of which were not included in the above calculations, but which would dramatically increase the ratio. Saveland (1987) used breakeven analysis to demonstrate that 'large financial gains' would accrue from the use of prescription fire to reduce fuels in critical areas. Dale et al. (2005) developed a case study for a large Colorado wildfire to demonstrate the substantial cost savings that would accrue from expanding the use of prescription fire. Mason et al. (2006) provided an economic assessment that showed significant net benefits from fuel removal investments to reduce the incidence of crown fire. Buckley et al. (2014) conducted an economic analysis on a large watershed in California after it was subjected to a damaging wildfire and concluded forest management, including prescription fire could save up to three times what it costs to suppress wildfire, as well as reduce the size and intensity of future fires by 75 percent. A quick glance at the economic and ecological benefits of prescription fire makes one wonder how Smokey Bear and fire exclusion advocates were able to sell a fatally flawed federal wildland fire policy for half a century; it defied common sense and ignored both a thousand-year natural resource management record and published peer reviewed science. The facts are obvious and straightforward: plants grow and die resulting in accumulating combustible material in most ecosystems; the longer the fire-free interval, the more fuel that accumulates; and the more likely a high intensity, damaging fire will be the result.

The 13 southern US states include more than 534 million acres (216 million ha). When European settlers arrived, perhaps as much as 80% of it burned every few years. According to Gramley (2005), 214 million acres (86.6 million ha) were classed as forestland in 1997. In 2011 these states treated 20.2 million acres (8.2 million ha), about 61 % of it for agricultural purposes which includes range management (Melvin 2012). Assuming a 3-year average fire return interval on forested lands, only 24 million acres (9.7 million ha) or about 11% is currently burned with prescription fire. This simple calculation reveals the enormity of the gap between the percentage that burned historically and that currently burned. This disparity is a major reason why southern states often report more wildfires than all other sections of the contiguous United States combined (Gramley 2005). The number of southern wildfires occurring annually varies widely depending upon the weather, but has recently ranged between ~30,000 and 65,000. Although these fires tend to be smaller (partially due to the extensive road system throughout much of the South that facilitates suppression compared to the western US and Alaska, and the fact that relative humidity recovers almost every night), the South still occasionally leads the nation in acreage subjected to wildfire as well (Gramley 2005). According to the US National Interagency Fire Center (NIFC) (http://www.nifc.gov/index.html), southern US wildfire acreage averaged 1.73 million acres (404,687 ha) between 2008 and 2012. The number of wildfires, acreage burned and fire severity are all predicted to increase in response to our warming planet and metastasizing wildland urban interface (WUI) (Stanturf and Goodrick 2013; Mitchell et al. 2014)). The only practical means to mitigate these predictions is to substantially ramp up the use of prescription fire.

North American Fire History

Early Western Hemisphere explorers often commented on the smoke-filled skies they encountered (e.g., Verarzanus (Hakluyt 1582) off the Carolina coast in 1524 stated "*...verie great fires*"; Sir Francis Drake (Bigges 1586) off the coast of North Carolina observed "*...one special great fire, which are very ordinary all alongst this coast*"; de Laudonniere (1587) near Jacksonville, Florida wrote "*...although the medows were at that season all greene, and half covered over with water, nevertheless the lightning in one instant consumed above five hundred acres therewith, and burned with the ardent heate therof all the foules which tooke their pastime in the medowes, which thing continued for three days space...*". Catesby (1731) mentioned it took him three days on horseback to circumnavigate a fire northwest of Charleston, South Carolina. Travelling through Florida in 1774, William Bartram (1791) made this notation "*...riding through high, open Pine forests, green lawns and flowery savannas....having lately been burnt...*".

The magnitude of the impact of our ancestors' use of fire cannot be overstated. Rather than ignoring historic fire regimes as our generation did, previous cultures manipulated natural fire regimes to suit their purposes and in so doing created and maintained the vegetation mosaic that greeted European explorers on both sides of the North American continent. They demonstrated that humans and fire can have a synergistic although fragile relationship sustainable over millennia.

Native American Use of Fire

While a parallel situation prevailed in Australia (Gammage 2011), the following discussion centers on North America. Many authors have suggested that both the amount and impact of intentional burning by Native Americans during the past several centuries has been woefully underestimated (Gruell 1985; Denevan 1992; Mann 2005). Williams (2003) lists several hundred references to Native American fire use. Lewis (1973) listed 70 reasons why Native Americans used fire. Williams (2003) grouped their use of fire into 11 major categories: hunting; crop management; improved growth and yield; fireproof areas; insect and nut collection; pest management (think ticks (*Ixodidae spp*), and chiggers (*Trombicula spp*)); warfare and signaling; economic extortion; clearing areas for travel; clearing riparian areas; and felling trees. We would add tool making and creating firewood to that list.

Human fires were in large part responsible for creating and maintaining grasslands, including the Great Plains (Komarek 1965; Bond and Keeley 2005; Scheiter et al. 2012; Noss 2013). Historically, relatively close-interval fires (1-35 years apart) shaped the vegetative mosaic of 60-70% of the contiguous United States. These fires resulted in savannas and woodlands as well the open forests of longleaf pine in the South and ponderosa pine (*Pinus ponderosa*) throughout the West. Fire created or at least maintained the numerous balds on Appalachian mountain tops (see Knoepp et al. 1998 for indirect evidence) and maintained dynamic fire climax ecosystems throughout the Americas. Flatley et al. (2013) analyzed fire scars in the southern Appalachians and concluded return intervals were generally less than a decade apart. Cherokee Indians and the European settlers that followed used periodic fire to manage the extensive chestnut and oak forests (DeVivo 1991), the mast from which was a staple for humans and other animals. Abrams and Nowacki (2008) examined the historical record regarding the importance of Native American management of mast and fruit trees in the eastern US and concluded that

> Native American land-use practices not only had a profound effect on promoting mast and fruit trees but also on the entire historical development of the eastern oak and pine forests, savannas and tall-grass prairies.

Fire manipulation of mast was also a common practice in Europe according to Mason (2000). Pyne (2010) quotes John Wesley Powell in his 1878 *Report on Lands of the Arid Region of the United States* as follows:

> In the main these fires were set by Indians … for the purpose of driving the game … on a scale so vast that the amount taken from the lands for industrial purposes sinks by comparison into insignificance.

Pyne (2010) goes on to state that Frank Hough, the US Forestry Agent, in his 1876 report upon forestry noted that:

> … the pervasive and casual use of fire among the general population. Fire was what made their world habitable.

Pyne (2010) sums up ecosystem management from a Native American standpoint as follows: "*A land unburnt was a land uncared for – a land that was uninhabitable*" and Mann (2005) states:

> At the time of Columbus, the Western Hemisphere had been thoroughly painted with a human brush. Native Americans ran the continent as they saw fit". Botkin (1995) stated: "It often seems that the common impression about the American West is that before the arrival of people of European descent, Native Americans had essentially no effect on the land, the wildlife, or the ecosystems, except that they harvested trivial amounts that did not affect the 'natural' abundances of plants and animals. But Native Americans had three powerful technologies: fire, the ability to work wood into useful objects, and the bow and arrow. To claim that people with these technologies did not or could not create major changes in natural ecosystems can be taken as Western civilization's ignorance, chauvinism, and old prejudice against primitivism – the noble but dumb savage. There is ample evidence that Native Americans greatly changed the character of the landscape with fire, and that they had major effects on the abundances of some wildlife species through their hunting.

Native Americans used fire extensively to improve their standard of living by modifying fire to fit specific ecosystems. Their use of fire fell within the natural range of both frequency and intensity. That said, we recognize they significantly increased the acreage treated with dormant season fire (Martin and Sapsis 1992, Archibald et al. 2012). European colonists on the other hand tended to use the meat cleaver approach to improve their standard of living through whole-scale conversion of existing ecosystems into artificial ones, often monocultures of desired plant species. Cronon (1983) described these differing fire goals between Native American and white settlers as follows:

> The choice is not between two landscapes, one with and one without a human influence; it is between two ways of living, two ways of belonging to an ecosystem. These two ways are 1) the modification of ecosystems to live with and enhance the natural abundance, or 2) develop or change the natural order to increase land production, such as farming and grazing. Generally, the American Indians used the first method where they burned parts of the ecosystems in which they lived to promote a diversity of habitats, especially increasing the 'edge effect', which gave the Indians greater security and stability to their lives. Their use of fire was different from white settlers who used the second method to create greater uniformity in ecosystems, which promoted an increased food supply and led to the development of towns and cities.

We reproduce two fire return interval maps below, one depicting the contiguous US (Figure 3) based on a digitized atlas of Kuchler's Potential Natural Vegetation Types (Hardy et al. 1998) and the fire regime types ascribed to the Kuchler types in Brown and Smith (2000).

The other map shows the predicted mean fire return interval for the eastern US (Figure 4) from a regression equation that combines mean maximum temperature with population density as described in Guyette et al. (2006). Major differences occur along the spine of the Appalachians and along major rivers, but both maps agree that most of the southeast burned at least once every 10 years and virtually all of it burned as least once every 35 years. In fact most of the Atlantic Coastal Plain burned several times a decade. Frost (1998) also constructed such a map using a somewhat different approach that agrees with the two shown.

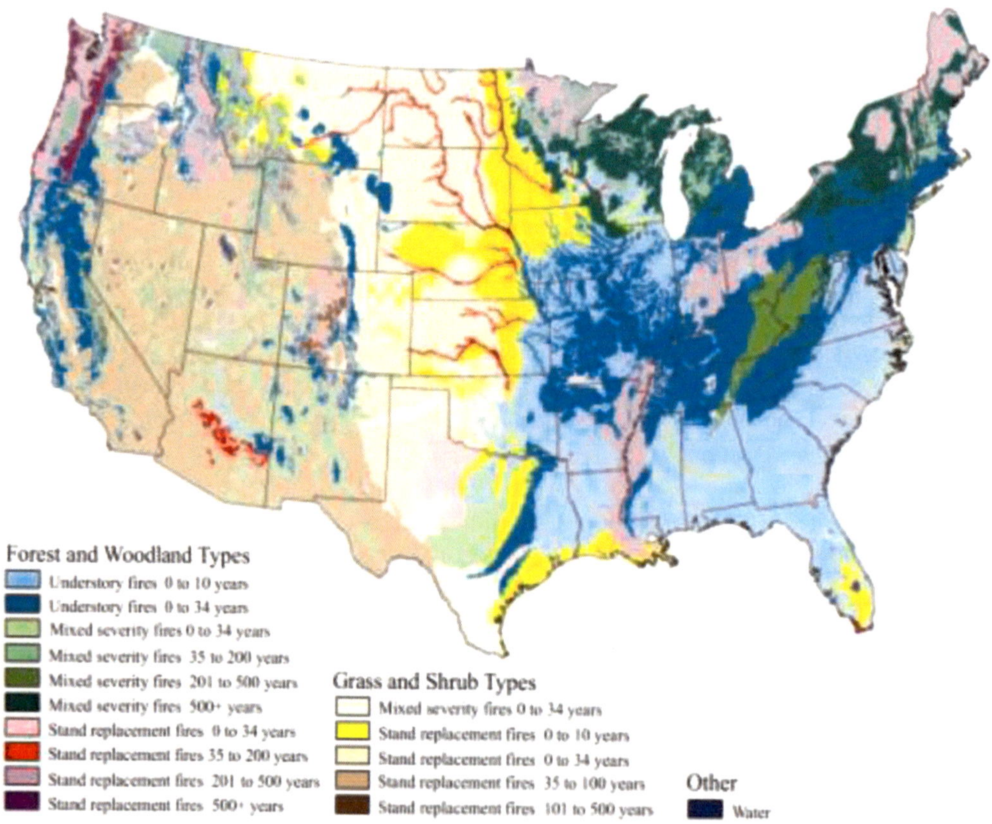

Figure 3. Pre-settlement fire types and return intervals (from Brown 2000). The term 'stand replacement' is used to denote fires that kill the complete plant including below ground meristems, as well as fires that just topkill plants such as graminoids and many hardwoods which resprout. Vegetation types placed in the 0-35 year bracket rather the 0-10 year bracket should *not* be interpreted as having fire frequencies between 10 and 35 years; this time range simply means there was not enough data or too much variation within the data to unequivocally place the system in the 0-10 bracket.

American Colonial Use of Fire

European colonists quickly adopted Native American use of fire; it was the tool of choice (the axe, plow and introduced livestock were not practical at the landscape level) to modify their surroundings prior to the industrial revolution. Stambaugh et al. (2014) reconstructed the fire, vegetation and environmental history of four oak stands in Oklahoma over the past three centuries and showed fire regimes changed spatially and temporally with cultural and land use changes.

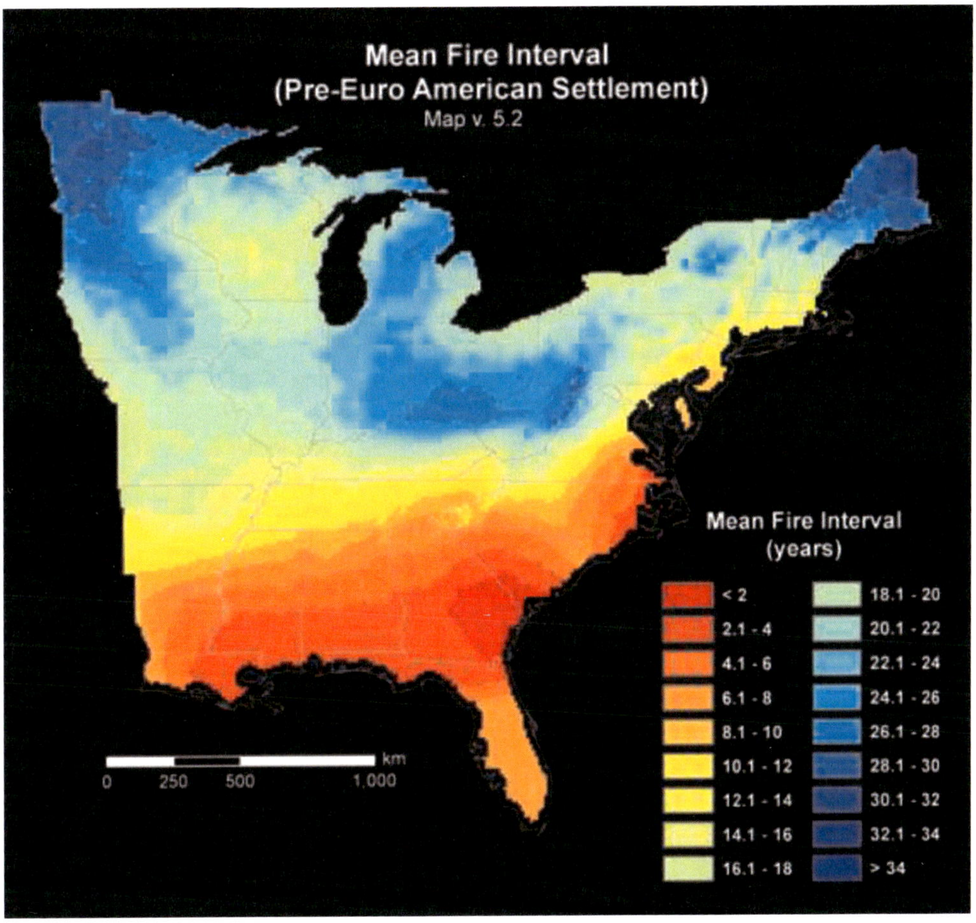

Figure 4. A first approximation of coarse scale mean fire intervals in the Eastern United States as predicted by mean maximum temperature and human population density based on Equation 1 in Guyette et al. (2006). The above figure is taken directly from that paper.

Huffman (2006) reconstructed southeastern pine savanna fire history by examining old-growth longleaf pine stumps. She found virtually all fires occurred during the growing season at a 2 to 3-year interval between 1679 and 1868. Additionally, she looked at fire scars on old-growth dead slash pine on a Florida barrier island; mean fire return intervals there averaged four years from 1864-1910 and from 1919-1947 (trees were raked around 1911-1918 during turpentine operations). A longer nine-year fire return interval occurred from 1959-2000.

Charles Sargent (1884) wrote an appendix to the United States census of 1880 that included a map of forested areas burned annually (Figure 5). Because he was mainly interested in forests, the Great Plains and other grasslands were generally left blank, as were areas where insufficient data were collected. We have no doubt most of these areas burned at least as frequently as forested areas from both natural (lightning), and anthropogenic sources (see Hough 1877).

The long growing season and plentiful rainfall across most of the South (except parts of Texas and Oklahoma) results in rapid plant growth. This produces a dense understory and heavy (10+ tons/ac [22.4_+ tons/ha]) fuel load in less than a decade on good sites - sooner than in other regions of the contiguous US.

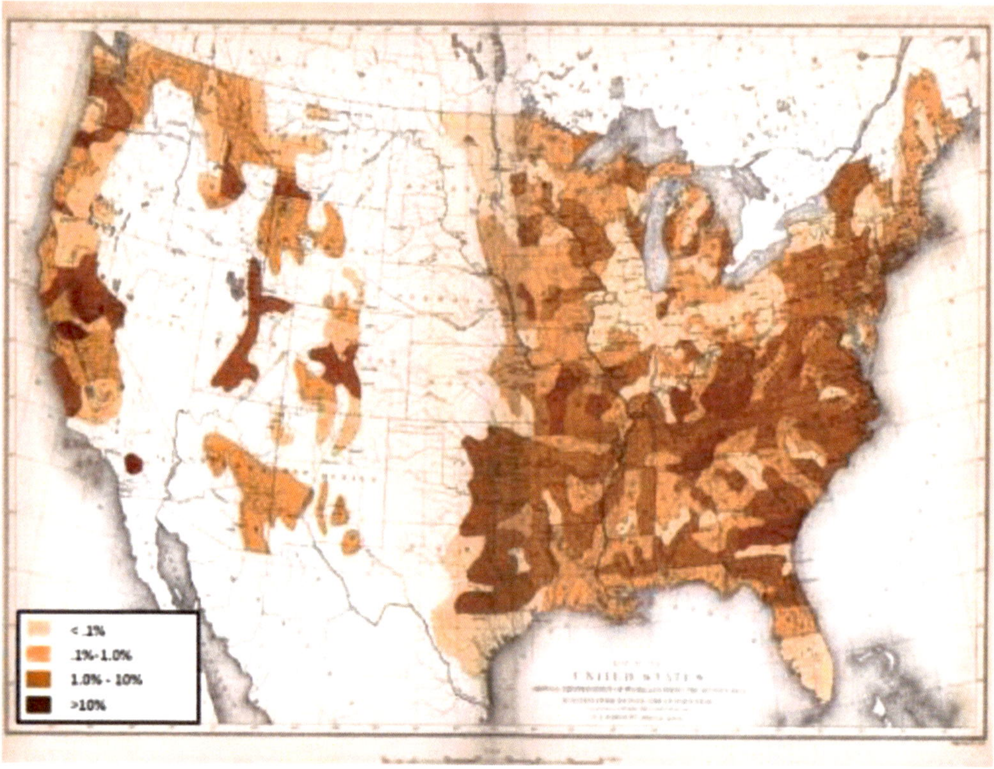

Figure 5. Woodlots within settlement areas that burned annually. This map was an appendix to the 1880 US census; the lightest color represents settlement areas where less than 0.1% of the woodlots burned annually, and the darkest color represents more than 10% of the woodlots burned annually (courtesy of the Forest History Society, Durham, North Carolina www.foresthistory.org).

Such conditions are conducive to extreme fire behavior. Rural southerners understood that they had to live with wildland fire – that it could not be purged from the landscape. The simplest and most rewarding approach was to harness fire to meet their needs and reduce the risk of wildfire to themselves and their homesteads rather than waiting for nature to strike the inevitable match, and then reacting in emergency mode as US federal policy dictated for over 50 years. The 1911 Weeks Act and 1924 Clark McNary Act allowed the USFS to use the carrot and stick approach to coerce all states to formally accept fire exclusion although some southern state natural resource agencies voiced the error of such thinking and its inevitable outcome. In 1935 the USFS instituted a policy that required every fire be actively suppressed with the objective of full containment by 10am the next morning for as many mornings as it took. In less than a decade, the inability to successfully exclude fire on southern national forests became so obvious that in 1943 the USFS quietly rescinded its ban on intentional fire on some of these forests. The official reason given for this exemption was the lack of available manpower during the war rather than the fallacy of the program, but the fact that they did not void this exemption once the troops returned home speaks for itself.

From the perspective of fuel mitigation, the major benefits of a prescribed burn program are: 1) much of the smaller accumulating dead fuel (duff and branchwood less than about 1 inch (2.5 cm) in diameter) on the forest floor is consumed; 2)Understory shrubs and hardwoods are top-killed and confined to the ground cover rather than allowed to form a midstory ladder capable of moving fire from the surface up into overstory tree crowns and; 3) Reaction (Rothermel) fire intensities (combustion rates) are typically lower (although Byram fireline intensities can be higher due to increased flashy fuels) (see Alexander 1982 and Wade 1987 for differences between various measures of intensity) for at least the next 3 to 4 years, making fire suppression considerably safer, easier and less expensive (e.g., Cochrane et al. 2014). Prescribed fire can be used in ecosystems world-wide to reduce hazardous fuel accumulations. Instances where wildfire suppression was facilitated and/or intensities and damage substantially reduced as a fire moved from an older to a younger rough are commonplace in the literature (e.g., Cumming 1964; Wagle and Eakle 1979; Biswell 1989; Omi and Martinson 2004; Raymond and Peterson 2005; Cram et al. 2006; Covington 2014).

Numerous publications describe the benefits of periodic intentional fire on many US Indian reservations that never bought into the fire exclusion policy and continued to use periodic fire, thereby providing examples of the benefits of periodic fire over time (Weaver 1947; Kallander 1958). U.S. President Abraham Lincoln said '*the first thing to do when you find yourself in a hole is to stop digging*', but the USFS took more than half a century to stop digging and officially change its fire management policy. More than a decade later, implementation remains lackluster (see Schiff 1962; Pyne 1982; 2010; Biswell 1989 for in-depth discussions). The USFS would like to forget the fact their fire exclusion policy is the major reason the US is now embroiled in catastrophic wildfires and astronomical fire suppression costs, but its leadership is obviously having trouble embracing the concept of landscape prescribed fire. USFS Chief Tidwell and his staff continue to espouse the need for landscape prescribed fire, but have yet to find the intestinal fortitude to make tough decisions and provide staff and funds to facilitate implementation. Our intent here is not to lambast the USFS, but as the late Florida State Forester John Bethea said:

> You can no more get to where you don't know where you're going than you got to where you think you are from where you don't know where you've been.

In other words, understanding how one arrived at a particular point will improve chances of turning things around. In-depth discussions of fire in North American ecosystems can be found in DeBano et al. (1998) and Brown and Smith (2000), world-wide in Scott et al. (2014) and in numerous books written by Dr. Stephen Pyne (a list of which can be found on his personal website www.stephenpyne.com), all of which contain extensive lists of references

Both the US and Australia are now dealing with mega-fires, a nebulous term whose definition fire scientists have yet to agree on (Kodas 2014), but which are basically massive fires that result in catastrophic economic damage and/or loss of human life, and which defy control until weather or topography changes, or they run out of fuel (Williams et al. 2011; San-Miguel-Ayanz et al., 2013). Work by Roos and Swetnam (2012) found that before the 20th century, this phenomenon had not occurred in the western US during at least the past 1400, years.

Littell et al. (2009) suggested area burned in the western US between 1916 and 2003 was controlled by climate rather than by fuels. Although weather, especially drought, may explain some of the variation in fire size, we submit that fire severity is much more important than size and that fires escaping initial attack are now much more difficult to control and more damaging because of unprecedented fuel loads as a result of seven decades of fire suppression (Williams et al. 2011; Stephens et al. 2014).

Scott et al. (2014) attribute the increase in mega-fires to climate change; we agree that as the summer rain-free period extends longer into the fall, fire size and intensity tend to increase. That said, large fires were historically common in the western US, while mega-fires (which by definition are more than just large fires) exist only because their behavior defies suppression efforts, no matter how many resources are committed – and in our opinion are a direct result of increased fuel loads arising from extended periods of fire exclusion.

Ecosystem damage and fire-related human mortality could have been drastically reduced if the USFS had only had an open mind when formulating its fire policy, or looked to India where fire exclusion was attempted more than half a century earlier. European foresters who arrived in India in the 1850's instituted a fire exclusion policy in spite of opposition based on the experience of untold generations of native land managers. The failure and eventual overturn of that policy was recorded as it unfolded and was synthesized by Underwood (2013).

Early Australian foresters where aware of the Indian example, but they ignored it as described in the same book. In the next section, we recount the Australian fire management story which unfolded half a century after the Indian experiment.

AUSTRALIAN EXAMPLE

The story of fire management in Australia closely paralleled that of America to the end of WWII, then diverged radically with Australia embracing the landscape use of fire to reduce fuels, while the US continued down the fire exclusion path with dramatically increased expenditures in the vain attempt to override nature. Then as the twentieth century drew to a close, both nations again reversed direction.

Australia is again dealing with mega-fires because it largely abandoned its successful hazard reduction program on the grounds that it was costly and unnecessary (a victim of its own success). Surprisingly enough, or maybe not so surprising, resource managers in both the US and Australia are now coping with the same barriers to the proven approach of using periodic fire at the landscape level (Underwood 2010). The only sure predictions are that fire will continue to be a dominant factor shaping ecosystem mosaics, and many politicians and policy makers will continue to act in their own self interest rather than for the good of the general public until forced to do so.

After an exhaustive search of the literature and in-depth discussions with Australian Aborigines about their fire practices and handed-down wisdom, both Gammage (2011) and Hallam (2014) concluded that at the time of European contact, the land was "*not as God made it but as the Aborigines made it*".

When Captain Cook arrived on the east coast of Australia in 1770 and made forays inland, he and other members of his party were puzzled over the fact the country had a wide-spaced overstory carpeted in grass with no 'under wood'; many equated it to a well maintained English park. Subsequent explorers often expressed dismay over their inability to explain the fairly static ecotones that existed between rainforest and scrub and grassland. Gammage (2011) thought this vegetation mosaic was the product of repeated deliberate fires designed to achieve that specific mosaic, and not the product of random ignitions. He devoted several chapters to developing three major aboriginal land management tenets that created and maintained the Australia landscape of 1788 which we believe would be equally effective today:

1. Unlike the Britain of most early observers, about 70 per cent of Australia's plants need or tolerate fire. Knowing which plants welcome fire, and when and how much, was critical to managing land. Plants could then be burnt and not burnt in patterns, so that post-fire regeneration could situate and move grazing animals predictably by selectively locating the feed and shelter they prefer.
2. Grazing animals could be shepherded in this way because apart from humans they had no serious predators. Only in Australia was this so.
3. There was no wilderness. The Law—an ecological philosophy enforced by religious sanction—compelled people to care for all their country. People lived and died to ensure this. The Law prescribed that people leave the world as they found it.

Over time the Australian Aborigines developed and implemented precise, fine-grained, local land management practices that varied according to climate, terrain and vegetation just as their contemporaries did throughout the world. The result was a well cared-for environment that gave them an idyllic fairly high standard of living; they lived in harmony with the land using fire to enhance the bounty it provided them. European colonists, however, dismissed these indigenous peoples and their age-old traditions as irrelevant. Within a century, the landscape was fuel chocked (including numerous introduced species), fostering mega-fire catastrophes with attendant loss of life. Land clearing contributed to the Black Thursday fire in 1851 (about 12.4 million acres (5 million ha) (just as it resulted in the massive 19th century fires in North America). The 1939 Black Friday Fire, burned roughly 5 million acres (2 million ha) and another 2.4 million acres (971,000 ha) were subjected to deadly wildfires in 1944 destroying over 500 homes. Up until WWII ended and men returned home looking for work, Australian fire history and policy closely paralleled that in the US. At that point fire policy diverged radically; in the US, military surplus was available and the USFS acquired surplus off-road vehicles and aircraft to substantially upgrade its suppression capabilities.

Australia did not have a vast inventory of surplus military equipment, so did not follow the post-war American model, but instead reversed direction and turned to the large-scale use of fire for fuel reduction (Underwood 2010). Based on past experience, they recognized it would be cheaper to manage fire than attempt to expunge it; and best of all, it worked while fire suppression had not yet been able to stop a single catastrophic fire without concomitant changes in the weather or fuel complex. The transition took nearly a generation, and during that time there were still occasional conflagrations. But by the 1980's, most forested areas were routinely underburned for hazard reduction. For example, Western Australia prescribe burned 12 to 14% of the entire forest estate annually (there are no federal forests in Australia). Troublesome fires still occurred at the wildland urban interface and during prolonged droughts, but the consensus was that the wildfire problem had been solved – or so it appeared. The solution unraveled in the 1990's for the following reasons:

1. Humans are conditioned to respond to threats. What the fuels management program achieved (especially in Western Australia) in the second half of the 20th century was to significantly reduce the threat of large, intense wildfires breeding complacency

and overconfidence. Budgets were cut, field stations closed, recruitment of professional and field staff slowed or ceased. Fuel reduction burning was no longer given the highest priority. 'Why do it when we don't have any bad wildfires?' was a recurring theme. In a nutshell, Australians became victims of their own success.

2. Forestry (or more correctly timber harvesting) became politically unpopular, and because of the way the two were linked on the ground through the forestry profession, so did fire management. By the 1980s, urban Australians had discovered 'the environment' and it became a new religion. Forests were no longer 'the bush', but fragile ecosystems, precious icons, nature's cathedrals, to be 'saved' from destruction by evil foresters. The old threat of the destructive bushfire was replaced by a new threat: fuel reduction burning! It was (and still is) routinely described by environmentalists as 'fire-bombing' or 'napalming', conjuring up images of Dresden, Germany during WWII or the Vietnam War.
3. The Australian political and electoral system allows small well-organized lobby groups disproportionate capacity to influence government policies. From the mid-1980s onward, environmental groups opposed to prescribed burning began to control election outcomes. Both of the major political parties, one or the other of which was always in power, became hostage to green lobbyists. Forest and fire management was one of the first things to suffer.
4. In the 1980s there was a profound change in the Australian public service. Up until that time, the so-called 'Westminster System' applied in which public servants were appointed on the basis of their know-how and experience, not their politics. Responsibility for public administration in areas such as forest management was in the hands of professionals who had come up through the ranks, not political appointees parachuted in from above. Under the new arrangements, the role of the senior public servant was no longer to give options and professional advice, but to toe the line of the political party in power. Departmental heads were placed on short-term contracts to ensure they could be easily dispensed with if they stepped out of line. This situation made it easy for environmentalists, operating through a political party which they controlled through the electoral system, to manipulate forest policy and land management practices. [if this sounds familiar to American readers, it should, as there are parallels with the American situation].
5. Finally there were the two great transfers. The first was a massive rededication of State forests as national parks and their falling under new administrators. The bulk of the national parks administrators in Australia (especially in the eastern States) at that time did not have backgrounds in forest or land management, but were zoologists, botanists or environmental scientists. They tended to fear fire, and did not want responsibility for it. They supported the second great transfer: this was shifting the responsibility for fire from those responsible for forest management to those responsible for emergency response. The traditional bushfire gatekeepers (fire managers) in their remote ranger stations were at an enormous disadvantage in all this. They were few in number, were politically naive, and had no-one to speak for them or support them politically. Right across Australia during the 1990s, they began to disappear, transferred to desk jobs in the city, retired and not replaced, and starved of resources as well as moral support. The once-routine programs of systematic fuel reduction collapsed Even in Western Australia where there was continued agency

> support for burning; the annual program began to wind down. By 2003 it had reached the point at which it began to become ineffective: large, intense forest fires began to re-occur for the first times since the 1960s. The downgrading of a prescribed burning program is worse than its closure; it is a double blow. A program of fuel reduction burning will not do the job if it does not meet a certain threshold in terms of area covered annually, and strategic distribution. Big fires which have worked up a head of steam in long-unburnt bush simply go over the top or around small, scattered or inadequately fuel-reduced areas. When this occurs it enables the opponents of fuel reduction burning to portray it as not only environmentally damaging, but worse, ineffective in preventing fires. The fact that fire managers had never said it would 'prevent' fires was conveniently overlooked.

Australia's fire response system was technologically advanced and efficient, but the decision to abandon or significantly reduce fuels management, and replace it with the suppression approach produced the inevitable result. Instead of improving, the situation rapidly deteriorated, culminating in the 2009 Black Saturday fires in Victoria. The highly vaunted suppression technology was useless against fires pushed by high winds through heavy dry fuels. There was no Plan B and many people died.

As in the past, this disastrous rash of fires was followed by a rash of inquiries, commissions, conferences, seminars, workshops *ad infinitum* including separate and independent inquiries by the House of Representatives, the Senate, the Victorian Parliament, State Coroners and a Royal Commission (Underwood 2010). None of these reports contained information that would surprise a veteran fire manager. Every study reached the same major conclusions:

1. Fuel levels in eucalypt forests must be kept below certain threshold levels and fuel-reduced areas thoughtfully distributed across the landscape.
2. The way to achieve this is through a program of periodic, planned burning under mild conditions that reduces fuels systematically over about 8-10% of the burnable forest every year, thereby treating 25-30% of the landscape in a 3 year period (for comparison, the southern US treats about 11% of the natural landscape over a 3-year period).
3. If we do this, and continue to maintain a sound bushfire suppression system, wildfires will be easier and safer to control, and will do less damage, even under the most severe weather conditions.
4. The community must have a Plan B, enabling evacuation to safe places when all else looks likely to fail.

These findings are *not* likely to be implemented for the following reasons (the last two are the very same barriers that prevent more widespread use of fire in the United States):

1. Opposition to fuel reduction burning remains virulent in environmental organizations and within academia in Australia. This opposition is incurable, because it is not based on science, but on ideology. However, neither the activists nor the academics are the main problem.

2. The main problem is the vacuum left by the removal of the gatekeepers and their supporting leadership; this has not, and is unlikely to be filled.
3. And the situation is made worse by the increasing bureaucratization of prescribed burning. So many constraints have been erected it is becoming very difficult to actually conduct a burn. This is a clever strategy as it allows those who oppose fuel reduction burning to claim publicly that they support it, while simultaneously making it nigh impossible for burning to take place through the imposition of bureaucratic processes that the public never sees.

Altangerel and Kull (2013) analyzed the submissions to a parliamentary inquiry after the 2009 fires and concluded that differences resulted from contrasting views about whether humans or nature are voluntarily or involuntarily exposed to wildfire risk; as stated earlier, we believe this is an emotional and not a science based issue. Correcting the bureaucratic morass in which Australian fire managers find themselves entangled will require major cultural changes including replacement of top management in the environmental agencies and emergency services, but it is not likely to happen in the present era of a political, rather than professional, public service. Without leadership from the top to establish the culture, and require subordinates to achieve annual fuel reduction targets, the job will not get done. Readers interested in learning more about large, damaging fires in Australia can access the Romsey Australia website at http://home.iprimus.com.au/foo7/firesum.html.

These institutional and cultural changes have occurred at a time when large numbers of Australian are moving out of the cities to live at the interface, and where it is becoming increasingly difficult for leaders to lead. Nothing can be done in Australia any more without calling for public submissions, consulting stakeholders, preparing draft plans, conducting public meetings, test cases, and pilot studies. Regrettably Australia is also seeing a growing threat of litigation the moment something happens that someone does not like. The same situation exists in America (Wilent 2014a). All of this has produced a sort of bureaucratic paralysis that hangs like a dead weight around the necks of the nation's bushfire managers. Finally, there is the inability of the community and media to see beyond short-term losses to long-term gains. The concept of the trade-off seems no longer to be comprehended. There are many good men and women still working as bushfire managers and in fire research in Australia. Their devotion to duty and their professionalism is unquestioned. The trouble is, increasingly they are being constrained, corralled, repressed, outvoted or stood up in witness boxes in courtrooms. The tolerance for error has disappeared, replaced by a fear of trial by media or of having to take the rap for an incompetent Minister. Expertise in the field is declining to the vanishing point, so that errors will become more commonplace. There is a new trend out in the bush, especially within the Australian national parks agencies. The refrain has become: "*why should I do a fuel reduction burn, and draw the wrath of the greenies and media down on me for being a villain, when I can sit back, wait for the inevitable wildfire and become a hero as a firefighter?*"

Both American coauthors have personally heard these same words uttered by natural resource managers in the US and see very close parallels to these institutional and cultural changes in the US. There must be a reason why educated and intelligent people cannot understand the basic cycle of fuel accumulation that leads to damaging wildfire, followed by several years with few damaging fires until fuel again accumulates past the threshold where damaging fires return. We fear the reason is that many people these days, including university professors, have almost no contact with fire in the real world. They all too often rely on computers rather than field studies/observations for their knowledge.

Other Continents

The story is similar throughout the world. Humans have used fire to shape the vegetation to their liking on all continents where wildland fire is common (see Silva et al. 2010 for the European situation). Even in 'more developed' nations, segments of the rural population often still use fire or at least remember their ancestor's dependence on, and respect for fire. In less developed nations, fire is still the tool of choice (Underwood 2007; Butz 2009; Scott et al. 2014). Travelers from more 'developed' nations to such countries, whether Costa Rica or Myanmar, are often surprised by the laissez faire attitude regarding fire, evidenced by the number of unattended fires burning around garden plots and houses (either constructed of cinder block or with yards swept free of surface fuel). Based on our travel to less developed countries, we agree with Pyne (1995) that "use of fire by native peoples to change ecosystems or portions thereof is almost universal".

Even in areas where the human footprint has been light, such as the primeval boreal forest of northern European Russia, fire has shaped the ecosystem. Kauhanen et al. (2014) used satellite imagery, field surveys and historical records to show this region is characterized by small fires (median size 0.3 ac (0.12 ha) with infrequent larger stand-replacing events akin to those of the early 2000s triggered by drought and bark beetle (Ips typographus) induced tree mortality.

The Changing View of Fire

Many 'advanced' civilizations of the past, including those of Greece, China and Mesoamerica recognized the inevitability of fire as well as the profound changes in flora and fauna it produces and therefore included fire as one of the four essential elements (fire, water, air and earth). 'Developed' countries today, however, don't share this vision, perhaps because of the lingering misconception that humans can mold this natural force however desired. Current thinking recasts them as atmosphere, biosphere, hydrosphere, and lithosphere, omitting fire (Robinson 1987). She presents evidence to show this omission is in error because fire radically alters climate-driven succession *"...thereby significantly altering the structure and balance of global flows of mass and energy and the distributions of global biota"*. Her statement continues to be bolstered as fire historians reconstruct Earth's historical fire record from hard evidence, as well as from handed down lore (both traditional and anecdotal) (Alvarado et al. 2011; Emery et al.2014; Scott et al. 2014).

During the past century historical fire regimes were cast aside denying ecosystems the rejuvenation they required, thereby forcing new, unprecedented species combinations; the process is ongoing and fire managers are trying to figure out how to best manage this new landscape fabric. A fire suppression policy will obviously impact fire's ability to nurture fire-adapted ecosystems, but in practice, politics is often the overriding factor. For example, Steelman and McCaffrey (2011) looked at the management of two large wildfires on national forests in the southwestern US and concluded internal pressures were as important as external ones in shaping fire suppression strategy. It should come as no surprise to rurbanites[1], exurbanites, and anyone with access to TV or the web, and be obvious to those who still routinely use fire, that we do indeed live on a planet of fire.

HOW PERVASIVE IS FIRE?

We are much better informed about wildland fire than we have been in the past, but do current fires differ from those our ancestors dealt with, or are we simply more aware of them, in part because of our accelerating home-building intrusions into wildlands? To answer this question, we need to quantify the current level of wildland fire and put it in historical context. This has yet to be done with any degree of accuracy, although the increasing use and sophistication of satellites continues to improve the 'guesstimates' (see Bowman et al. 2003; Giglio et al. 2010). Even the US has yet to devote the resources to develop an accurate measure of current burn acreage.

United States Statistics

The National Interagency Fire Center (NIFC) compiles annual numbers and acres burned for both US wildfires and prescribed fires. We have no problem with their wildfire statistics, but the posted prescribed fire statistics are worse than misleading. NIFC makes no mention of the accuracy of these annual prescribed fire summaries nor do they include explanatory notes stating they are incomplete. Without such a disclaimer the reader is led to believe these are accurate totals, rather than just the tip of the iceberg. In reality, Florida and Georgia together typically treat more acres with prescribed fire than the NIFC combined total for all 50 states (e.g., 3+ million acres (~1.22 million ha) versus a little under 2 million acres (~800,000 ha) for all 50 states in 2012).

The information included in NIFC prescribed fire accomplishment statistics also differs among states. For example, prescribed burn acres reported by Florida are typically more than twice those reported by Georgia, albeit these apparent differences are not real because Florida includes agricultural burns in its prescribed fire database while Georgia does not; in recent years such burns in Florida have ranged between about 700,000 and 1 million acres (~283,000 to 405,000 ha) per year.

To us, the most troubling aspect of the NIFC statistics is not the fact that they are wrong, but that NIFC is the 'go-to' source for fire information and the layperson or policy-analyst has no idea how misleading these statistics are.

We have no reason to suspect this is the case, but a convincing argument could be made that the government is again suppressing the truth (see Pyne 2010 for an earlier example). A more detailed discussion of the NIFC reporting errors can be found in Wade (2014).

An independent nationwide prescribed burn summary (Melvin 2012) exists for calendar year 2011 which confirms the enormity of the NIFC under-reporting. Actual acreage treated during 2011with prescribed and controlled fires (prescribed fires are executed following a

---

[1] Matthews (1992) defined 'rurbanization' as '*the invasion of affluent urban and suburban-oriented people into rural areas, looking for a self-defined 'country' lifestyle, while importing urban attitudes and values and expecting urban amenities.*'

written plan while controlled burns are not) based on input from all 50 states was 7.9 million acres (3.2 million ha) for forestry objectives and 12.3 million acres (4.98 million ha) for agricultural purposes yielding a combined total of 20.2 million acres (8.17 million ha). In comparison, the 2011 summary of acres treated with prescribed fire for all 50 states posted on NIFC was 2.1 million acres (0.85 million ha), off by a factor of 10! In our opinion that is more than misleading.

World-Wide Statistics

Currently, intentional fires are undocumented in most less-developed countries. Satellites cannot currently differentiate between planned burns and wildfires, but it is clear that mega-fires are becoming increasingly common (see NIFC website; Flannigan et al. 2013; Scott et al. 2014); all continents except Antarctica now have their share (Williams 2007, 2013; Williams et al. 2011).

The largest single fire recorded in the 20th century was the Great Black Dragon Fire which burned roughly 18 million acres (7.3 million ha) in China and Siberia in 1987. Although this record will be hard to surpass, the 21st century has already had its share of mega-fires: Botswana - 9.8 million acres (3.6 million ha) in 2008; Australia - 1.1 million acres (445,000ha) in 2009; Bolivia - 3.7 million acres (1.5 million ha) in 2010; Russia - roughly 27,000 wildfires which burned more than 2 million acres (832,000 ha) during the summer of 2010; and the US which has exceeded 5 million acres (2.02 million ha) in 10 of the last 14 years. These and other fires were discussed at the 5th International Wildland Fire Conference held in Sun City, South Africa in May 2014 (http://www.workingonfire.com/index.php/2013-06-26-11-01-07/latest-news/71-5th-international-wildland-fire-conference). One of the take home messages was that wildfires have a much larger deleterious economic impact in developing countries than in wealthy ones.

Satellite Products

A global early warning system for wildland fire is now operational and can be accessed at http://www.fire.uni-freiburg.de/gwfews/. A USGS satellite records burn scars, and a NASA satellite has been recording wildland fires around the globe since 2002; this information can be accessed at http://www.nasa.gov/mission_pages/fires/main/climate/. Another website with active worldwide fire and burned area products is http://modis-fire.umd.edu/News.htm.

Figure 6 shows the number of fires worldwide over the 7-day period 9/22-29/2014. Although the preponderance of data is from wildfires, prescribed fires and agricultural burns, all open flames that can be seen from space are shown including camp/cooking fires, structural fires, oil well gas flares etcetera. Averaging fire numbers over a 10-year period helps sort out locations where fire occurrence is chronic and were it is infrequent (Figure 7).

The average acreage burned annually between 1995 and 2013) is shown in Figure 8. Although actual acreage burned annually can only be roughly estimated from Figure 8, the map does show where most of the acreage burned occurs - sub-Sahara tropics and sub-tropics of Africa and Australia.

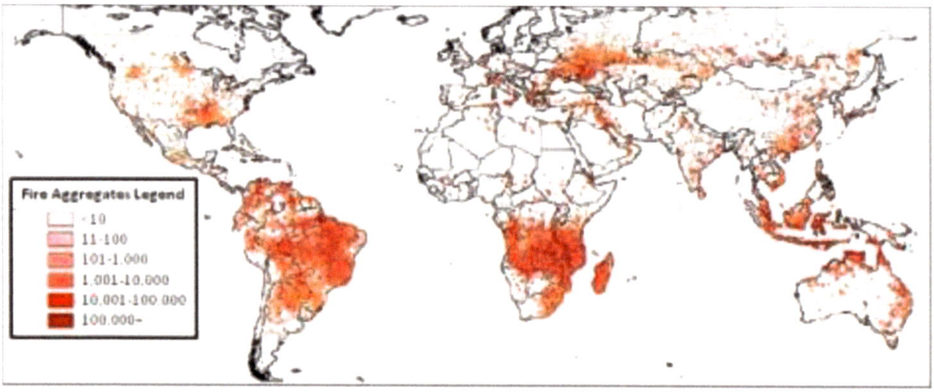

Figure 6. Number of fires world-wide for the period Sept 22-29, 2014. Taken from NASSA's FIRMS Web Fire Mapper https://firms.modaps.eosdis.nasa.gov/firemap/?x=-5.400000000000006&y=5.700000000000003&z=2&g=g&v=2&r=5&i=nw&l=ad,ct.

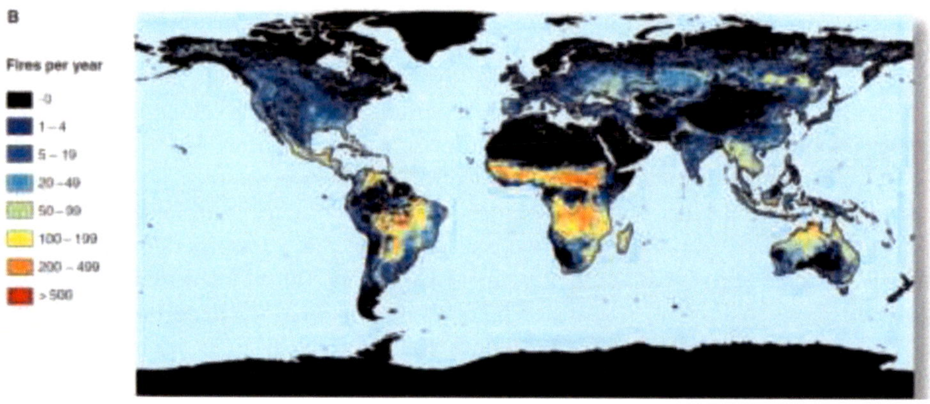

Figure 7. The annual average number of fires observed by satellite earlier this century as derived by Giglio et al. (2006) (adapted from Figure 6.5 in Scott et al. 2014).

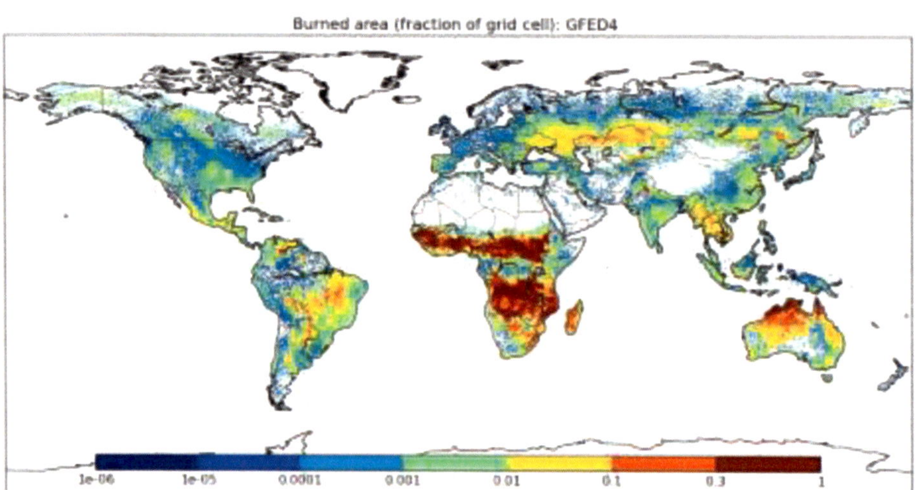

Figure 8. Fourth generation of the Global Fire Emissions Database (FED4) burned area taken from the Global Fire website (http://www.globalfiredata.org/) constructed from data in Giglio et al. (2013).

As an example look at the Australian Northern Territory (ANT), an area of over 10.8 million acres (~4.4 million ha), which shows a high level of annual fire during the 13 year period. This is no recent anomaly as the Aborigines historically burned roughly 50-60% of that area every year early in the dry season with low intensity fire. European colonists displaced the Aborigines and abandoned the pastoral use of fire, locally called fire-stick farming, but frequent late-season lightning fires resulted in about the same acreage burnt every year – the difference was that flame lengths went from less than 2 ft (0.6 m) long to 15-to 20 ft (4.5 -6 m) long with an estimated additional annual release of 24,000 metric tons of carbon.

Within the last decade some of these lands have been converted to Indigenous Protected Areas and are again managed with fire-stick farming. The Nature Conservancy has been a catalyst in this transformation and estimated that although about the same acreage is burnt each year, wildfires now comprise less than 5% of the total area burned with a concomitant reduction of almost 26,000 metric tons of carbon emissions (Rogers 2014). ANT is characterized by a very pronounced wet season, followed by a very pronounced dry season. It can thus be inferred from both figures that once the rain stops and the water table drops below the surface, every day is a burn day until the desired acreage is treated following traditional knowledge passed down from generation to generation.

FIRE AS A PROCESS

Wildland fire has been, and continues to be a ubiquitous global force, but exactly what does fire do? How does it sculpt the Earth's vegetation? As a general concept, fire is the oxidation of organic matter, the same basic process as decomposition, albeit it takes place at a much faster rate. Fire is neither innately destructive nor constructive; it is simply a natural process that, like all others, produces change – a basic premise underpinning the field of ecology. Whether these changes are viewed as desirable or not depends upon their compatibility with human values. When within the bounds of ecosystem resiliency, these changes ensure plant community stability; outside these natural boundaries, dramatic unexpected changes take place and fire alone (or lack thereof) is not likely to restore the previous plant communities without chemical or mechanical pretreatment. Fire is the catalyst that starts, facilitates, accelerates, decelerates or stops the myriad of natural processes necessary to ensure the health of, and perpetuate fire-adapted ecosystems. Sooner or later, fire is required to rejuvenate and maintain most terrestrial ecosystems and the various ancillary loops and cycles that keep them fully functional (Vogl 1976). Virtually every function involves complex webs of interactions and feedback loops. Most large geographical regions have summarized fire effects for their locale. A sampling includes the USFS updated state-of-the-art reviews of fire effects on fauna, flora, cultural resources and archeology, soils and water, air, and nonnative invasive species, collectively called the 'Rainbow Series' (Rocky Mountain Research Station 2000 – 2012); Wright and Heinselman (1973) emphasized the Great Lakes region; Wade et al. (1980) emphasized South Florida; Higgins et al. (1989) did the same for the Northern Great Plains; and Sugihara et al. (2006) emphasized California; Keeley et al. (2011) did so for Mediterranean ecosystems; Alexander and Euler (1981) for the boreal mixedwood forests of central Canada; Silva et al. (2010) for Europe; and Whelan

(1995) and Scott et al. (2014) world-wide perspectives. Rather than attempt to discuss fires' suite of impacts, we instead refer the interested reader to the above references. We do however, include examples throughout the chapter, but keep in mind that many are negative impacts resulting from inappropriate fire regimes and produced by relatively uncommon large intense wildfires.

Fire Management and Regimes

Ecosystem health is declining throughout much of the world (Figure 9) for a number of reasons (Shlisky et al. (2007). Destruction of native plant and animal associations in the name of progress, or simply for greed, is accelerating, if for no other reason than because we are getting better at it. According to Shlisky et al. (2007), major human actions contributing to globally altered fire regimes are urban sprawl, livestock production, ranching, farming, resource extraction, climate change, inappropriate fire use, and fire suppression. We add the following to this list: damming or diverting rivers; draining wetlands and aquifers; large-scale chemical use (reduced numbers of insect pollinators) and; introducing/facilitating the spread of exotic plant and animal species such as kudzu (*Pueraria lobata*) in the southeastern US and the European rabbit (*Oryctolagus cuniculus*) in Australia.

Shlisky et al. (2007) reported that over 60% of the ecoregions assessed world-wide were degraded or very degraded. We suggest this decline in ecosystem health has a lot to do with changes in fire intensity, frequency, and/or timing: too much, too little, or under the wrong weather/season. When the problem is too much fire (increased frequency, intensity or severity), it is almost always because humans, at least initially, did so for economic reasons; for example to convert dense forest to pasture or cropland (Cochrane 2003; Brando et al. 2014).

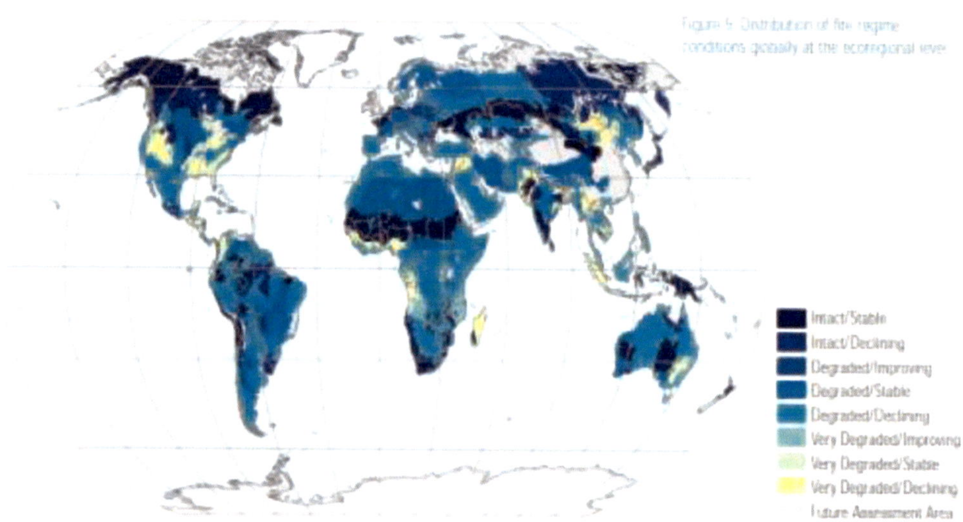

Figure 9. The distribution of fire regime conditions globally at the ecoregional level (from Shlisky et al. 2007).

Where people are tied to a particular plot of land and use fire more often than their ancestors did in an effort to increase their standard of living, ever-tightening feedback loops can result, and peoples standard of living tends to slide back below where it was when they started the process (Pyne in Scott et al. 2014 calls this 'biotic strip mining'). Moreover, the situation is exceedingly difficult to correct because the people responsible are now even more dependent upon fire for survival. They typically are very motivated to continue, or even further increase fire frequency and are vehemently opposed to any reduction in its use (Komarek 1981; Kull 2002).

This was the situation many rural southerners in the US, especially those in the Piedmont and southern Appalachians, found themselves in during the first few decades of the 20th century. Virtually all were trying to eke out a living on soils worn out by years of poor farming practices. They annually burned their lands as well as all surrounding property to provide forage for their free-ranging livestock and to protect their homesteads from wildfire. Most of the land was owned by wealthy absentee urbanites who wanted a continual income flow from timber, but this was impossible as long as subsistence farmers prevented regeneration with their fire practices. These wealthy landowners successfully lobbied federal and state agencies to ban the use of intentional fire (Komarek 1981).

Rural southerners across the southern US recognized that fire exclusion would both eliminate the forage that sustained their livestock through the winter and significantly increase the likelihood that they would lose their homes and even their lives in subsequent conflagrations if fuels were allowed to accumulate. There was no middle ground (which is where the answer was). As long as people accepted the gift of fire and its inevitability on nature's terms, they could live in relative harmony with this natural force, but when governments intervened, thinking they could manipulate natural processes like a spigot to control water in a pipe, the situation invariably deteriorated, sometimes very quickly.

Too Much Fire

Numerous current examples can be found worldwide where humans have shortened the natural fire frequency cycle. Burning the same area too often during the past several decades is the reason the Siberian taiga is degraded (Hardesty et al. 2005), but this ecosystem is still within its historical range of variation and can thus be restored by simply increasing the fire-free interval. According to Mouillot and Field (2005), about a third of the total annual burn acreage worldwide is in the tropics where fire every year or two is required to perpetuate the vast grasslands, savannas and woodlands. Here the problem often is fire during the wrong season. Kilahama (2011) and Maxwell et al. (2011) describe the fire situation in Tanzania and Zimbabwe respectively, and the successful integration of rural residents into improved fire management scenarios including the use of low intensity fire early in the dry season to prevent more intense and damaging fires later. For example, Myers (2006) quotes an FAO official who reported that in Mozambique, annual fires in degraded Miombo woodlands destroy thousands of houses and 15 percent of the crops each year including occasional human fatalities.

In Ghana, 29% of the gazetted forest estate, about 791,000 acres (320,000 ha) has been deforested and an additional 55% partially degraded, in large part due to repeated fire. The annual loss from these fires is estimated to be roughly US$100 million, an amount equivalent to about 2% of Ghana's gross domestic product (Hardesty et al. 2005). The situation was much the same in Zambia. As much as 80% of Kafue National Park (one of the largest in the world at 5,555,000 ac (2,248,000 ha)) burned annually during the height of the dry season mainly from fires started outside the park by poachers and from other illegal activity (Anderson 2014). Since 2011 park rangers have been burning earlier in the dry season to prevent large damaging fires over the ensuing months.

Too Little Fire

Too little fire is also a common problem worldwide, but depending upon how long it has been withheld, it can be technically more difficult to rectify. Although the cause may be obvious to experienced burners and the end result from a fire management standpoint may be the same, different combinations can yield a wide variety of ecological outcomes. Where the problem is too little fire coupled with intensity and severity that are also too low, tree cover will increase, shading out herbaceous fuels (Glitzenstein et al. 2003). The longer the fire-free period, the more likely the eventual fire will result in substantial overstory mortality as well as being considerably more difficult and dangerous to control.

Many plant communities once burned primarily with low intensity fires interspersed with occasional higher intensity fires that typically result in partial or complete overstory mortality and trigger a stand regeneration event. This scenario is often called a mixed fire regime (Brown 2000). Eliminating the low intensity fire means that when the area does eventually burn, the increased fuel load will likely result in greatly increased fire intensity and severity disrupting historical succession patterns. A look at Figure 9 shows very degraded/declining ecosystems in the southeastern US which overlap the range of longleaf pine. Occasional higher intensity and higher severity drought fires superimposed on the chronic lightning and anthropogenic fire regime maintained the diverse vegetal mosaic that ranged from swamps and bogs, to prairies and open forest (Wells 1942; Garren 1943; Myers 2000; Wade et al. 2000).

Currently many southeastern ecosystems suffer from a lack of low intensity fires, but have not yet passed the point of no return and can thus be restored by simply decreasing the fire-free interval. Returning fire to this landscape, however, gets more difficult as the fire-free interval lengthens and a miscue during the process can result in considerable overstory mortality (Varner et al. 2007; Hermann and Kush 2010). They discuss reintroduction techniques and how to avoid associated unwanted longleaf pine mortality. In ecosystems where plentiful rainfall and warm temperatures result in rapid accumulation of duff, and that were historically maintained by chronic fire such as these, once the fire-free interval exceeds 5-6 years on good sites, care must be used to limit duff consumption during the first burn. As the fire-free period lengthens, the initial re-entry burn is just the start of a slow decades-long process of decreasing the thick duff layer in small increments to avoid overstory tree root damage. One slip in this process can negate all progress. For example, one of a handful of remaining old growth longleaf stands was lost during a 'routine' burn more than a decade after this process had begun because too much duff was inadvertently consumed (Kush 2009). This can be a problem worldwide. When reintroducing fire in western conifer stands, managers typically have to contend with a developing midstory comprised of flammable conifers rather than hardwoods (see Hood 2010). One should keep in mind that even the most fire dependent ecosystems can foster catastrophic fire when just a couple of required return-interval fires are missed. Campbell (2011) described the process of reintroducing fire to the jarrah (*Eucalyptus marginata)* forest in Western Australia in the late 1950s after an extended period of fire exclusion; the process was described as "hair cutting". Low intensity burns on calm, sunny days in winter were used to progressively burn off layers of accumulated litter over a period of several years.

Resource managers are often reluctant to reintroduce fire after an abnormally long fire-free interval, but doing nothing should not be an option as the eventual wildfire will almost certainly wipe out the stand. Managers will never have complete knowledge, and quantitative information may be nonexistent, but experienced fire managers know enough to get started in every ecosystem we are aware of. They can localize general guidelines to begin restoration in fire-starved, fuel-clogged stands including those requiring stand replacement crown fires (Wade et al. 1998; Hood 2010; Reynolds et al. 2013). After several burns, managers are often surprised to discover native ground cover species that have not been seen on the site in decades, present and flowering (Varner et al. 2000). Success stories and not-so successful fire restoration attempts are often topics at fire council meetings (discussed later in this chapter) and are a compelling reason for local fire managers to attend.

Fires of the Wrong Kind

Fires of the wrong kind result in altered, undesirable fire regimes, defined by Hardesty et al. (2005) as those "*that have been modified by human activities, such as fire suppression and prevention, excessive burning, inappropriate burning, ecosystem conversion or landscape fragmentation, to the extent that the current fire regime negatively affects the viability of desired ecosystems and the sustainability of products and services that those ecosystems provide*".

The wrong kind of fire (depending upon ones perspective) is why Madagascar and Malaysia are shown in Figure 9 to have very degraded/declining ecosystems. Fire in these areas is being used at the landscape level to convert natural ecosystems to palm oil plantations. In both examples, the federal governments banned the use of fire by rural people who depended upon it for their livelihood while at the same condoning or facilitating its use to convert forests, often in nature reserves and other 'protected' lands (see below and Kull 2002). The wrong kind of fire is perhaps epitomized by the 1997–1998 mega-fires in Southeast Asia that burned more than 24 million acres (9.7 million ha), much of it fire-sensitive tropical forest. These drought-exacerbated organic soil fires elevated Indonesia to one of the top greenhouse gas producers worldwide and affected the health of more than 100 million people. The fires released enough iron to create an extraordinary red tide that devastated coastal marine ecosystems in the region, which in turn led to reef death by asphyxiation (Abram et al. 2003); (see also Ito 2011). Smoke from the fires reduced visibility and severely impacted shipping lanes between the Indian and Pacific Oceans for months resulting in economic losses exceeding 9 billion US dollars (Bowman et al. 2003). None-the-less destruction of both gazetted (areas a government officially sets aside as permanent reserves, which often includes eviction of indigenous people) and un-gazetted forest reserves to allow conversion to palm oil plantations continued unabated resulting in extirpation of species as chronicled by a July 2013 newspaper report stating the last stands of keruing paya

(*Dipterocarpus coriaceus*) in peninsular Malaysia were harvested when a Forest Reserve in Perak was cleared for oil palm plantations (http://news.mongabay.com/2013/0730-keruing-paya-extinction-palm-oil.html?fbfnpg%23itfT3cAs4gMDUR4q.99).

The expanding body of literature shows humans have, and continue to provide the ignition source for most fires. At the 2004 World Conservation Congress, the Global Fire Partnership presented overwhelming evidence that altered fire regimes should be a high-priority global conservation issue (Shlisky et al. 2007). TNC identified altered fire regimes as one of the key threats to biodiversity (Hardesty et al. 2005). The situation is much worse than many people realize; according to Shlisky et al. (2007), fire regimes in 13 of the 14 major habitat types they listed are threatened by inappropriate human fires or attempted fire exclusion. Pyne (in Scott et al. 2014) discusses these issues in more depth and provides examples.

In the case of too little fire, as an ecosystem approaches the boundary condition from which it cannot recover, increasing natural force is unleashed in the attempt to correct the imbalance. This process plays out with increasing vigor as the period of fire exclusion increases. These pyrotechniques are spectacular from a distance, ominous as one gets closer, and potentially fatal when humans are directly involved. The original flora and fauna is replaced by new communities, often with a high proportion of exotics that will be less diverse and tend to burn with higher reaction intensities, often under more severe weather conditions; the result being more damage to the site and to anthropogenic improvements, as well as increased human exposure to risk.

For example, if fire is withheld from most eastern US temperate ecosystems for as little as a decade, invading hardwoods, previously confined to more mesic sites by recurrent fire will aggressively spread uphill, shading out herbaceous ground cover and virtually eliminating reproduction of more pyric oaks and pines. Within a few decades, overstory species composition will radically change and fire will be relegated to periods of drought or in the autumn before precipitation packs down the newly fallen carpet of leaves. Nowacki and Abrams (2008) call this process 'mesophication'. See Wade (1991) and Wade et al. (1998) for examples of exacting conditions sometimes necessary to reintroduce fire into such ecosystems.

The increasing magnitude of damaging wildfires in the US over the past several decades is simply nature trying to rectify the results of attempted fire exclusion and our reluctance to implement the only practical solution (prescribed fire) on a landscape scale. Current levels of both intentional fire and wildfire should keep us mindful of the pivotal global role of fire as it and humans continue their ongoing love-hate relationship, whether it be fire exclusion, or wholesale land use conversion as in the case of Brazil for ethanol (Filho and Horridge 2011) or in the Pacific Islands to meet the demand for palm oil (Morgan 2013).

Appropriate Fire Regimes

Appropriate fire regimes result in ecosystem stability, albeit they are typically the product of centuries of human manipulation to create and maintain desired outcomes. Most ecosystems exhibit wide amplitude within which they are markedly resilient in spite of our current fire management decisions, but they do have limits. When a threshold is passed because of too much fire, restoration will take more than simply reducing burn frequency and may involve considerable effort such as the reintroduction of extirpated species, but the blueprint required to restore the ecosystem is typically straightforward compared to human ramifications and restoration efforts when the problem is too little, or the wrong kind of fire. In these cases chemical or mechanical pre-treatments are often needed.

Because of concerns voiced about the carbon footprint of fire, the Association for Fire Ecology, International Association of Wildland Fire, Tall Timbers Research Station, and TNC joined together to conduct an in-depth review of the literature on the subject. The result was a "white paper" titled *The Merits of Prescribed Fire Outweigh Potential Carbon Emission Effects* (Association for Fire Ecology et al. 2013). Their results are in sync with Underwood et al. (2008) which explains why frequent low intensity fire is a carbon sink, while higher intensity fires, especially when moisture levels are lower will always result in more atmospheric carbon emitted than used in plant regrowth.

How We Arrived at This Untenable Global Fire Situation

The details differ by location, but the stories have a common thread. In Australia the primary culprits are the ideologically driven academics whose fire experience is confined to their backyard barbeques. In North America the reasons include: 1) misguided government actions to 'protect' their constituency, 2) self-serving agendas, 3) simple human greed, and in some cases all three. Jim Brenner with the Florida Forest Service once said "*Removing the option of controlled burning does not stop the burning, it just removes the control*". But remove it they did, or at least the federal government tried its best to do so and succeeded in stopping human ignitions in all but the southern US and on American Indian reservations. See Putz (2003) for a humorous but accurate account of why historic landscapes have been maintained in many parts of the South despite federal and state laws that banned the age-old practice of burning the woods. World-wide, laws or policies that disregard or ban local traditions and culture, or jeopardize ones standard of living will likely never be peacefully accepted unless a better future can be guaranteed.

Indian reservations in the US and Canada provide examples where accumulated fire wisdom has not been lost because federal and state fire exclusion policies could not legally be unilaterally implemented on them. These reservations serve as long-term demonstration areas where the effects of natural fire augmented with anthropogenic ignitions can be observed. A significant practical benefit is that the fire practices that created and maintain these desired landscapes can be emulated elsewhere without extensive research or trial and error. Numerous authors have (e.g., Weaver 1947; Kallander 1958) and continue to (e.g., Blank 2006; Diekmann et al. 2007) point to the long-term results of fire-based ecosystem management practiced on reservations. Testimony before the US House of Representatives Natural Resources Committee by Carole Lankford, Vice Chairwoman of the Confederated Salish and Kootenai Tribes, in April 2014 stated, and we concur: "*... the nation would not experience the devastating wildfires it does if US forests were managed the way forestland on the Flathead Indian Reservation* [and most other reservations] *is* ..." (Devlin 2014).

Unlike the Flathead Reservation where fire is routinely used, lands where fire exclusion has been temporarily successful, are now fostering mega-fires with their attendant suppression costs and loss of life and property, not to mention ecosystem damage. Attempted fire exclusion was a misguided effort, doomed to failure from the get-go because it violated natural laws; its advocates viewed the accumulating peer-reviewed scientific knowledge to the contrary as flawed; the handed-down wisdom and experience of the people who had been managing fire-adapted ecosystems for thousands of years was ignored and resource managers were literally forced to accept the early 1900's European forest management template and fallacious Clementsian ecological theory.

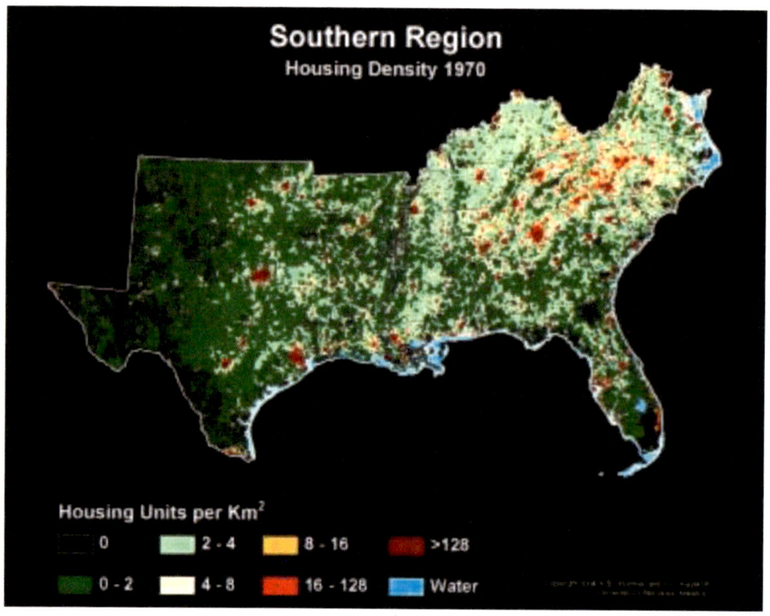

(a)

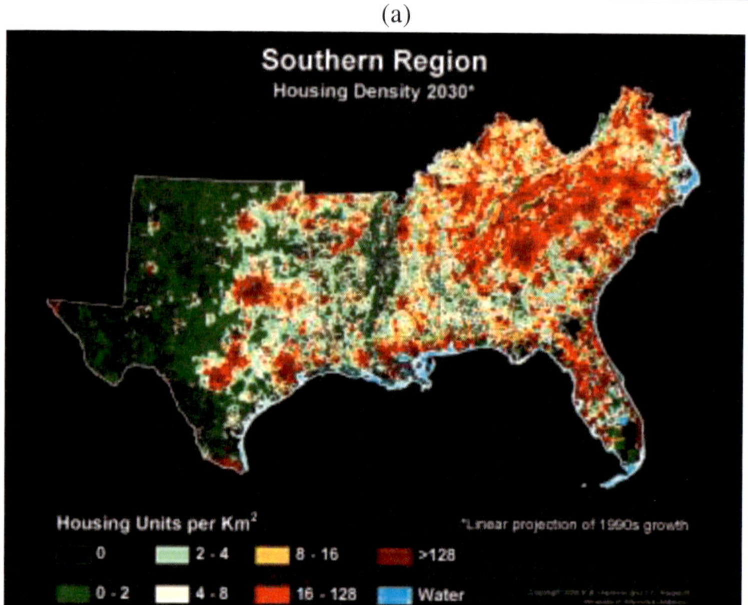

(b)

Figure 10. Maps of (a) the 1970 housing density and (b) projected 2030 housing density in the Southern Region (R-8) of the US Forest Service. Curtesy of the Silvics Lab, Department of Forest Ecology and Management, University of Wisconsin.

Southerners had been warning of the dangers of accumulating fuel associated with fire exclusion since at least 1875 (Blocker 1875); see Wade (2011) for a broader discussion. The situation was the same in the western US where landscapes had also been shaped by fire for thousands of years (Anderson 2005) and reflected the dynamic mosaic of fire intensity, severity and timing. Here to, managers such as Ogle (1920) warned that removing fire would increase fuel loads over time resulting in destruction of the forest landscape. And again the warnings were ignored by newly arriving federal foresters armed with their erroneous text book knowledge and government-sponsored propaganda (Miller 2010).

Schiff (1962) and (Pyne (1982, 2004, 2010) describe the battle between advocates of fire use and those promoting fire exclusion. Perhaps the most discouraging aspect of this controversy was that the burners had a success story thousands of years in the making and a body of science to back them up which the USFS suppressed (Pyne 2010), while the federal government had nothing to offer but a jaundiced, albeit well-meaning vision. The US federal government won the battle, but eventually lost the war because the public finally demanded accountability. One can only wonder why people in pivotal positions refuse to learn from the past and continue to believe they can exert their power to force natural processes to adhere to some idealized outcome.

World-wide, the increasing frequency of catastrophic wildfire is primarily due to long-term consequences of fire suppression in fire prone ecosystems (National Commission on Wildfire Disasters 1995; USDA Forest Service 2000). These fires are difficult and expensive to control and the situation is exacerbated by the increasing number of people building their homes in these fire-starved forests. The southern US situation is shown in Figure 10 and is the same in Australia and likely wherever people can afford to move out of congested cities.

Urban sprawl adds complexity to burn planning and execution and has had a negative impact on the ability to burn many landholdings which will only worsen as the mass influx of people into exurban areas continues. These hurdles can, however, usually be overcome if resource managers are willing to devote the extra effort to work with affected residents. For example many Florida state agency lands are ringed by homes, and yet considerable acreage is intentionally burned each year.

The Path Forward in the US

Nationally, fire management suffers from too much politics, too much focus on fire suppression, and leadership that appears to be curtailing momentum towards implementation of a real solution (Williams 2007; Pyne 2010; Wade 2011; Cochrane et al. 2012; Mutch 2014). Any remaining advocates of fire exclusion are not speaking out publicly and the press is very supportive of the concept of prescribed fire, but that said, fire on USFS lands continues to be mired in complex (unreasonable in our view) planning requirements and bureaucratic gobbledygook. We are hopeful, but not optimistic that The Cohesive Wildland Management Strategy (CWMS or CS for short) unveiled in 2013 will result in measurable increases in the use of prescribed fire in the US. The strategy is national in scope and based on three premises:

1. Restore and maintain landscapes - Landscapes across all jurisdictions are resilient to fire-related disturbances in accordance with management objectives.

2. Fire-adapted communities - Human populations and infrastructure can withstand a wildfire without loss of life and property,
3. Wildfire response - All jurisdictions participate in making and implementing safe, effective, efficient risk-based wildfire management decisions.

This initiative can be accessed at: >http://www.forestsandrangelands.gov/strategy/<. The CS vision is to *"safely and effectively extinguish fire when needed; use fire where allowable; manage our natural resources; and as a nation, live with wildland fire"*. It is a clear, concise attainable policy that establishes goals and a broad operating fire management framework which, rather than a 'one shoe fits all' approach, provides the latitude for USFS regions to develop guidelines and plans of action to match local conditions. The CS has resonated well with natural resource managers, but there was surprisingly strong opposition to allow regions the above leeway so it remains to be seen whether upper level agency personnel will facilitate or obstruct regional plans. It is a bad omen that rather than immediately funneling money to fast track writing and implementing regional plans, funding was instead used to produce state-of-the-art software that allows large fires to be broken into sectors which can be managed individually with different tactics. This has long been the case on the ground, but now slick smoke and mirrors presentations can be used to impress politicians and the lay public; in our opinion this just obfuscates the root of the problem which is fuel-chocked tinderboxes. If progress to date is a harbinger of things to come, it appears the emphasis will be placed on better managing wildfires rather than in ramping up the landscape use of prescription fire - the only practical option to solve the mega-fire problem.

The National Strategy, the final phase of the CS was released in April 2014 (refer to http://www.doi.gov/news/loader.cfm?csModule=security/getfile&pageid=526008). This strategy, developed by federal, state, tribal, local and community partners, and public stakeholders, outlines approaches to coordinate and integrate efforts to restore and maintain healthy landscapes, prepare communities for fire season, and better address the nation's wildland fire threats. We thus thought The Large Fire Conference held in Missoula Montana in May 2014 would have focused on solutions but we were wrong. Many of the speakers and panels described the acute need to restore landscape level fire. The USFS Chief again made it clear that the agency knows what the problem is and is going to reduce fuels to solve it, but he was silent on what steps he is taking to reverse the downward spiral in experienced prescribed burners through retirements, employee reductions and transfers to non-fire positions; there was no mention of administrative changes to facilitate a solution. There is a huge difference between standing up and telling people you are going to do something and actually doing something to facilitate accomplishment; it appears to some observers that his actions to date are masked in bureaucracy and instead intended to hinder accomplishment.

The ecosystems at peril need action not more obfuscation. There was a whole session devoted to Canadian landscape fire restoration, but no companion suite of presentations pertaining to the US. No speakers provided guidance to show how to cut through agency red tape, mitigate administrative constraints and marginalize unsupportive/timid supervisors. Coauthor Miller provided a template (the same one used at the end of this chapter), but not a single federal presentation described creative methods to circumvent administrative bottlenecks and hurdles to forge ahead and accomplish landscape level burns. There were however, numerous talks on large fire management, the repercussions of climate change, advancements in suppression tactics and equipment, and on planning to plan. This is another indication that the USFS has decided not to make landscape level use of prescription fire their primary focus - a grave mistake in our opinion that treats the symptoms rather than the disease.

The USFS Chief could dispel such cynicism with a simple directive stating landscape level fuel reduction is a national priority and as such will be a performance element for Regional Foresters. This would quickly translate into significant increases in acreage treated. Many field-level fire managers know what has to be done and how to do it; they just need the agency to facilitate rather than hinder their efforts and to reassure them that the agency will stand behind them if a bad outcome occurs as long as they were operating within the scope of their position.

The federal fiscal year begins October 1st so significant acreage could be treated each year during the dormant season and early growing season before the western wildfire season siphons money from the program to pay for suppression costs that exceed budgeted funds. Fire suppression costs rose from 16% of USFS budget in 1995, to 42% in 2014. The USFS has run out of appropriated fire suppression funds in 7 of the last 14 years forcing the agency to transfer money from other programs such as prescribed fire use. (Wilent 2014b). In our opinion, fire use is the ONE program that must be expanded rather than curtailed. In June, 2014, the USDA released a state-by-state impact of projects that had been postponed or canceled because funds were redirected to fire suppression (http://www.fs.fed.us/news/releases/usda-releases-state-state-impacts-limited-wildfire-suppression-recent-years). The list includes numerous ecosystem restoration projects.

The chart depicted in Figure 11 superimposes suppression costs over the number and size of wildfires over several decades. Note that even in years with below average number of wildfires and acreage burned, such as 2013, just a few mega-fires can result in astonishingly high suppression costs. According to Williams (2013) mega-fires account for about 0.1% of all wildfires, but roughly 95% of the total area burnt and 85% or more of total suppression costs. One should also remember that suppression costs are typically just a small percentage of the total cost when property improvements, timber, range, soil, watershed damages, and remedial measures are included.

The following is a cautionary tale that we hope moves natural resource managers in developing nations to avoid the mistakes the US has made, and continues to make. Many federal natural resource managers are polished examples of political correctness and risk aversion. Once humans began making fire management decisions, they assumed responsibility for maintaining ecosystem health, but when people consider every decision based on 'what effect will it have on my promotion potential' and agencies are staffed with political lackeys that have no backbone or moral fiber, ecosystems suffer. Such managers are experts at blaming others, or climate change for deteriorating ecosystem health during their watch. They are masters at advocating the need for landscape size fire treatments, but never actually having the 'right' conditions to put fire on the ground. Upper management gives speeches lauding plans and accomplishments while our nations ecosystems continue to deteriorate under the weight of unprecedented fuel buildup and resulting mega-fires. Many state legislatures have become so disenchanted with bungled federal land management that they are trying to wrest management of lands within their jurisdiction from federal agencies. This movement has grown to the point where it has been given a name - The Sagebrush Rebellion.

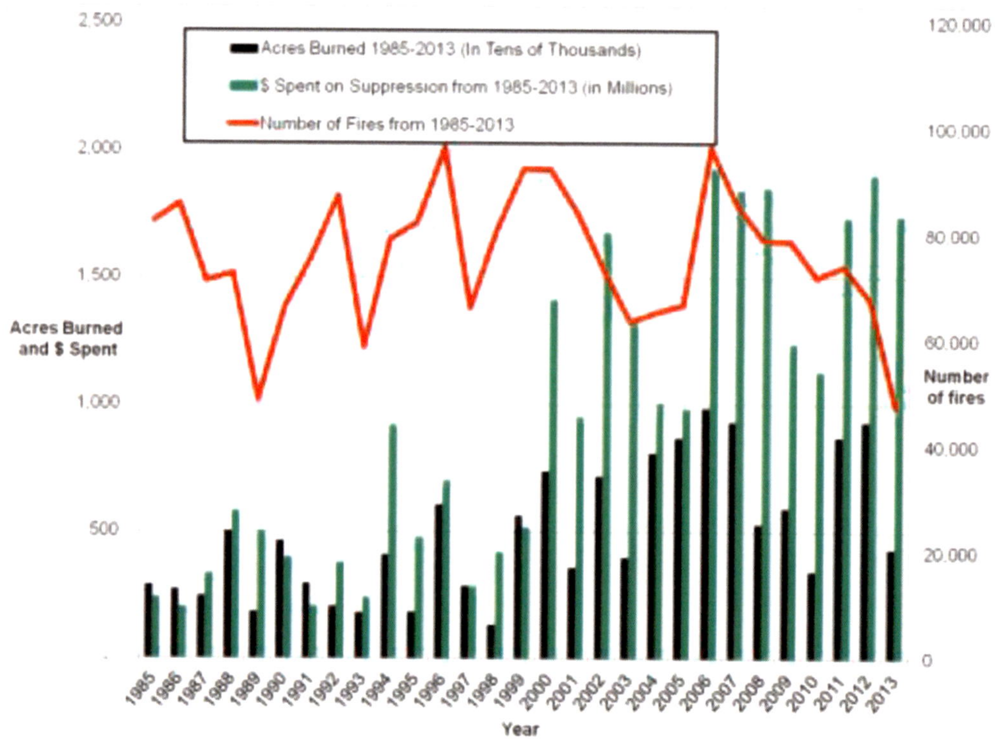

Figure 11. Federal fire suppression costs, number of wildfires, and area burned in the US between 1985 and 2013. Figure constructed from information on the NIFC website.

Federal agencies have many dedicated and competent people at the field level, but their hands are tied by bureaucracy, inept CYA (cover your a..) supervisors, and litigation (Wade 2011). The National Environmental Policy Act (NEPA) became law in 1970 and its goal was laudable, but environmental fringe groups have learned how to invoke this Act to forestall, and in many cases stop, scientifically based ecosystem management plans that have been approved after passing multiple levels of internal and external review. Groups filing these lawsuits cannot be educated because their agendas are based on some fuzzy utopian ideal, not science. These lawsuits take an enormous amount of scarce federal resources to fight in court, often over a period of decades.

Politically difficult compromises need to be reached to rescind or put on hold socially appealing legislation that drains natural resource agencies of funds that have little to do with actual ecosystem management. Prescribed fire is a safe way to apply a natural process, ensure ecosystem health and reduce wildfire risk and as such, funds should be allocated to substantially increase its landscape footprint. We use the WUI as an example where reforms could result in substantial savings. Suggestions include:

1. There is no WUI on federal wildland. This zone results from decisions made outside federal land boundaries, so federal fire managers should not be saddled with their protection. As Stephens et al. (2013) state "*Because the federal government has no jurisdiction in development policies in the privately owned urban-wildland interface, state and local jurisdictions could pay for fire suppression in the interface. This*

would enable a significant increase in critical forest restoration funding and would probably reduce building in the interface". A look at past wildfires shows innumerable cases where many more houses could have been saved than were lost if the task of attempting to defend those situated in marginally defensible locations was left to the discretion of structural fire departments or to the occupants.

2. If one chooses to build a home down slope from an active volcano, on the outer most sand dune, or within or directly above a fire dependent ecosystem, that is a personal decision and neither state nor federal firefighters should be expected to make heroic efforts to protect such structures and needlessly place themselves at risk in so doing. People that make such short-sighted decisions should be denied reimbursement for losses from public funds collected from more prudent tax payers.
3. Require that a letter outlining fire risks, and formalized smoke corridors be a part of every exurban and rural real estate transaction to ensure potential buyers are forewarned. See Wade et al. (2006) for examples of ordinances and items to consider when drafting such an ordinance.
4. Change tactical priorities so federal initial attack forces can concentrate on stopping the fire rather than being diverted to protect structures on in-holdings, often at considerable risk to themselves.
5. Convince insurance companies to require exurban and rural property owners to pay the true premium, rather than spreading the cost over as many policy holders as the law allows. This simple change would be strong encouragement for such property owners to take the necessary Firewise steps that could qualify them for premium discounts.
6. Require stricter building codes for structures on rural and interface properties including an outright ban on untreated wood shingles and plastic soffits, and provide an option for installation of dry hydrants or other water sources not dependent on municipal water mains.

Progress in the Southern US

Even though many southern state natural resource agencies, non-government organizations and private landowners are struggling with declining budgets, they have a better appreciation for, and dedication to both facilitate and actively engage in the only known solution to minimize the occurrence of conflagrations; a solution backed by hard science and a track record that has stood the test of time – applying appropriate fire regimes at the landscape level. Many have enacted statutes that promote the use of prescribed fire while safeguarding the public. The Florida Prescribed Fire Statute (Statute 590) is one of most comprehensive and was updated in 2014 (http://www.leg.state.fl.us/Statutes/index.cfm?App_mode=Display_Statute&URL=0500-0599/0590/0590ContentsIndex.html &StatuteYear=2012&Title=-%3E2012-%3EChapter%20590). The section that pertains to prescribed fire is called 'open burning' chapter 590.125. The Florida Administrative Code 5I 2 provides rules and regulations pertaining to Statute 590 (Refer to https://www.flrules.org/gateway/ChapterHome.asp?Chapter=5I-2).

As soon as landscape fires are thoughtfully placed, the probability of damaging wildfires will diminish and once substantial acreage is brought into compliance, the specter of mega-fires will become a historical footnote. Prescription fire continues to flourish in the southern US. Although most of the South is classed as WUI, these states are none-the-less in an enviable position from an environmental regulation standpoint because both the regional US Environmental protection Agency (EPA) and many state environmental protection agencies understand the deleterious consequences of uncontrolled fire in heavy fuels on public health and safety; these agencies thus tend to be sympathetic toward the use of prescription fire.

Southerners have been able to turn the scarcity of some flora and fauna into opportunities to use fire. Over the past 150 years, longleaf pine has disappeared from over 95 % of its range mainly due to type conversion to faster growing species. This situation has resulted in several federal and state incentive programs, some of which require the use of fire. The red cockaded woodpecker, (Picoides borealis) is a federally listed threatened and endangered (T&E) species which requires an open pine habitat maintained by frequent fire; wherever colonies exist, they must be protected and their habitat maintained by federal statute.

Expanding the Use of Fire

Melvin (2012) surveyed state agencies responsible for prescribed fire in all 50 states in the US regarding an array of fire management issues; all 50 completed the questionnaire and ranked the impediments to increased fire use in their state, the results of which are shown in Figure 12. The lack of trained prescribed fire managers, training opportunities, private contractors, and partnerships was the major bottleneck to increased use of prescription fire.

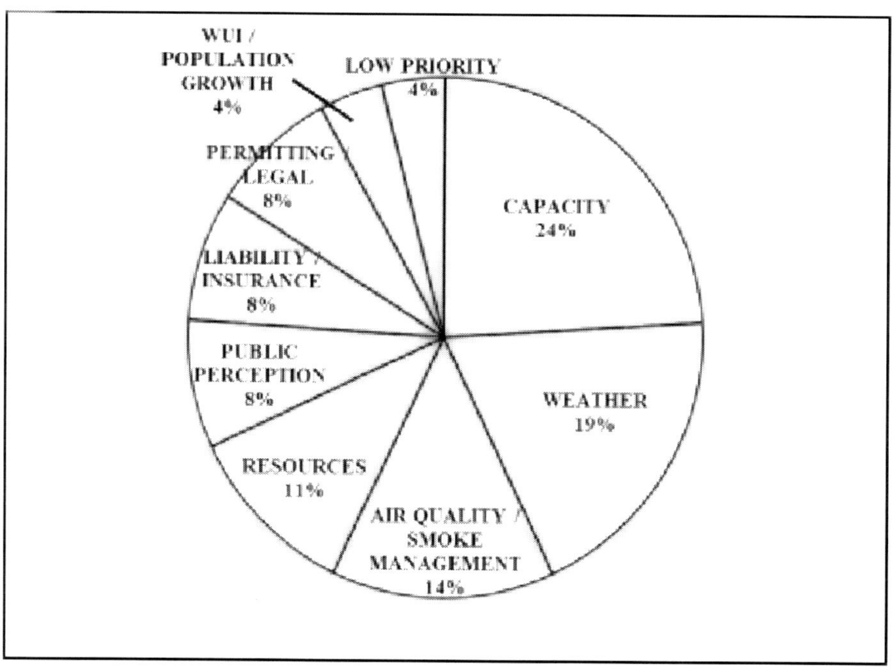

Figure 12. Impediments limiting prescribed fire use as ranked by all 50 state fire management agencies (from Melvin 2012).

Weather was the second greatest impediment in the South and is a politically astute answer because it shifts responsibility for lack of progress to nature; however, we suspect an analysis of the number of burn days would show numerous opportunities were forgone because burn prescriptions were too restrictive, other priorities were judged more important on a given day, or because favorable conditions occurred on weekends. In the South, liability/insurance ranked second, likely because most land is in private ownership and classed as WUI.

Southern states facilitate the use of fire in many ways. Every state that requires a permit or authorization to conduct a prescribed burn does so without charge and allows it to be requested the day of the burn; in some cases authorization can be obtained the day before the planned burn. In other regions, however, states may require several weeks lead time as well as charge a fee, both of which constrain rather than facilitate fire use. The nine southern seaboard states all offer certified prescribed burn manager training to make burners more knowledgeable and proficient; many offer incentives to certified burners that expand the legal burn window. Many also allow nighttime burns which expand the burning window. Some states and private organizations have dedicated burn teams ready to assist where needed which is a great way for landowners to get on-the-job training. When fire managers start each morning thinking it is a burn day and their highest priority, annual burn goals are typically met.

Florida enacted a law in 1977 that allows the Florida Forest Service (FFS) to use fire to reduce dangerous fuel accumulations on private property (see Wade and Long 1979); following specific protocol, a notice is sent out with the annual tax bill and if the landowner does not object in writing within a given time-frame, the FFS is authorized to use fire on the property. If a landowner refuses and a wildfire subsequently starts on, or crosses that property and causes damage to adjoining lands, litigation is likely. Landowners who do nothing to mitigate fuels on their property act irresponsibly and should not expect to be absolved of contributing to damage on adjacent actively managed lands.

Mark Melvin has developed a list of '*Realities of Practice into the Future*' with which we wholeheartedly agree. He defines the current dilemma as '*ecological imperatives versus societal attitudes and regulation*' and the solution as '*the integration of science, art and policy with input from all stakeholders.*' His guiding philosophy is: "*Prescribed fire is a critical component of fuel management, wildfire mitigation and maintaining forest health*". It embraces four challenges:

1. All regions need to find creative means that restore ecological fire at landscape levels
2. Balance public health/safety and the ecological role of fire
3. Train the next generation of more sophisticated [better equipped] fire practitioners
4. Repair the broken human – fire linkages.

THE COALITION OF PRESCRIBED FIRE COUNCILS (WWW.PRESCRIBEDFIRE.NET)

Prescribed burners are doing their part. The Coalition of Prescribed Fire Councils (CPFC) was incorporated in the United States in 2009 as a 501c3 non-profit organization. Member councils currently stand at 31 representing 27 states and British Columbia with council development formally under discussion in two additional states (Figure 13).

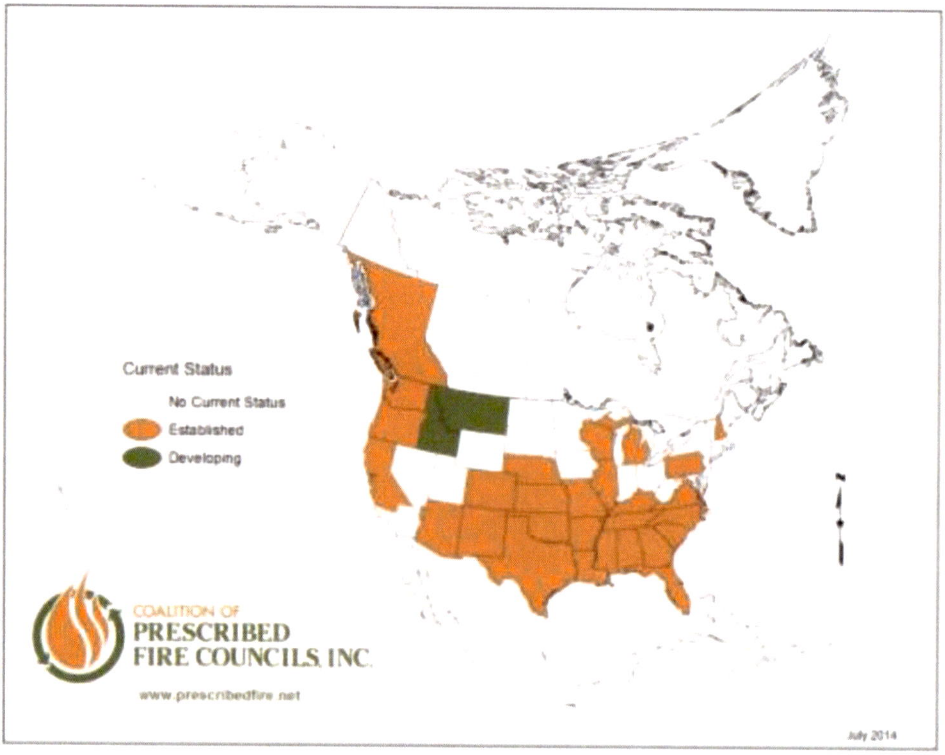

Figure 13. Map showing status of prescribed fire councils in North America as of November 2014. Some states such as CA and FL have regional councils. Two non-government organizations also belong.

The mission of CPFC is to promote the use of prescribed fire to: 1) enhance public safety; 2) manage natural resources; and 3) sustain environmental quality. Their overarching goal is to create one voice to assist fire practitioners, policymakers, regulators, and citizens with issues surrounding prescribed fire use; in other words to protect the right of, encourage the use of, and promote public understanding of prescribed fire. Their primary focus is to build partnerships around fire and they are hard at work doing so.

Member councils have many success stories. At least seven southern states have efforts underway to convince state and county governments to pass resolutions supporting the use of prescribed fire and acknowledging that it is a property right. To date all 67 counties in Florida have passed such a resolution and all but 2 (metropolitan Atlanta) of the 159 counties in Georgia have followed suit Appendix C lists items to consider when drafting a statute to protect and facilitate the intentional use of fire.

A number of states have also passed 'Prescribed Fire Awareness Week' proclamations. Each year, council members join the respective Governor in a press conference kicking off the week and, depending upon the state, put on equipment demonstrations and man a 'Q&A' desk located in the capital to provide information and answer legislator questions. Many Councils are involved with field days where students and the public are invited to watch a burn. Councils are proactive in the smoke management/ air quality arena. Many have met with federal, state, and regional air quality regulators and hosted 'regulator' field days to demonstrate first hand, the work that goes into planning and executing a burn; in fact they often involve their guests in conducting a burn, letting them use a drip torch or ATV-mounted igniter. They do the same for the press. CPFC and member councils are directly involved in helping states develop/revise smoke management plans as mandated by the federal EPA. CPFC regularly organizes input to EPA regarding policy and rules. They also oversee After Action Reviews (AAR) when things do not go as planned on a burn.

When requested, CPFC works with states to secure liability protection, becomes a stakeholder in burn ban declarations during severe drought, and provides advice regarding Burn Manager Certification programs. Other outreach activities include helping state councils develop websites, assisting in message campaigns, hosting media events, assisting in development of state fire strategies when invited, disseminating information on safety and training opportunities, and in general, providing expertise on fire-related issues.

CPFC believes collaboration between all stakeholders is the key to success. Every fire council is a passionate, professional, committed group of fire practitioners. When a council becomes involved in solving a fire management issue, they act as a catalyst to resolve the issue at hand. Their approach is to develop seamless partnerships; they will not accept failure. They are committed to using fire to meet the demands of ecosystem health.[2]

SUGGESTIONS TO FACILITATE RETURNING FIRE TO THE LANDSCAPE

Some technocrats and bureaucrats use their position to ensure prescribed burning moves forward at a barely perceptible pace. We think that is a mistake. Burn managers will never have all the information they desire, but there is more than enough readily available knowledge to rapidly proceed in most ecosystems. What is needed is increased funding for prescribed fire so state and federal natural resource agencies can build capacity, both staff and equipment. Fire managers in an area should have paperwork in place to facilitate borrowing/renting or loaning both from each other as needed.

Managers at all levels need to stop emphasizing potential negative side effects and 'what ifs', and get rid of senseless '*one shoe fits all*' policies and constraints to the judicious use of fire. Everyone needs to recognize that restoring appropriate fire regimes is a win – win situation for both humans and ecosystems.

Liability protection for burners varies widely across states (Sun and Tolver 2012), but should be adequate to fend off frivolous litigation (see Wade et al. 2006 for an in-depth discussion); it should rank near the top of priorities in all states yet to have such coverage in place.

Prescribed burning is as much art as science and fire behavior continually responds to changes in topography, fuel, and weather as a fire spreads through the unit and as the day progresses. Conscientious burners attend training to acquire the knowledge to anticipate changes in fire behavior and modify firing technique and ignition pattern to maintain desired fireline intensity and achieve burn objectives.

[2] Personal correspondence with Mark Melvin, Joseph W Jones Ecological Research Center, June, 24, 2014.

Occasionally burn objectives will not be met, but such instances can be minimized by giving burners the latitude needed to properly execute burns when desired burning conditions arrive. When unacceptable outcomes (e.g., escapes or smoke intrusions into a smoke-sensitive area) result from circumstances beyond human control such as unforecast weather changes, burners should be protected. Mistakes in judgment do occur and need to be addressed, but the witch-hunt some federal agencies and journalists engage in to find a scapegoat for a decision made (or not made) in an emergency situation is counterproductive. Have good managers in place and empower them to discipline unacceptable behavior quickly and efficiently as private companies do, rather than getting mired in agency bureaucracy and equal opportunity minutia that destroys employee pride and Esprit de corps.

Herbert Stoddard (1961), the founder of Tall Timbers Research Station said "*Fire may well be compared to a two-edged sword which requires judgment, care, and experience to properly handle...*". Organizations need to allow people to learn from their own, as well as the mistakes of others and facilitate the use of mentors. It is the only way a person can grow into a knowledgeable, expert practitioner. Most fire council meetings have a 'lessons learned' time slot on the agenda. There is also a national website dedicated to just this purpose called the Lessons Leaned Center accessible at http://www.wildfirelessons.net/home.

Dwight Eisenhower said '*plans are nothing; planning is everything.*' We agree and highly recommend a written plan be prepared prior to every burn. It forces the planner to think about a number of factors that might otherwise be overlooked and serves as a checklist to make sure all pertinent factors have been considered. A one or two page plan plus maps and a brief contingency plan is usually adequate; the more complex the burn, the more involved the plan should be. An integral part of any plan is a post burn evaluation within a few days describing any problems in burn execution, crown scorch etcetera, and suggestions of what to do different, if anything, next time. A follow-up assessment should be conducted after the next full growing season to record how the unit looks and whether it appears as though the burn objective(s) were met (photos are very helpful). We do not recommend using the USFS template which can easily surpass 40 pages, but it is worth looking at as a reality check for your plan. Examples of simple plans can be found in Waldrop and Goodrick (2012), on several state agency websites and perhaps at your local wildland fire agency office. We recommend that someone in a supervisory potion sign off on the plan as well, but see no need for a profusion of signatures. Proactive state agencies tend to approve or deny plans very quickly, although some require much more lead time to mull then over and get approval of other regulatory agencies; this constraint can severely hamper quality burns, particularly those that require a narrow range in fuel moisture conditions as they cannot be reliably predicted more than a few days in advance.

Conducting prescription fires is dangerous, arduous work and is no place for folks that aren't willing to stay in top physical shape, that aren't willing to work outside normal office hours or that are not receptive to constructive criticism. Everyone should strive to become more proficient and pitch in where needed to accomplish the task at hand regardless of their job description; anyone that thinks it should be unionized should find another job. The US coauthors are aware of federal fire managers that have retired in frustration because they could not field a crew to burn outside of normal business hours.

Although there are many challenges to putting fire on the ground, both the barriers and catalysts are human in nature rather than technical. Thus if an organization makes the decision to institute a prescribed burn program, makes it a high priority goal with top management setting the tone, gives staff the equipment and training needed, and then requires attainable annual burn acreage accomplishments, the results will be very positive. Read on for an example of how this is being accomplished in a very challenging ecological and human environment.

A Success Story

Noted fire historian and author Stephen Pyne (In press) has said "*They [Floridians] know that in the public eye, if one of them fails, they have all failed. Elsewhere, fire divides. In Florida, it joins*". In 1975, fire managers in South Florida representing federal, state and local agencies banded together to form the South Florida Wildland Fire Council, in large part to create a forum where they 'checked their weapons at the door' and then get together to discuss their widely differing views of the fire 'problem'. Irrespective of agency policy, every member recognized that fire and water were the two basic components that drove these sub-tropical pine and grassland ecosystems, but their ideas about how to handle the acute arson problem and ubiquitous lightning fires covered the spectrum. They were able to agree on the core premise that fire was not only inevitable, but beneficial under a range of weather conditions and application methods, and then share what worked under what weather and fuel conditions and what did not. It was an amazingly successful process to argue their views, but eventually agree to tweak their fire management decisions and stretch their operating policy guidelines in order to try solutions that might help achieve their agencies objectives.

Members began sharing equipment and personnel, often without the requisite paperwork and endless hoops agencies are throttled with, thus putting their careers at risk - simply because it was the right thing to do. In just a decade, these dedicated public servants turned a seemingly intractable situation into such a success story that two additional fire councils were formed: the North Florida Prescribed Fire Council and the Central Florida Prescribed Fire Council. These groups unite fire practitioners across agencies, ownerships, and missions to promote and protect the use of fire as a management tool. Collectively they worked to get all 67 counties in Florida to pass county resolutions promoting prescribed fire. Even counties where the population is composed mainly of retirees from northern states where their ancestral ties to wildland fire had been forgotten, passed resolutions. The fire councils encourage the exchange of information, techniques and experiences among practitioners of prescribed fire and work with state and local agency fire management and regulatory staff to improve laws and administrative rules to facilitate appropriate use of fire while protecting the safety and health of citizens, and protecting a landowner's right to burn.

When land managers outside the Deep South talk about the need to increase the number of burns and area treated, they frequently dwell on the reasons it cannot be done. When someone raises the point, 'well its being done in Florida' many respond that 'Florida is a different case. They don't have this or that challenge.' That is simply not true; Florida has its own challenges that are every bit as complex as the challenges faced elsewhere in the country and has very difficult WUI issues. One big difference is, as Stephen Pyne said, and we paraphrase: In Florida when fire managers are challenged, they work together to solve the challenge; they do not hide behind their organization skirts. These fire managers do not all think and operate the same, but they *all* recognize that Florida ecosystems not only will burn, but that they need recurrent fire to thrive.

We present two of many successful land management programs that show while there are some things about Florida that make it easier to burn; there are many things that compensate to make it every bit as challenging. Both are managed under the jurisdiction of the Saint Johns River Water Management District.

The St Johns River Water Management District (SJRWMD) is located in northeast Florida. Its mission is to "*ensure the sustainable use and protection of water resources for the benefits of the people of the District and the state of Florida*". (Refer to: http://floridaswater.com/overview.html). To accomplish this mission, the District is expected to strike a balance in water needs by educating the public about water conservation, setting rules for water use, conducting research, collecting data, managing land, restoring and protecting water above and below the ground, and preserving natural areas. Regarding land management, Florida Statute 373.1391 states

> "Lands titled to the governing boards of the districts shall be managed and maintained, to the extent practicable, in such a way as to ensure a balance between public access, general public recreational purposes, and restoration and protection of their natural state and condition".

Currently the District manages a multiple use program on 420,000 acres of land. An active fire management program is essential to meeting this legislative charge.

Julington Durbin Preserve

The Julington Durbin Preserve is located within the city limits of Jacksonville, FL (2012 population 840,000) with portions adjacent to Interstate highway 'I-95' (one of the busiest Interstates in the US) (Figure 14).

The preserve is jointly managed by SJRWMD and the City of Jacksonville. It shares a common boundary with hundreds of individual condominiums, each with a picture window facing the lands being managed. Figures 14 and 15 show burns are literally conducted in their backyards. Management of Julington Durbin is not just a fire management story; it is a forest health story. The property is about 2,100 acres (850 ha). Twelve hundred of those acres (~490 ha) are wetlands and swamps along miles of two creeks, but down the middle of the peninsula, there are 900 acres (365 ha) of fire-dependent sandhill plant communities. Sandhills are rare communities dependent upon very frequent fire - once every two to five years. In this case, nearly half of the sandhill had been planted to Ocala sand pine (*P. clausa* var. *clausa*) by the previous owner. One of the primary ecological values of sandhills is the high species diversity of the groundcover. Sand pine typically forms dense stands, the deep shade diminishes groundcover substantially, degrading the conservation value of these sites. In addition, Ocala sand pine is a species that has a crown fire stand replacement regime on a 20-60 year return interval (Cooper 1973), threatening the safety of adjacent human communities with high intensity crown fires.

In 2004, SJRWMD harvested the sand pine. In May 2005, they began burning the sandhills. During that timeframe, the condominiums were under construction and SJRWMD was trying to get as much done as possible before they were occupied. The burns triggered a response from the groundcover that had been suppressed under the sand pines.

(a)

(b)

Figure 14. Aerial view of Julington Durbin Preserve burn units, Jacksonville, Florida. Note proximity of burn units to major interstate highway (a) and condominium development with corresponding recent prescribed burning history (b).

Following burning there was abundant wiregrass recovery and many wildflowers. SJRWMD followed up with burns and mechanical sand pine control in 2008, 2009, and 2012. Because the area is so close to a large human population (nearly one million residents live in Jacksonville), the area is used for recreation by hikers, bikers and horseback riders. Three public schools use the area to train their cross-country teams.

During a burn in May of 2012, fifteen separate people came up to fire staff and thanked them for burning the area. In fact, during one recent burn, one of the neighbors approached the Incident Commander and said they were glad the fire mangers were there and that they were beginning to feel neglected because an adjacent unit next to other residents was being burned more often than the one in his backyard.

District fire managers have established a reputation and an effective relationship with the urban neighbors. SJRWMD staff protect that relationship as if it was gold. During one burn the wind direction began to shift to the point where if it continued, smoke would have begun blowing on one of the condominiums, so a shallow fireline was created through intact groundcover to cut off the burn. The line was rehabilitated and the burn team came back and finished the burn on a day when the smoke would not blow toward the homes. The relationship with the neighbors is too important to tarnish it by burning more acres when the smoke might affect them.

Because the more you burn the easier it gets, SJRWMD staff are trying to burn this area on a two-year rotation. A two-year rotation produces less smoke and has reduced fire behavior, which makes the fire easier to control, and produces positive ecological effects. Figure 15 gives one an idea how fire managers have been able to repetitively burn units near the neighbors.

Figure 15. Proximity of St. Johns River Water Management District fire management personnel to a prescribed fire ignited near condominiums in Jacksonville, Florida.

What Are Our Tools for Success?

In order to be successful, fire managers at SJRWMD use the following seven tools: they: communicate; cooperate with other fire managers and with the neighborhoods; build confidence; manage their smoke; prepare; train and; empower burn teams.

Communicate

SJRWMD uses every tool available to communicate the prescribed fire message. They have general messages on their website (http://www.sjrwmd.com/) about fire-related land management and specific pages about fire management. They have messages about restoration projects that include segments on how important fire is to restoration. They announce individual burns on their website on the day of the burn. They send out annual fire plans to neighboring homeowner associations, and they inform them of individual burns through email and phone trees.

Great Outdoors has their own closed circuit television; they put fire messages on their television system on days when prescribed burns occur. Figure 17 shows one of the signs SJRWMD puts on the property boundary line that announce "*This property is managed with prescribed fire in order to maintain its ecology and to reduce the risk of wildfire*", so that new neighbors are aware the neighboring land is managed with fire when they have not lived there long enough to observe a burn. Kiosks are located at the entrance of the property and adjacent to restoration projects that inform visitors about the importance of fire management and the role it plays in restoration.

Fire managers work with the media during Florida's annual 'Prescribed Fire Awareness Week'. They joined together to hold a mini fire academy for members of the media. Figure 18 shows a reporter carrying a drip torch during that event. Once a reporter has had the chance to carry a drip torch and participate in a prescribed fire, he/she has a completely new perspective on the prescribed fire message.

Figure 17. Prescribed fire sign on boundary.

Figure 18. News media participating in a prescribed burn during prescribed fire awareness week.

Cooperate

Fire managers collaborate with all federal, state, and local agencies in the area who work with fire. They regularly trade personnel and equipment to complete prescribed burns. Typically, SJRWMD fire management does not request reimbursement because everyone knows that they may loan another agency an engine and two people today, but next week they will borrow a tractor plow and an operator.

Fire managers also cooperate with agency partners and our neighboring communities when a community wants to become *Firewise*. Initial time and energy spent helping a neighborhood become *Firewise* is a sound long-term investment for the District specifically and fire management more broadly.

Build Confidence

When fire managers determine the resource needs for an individual burn, the number of suppression resources is determined by two things: what is needed to catch a potential escape and what is needed so every neighbor who looks out their window and sees fire, also sees someone managing it. In the WUI, fire managers often include big red structural engines, not because they will actually be used to attack fire, but because residents feel safer when they see a big red truck. The big red fire trucks are one of the reasons our efforts to cooperate are so important.

Manage Smoke

Proper smoke management is crucial, especially when burning in the WUI. Fire managers should not compromise on dispersion, or wind forecasts. Skimping a few points on the dispersion index (Lavdas 1986) (A numerical index that estimates the atmosphere's capacity to disperse smoke from prescribed burning), or a few miles per hour on the wind, may let you burn on an extra day or two, but the potential problems outweigh the benefits. SJRWMD fire managers also mop up aggressively when necessary in the WUI as a demonstration of their commitment to minimizing impacts to neighbors.

After several burns, large fuels such as old stumps are fewer, making mop up easier. Smoke-related complaints or accidents would erode that public confidence that SJRWMD has worked so hard to build. This level of care (and corresponding costs) may sound like 'pie in the sky' to more cavalier burners, but the public record attests to the fact there have been *no* complaints and *no* accidents to date on any SJWND properties where 30,000+ acres are treated annually – an enviable record indeed!.

Prepare

SJRWMD fire mangers prepare perimeter fire boundaries. If residential neighbors have heavy fuels, fire managers will mow or drum chop a 30 to 60 foot wide firebreak inside the boundary to reduce the risk of escape. SJRWMD fire managers always have a hard fire line (a line down to mineral soil) next to a boundary, and in the WUI, we modify the adjacent fuels to convert the fuel from a shrub fuel model to a litter or grass fuel model in order to reduce fire behavior at the edge.

Train

A great American football coach once said, '*We practice like we play*'. Fire managers everywhere need to train like they work. If we are going to pool agencies for fire management, we need to pool agencies in training so we have commonality before we reach critical points. SJRWMD is blessed with very little turnover[3], so costs otherwise associated with repeated initial training can be spent on more advanced training. SJRWMD fire managers are encouraged to obtain additional fire training each year in addition to the required refresher training.

Empower Fire Managers

Because SJRWMD has so little turnover, their fire managers are experienced and are therefore empowered to make decisions. A person managing a fire next to a subdivision or an interstate highway should not have to call someone in an office fifty miles away in order to

[3] Coauthor Miller will be happy to discuss the reasons for low staff turnover or other aspects of the above examples. He can be reached at srmiller@sjrwmd.com.

implement a decision during a fire. More process does not necessarily make for better or safer burning. Hiring, training, retaining and empowering effective fire managers makes for more, better, and safer burns. There is significant benefit resulting from employees who have been working in a position through multiple fire rotations. Fire managers who have burned the same unit previously are much more effective at recognizing and responding to anomalies, which if left unattended often grow in significance.

CONCLUSION

We have described the pivotal role fire has, and continues to play in shaping Earth's vegetative mosaic. Currently many ecosystems are suffering from inappropriate fire regimes. We give some examples, but deliberately chose not to showcase the myriad rare flora and fauna worldwide that are in trouble, the overwhelming majority because of a lack of appropriate fire. Much of our discussion hinged around the use of fire for hazard reduction because upper management and elected officials are focusing on wildfire suppression to protect public health and safety rather than on ecosystem restoration even though that strategy will not solve either issue. But embarking on a policy to annually treat a substantial portion of the landscape with fire will solve both. If society does not want fire-prone ecosystems to disappear, appropriate fire regimes will have to be returned to the landscape in meaningful doses. To facilitate this undertaking, we urge the public to insist that policymakers and upper-level federal and state natural resource managers facilitate the application of prescribed fire by making adequate funding available and backing on-the-ground staff decisions about where, when, and how to execute burns and any necessary pretreatments. Enough knowledge exists to immediately begin the process and we present a fire management program that shows fire can be successfully reintroduced in long unburned ecosystems within a very complex and challenging location, the WUI. Gaps in the knowledgebase can be closed through adaptive management as practitioners develop expertise. Fire managers have a challenging future to put appropriate fire regimes on the ground at the landscape level in an ever expanding WUI while at the same time complying with regulatory restrictions intended to safeguard public health. To that end, regulatory agencies need to recognize fire is inevitable and that facilitating its use will help achieve public health and safety goals while restricting its use will have the opposite outcome.

The longer we procrastinate, the more difficult the task. If all of us would just stop making excuses, stop waiting for someone else to make decisions for us, be proactive, get off our derrieres, and put 'getting fire on the ground' at the top of our daily priority list, society would soon see huge strides in ecosystem restoration and a concomitant decrease in damaging wildfires. It is incumbent upon all natural resource managers, city, state and federal administrators and elected officials that they leave the land in better shape than they found it.

ACKNOWLEDGMENTS

The coauthors have modified presentations given in the past for use in this chapter. Several Mark Melvin slide presentations were also modified for use herein with his permission. Authors including Jeff Hardesty, Ron Myers, and Steve Pyne who, although we cite them, deserve much more credit as they have shaped our thinking. We are indebted to Jim Brenner, Fire Management Administrator for the Florida Forest Service and Mark Melvin with the Joseph W. Jones Ecological Research Center in Newton, GA for thoughtful draft review comments and especially to an anonymous reviewer. We thank Neal Edmondson, Georgia Forestry Commission and Mark Melvin, Joseph Jones Ecological Research Center, for their efforts that resulted in the Georgia Fire Council underwriting the cost of color figures.

REFERENCES

Abram NJ, Gagan MK, McCulloch MT, Chappell J, Hantoro WS (2003) Coral reef death during the 1997 Indian Ocean dipole linked to Indonesian wildfires. *Science* 301, 952-955.

Abrams MD, Nowacki GJ (2008) Native Americans as active and passive promoters of mast and fruit trees in the eastern USA. *The Holocene* 18(7), 1123-1137.

Alexander ME (1982) Calculating and interpreting forest fire intensities. *Canadian Journal of Botany* 60, 349-357.

Alexander ME, Euler DL (1981) Ecological role of fire in the uncut boreal mixed wood forest. pp 42-64 In Proceedings Boreal Mixedwood Symposium, 14-18 September, 1980 Thunder Bay, Ontario, Canada. Canadian Forest Service Great Lakes Forestry Research Center .COJFRC symposium proceedings 0-P-9 (Sault Ste. Marie, Ontario)

Alvarado E, Mason L, Leighton A, White G, Morishima G, Durgl J, Hardy C, Erickson J, James L, Isaac E (2011) Integration and application of traditional ecological knowledge and modern science for contemporary wildland fire management in tribal lands of North America. FAO at the Vth International Wildland Fire Conference, 9-13 May, 2011, Sun City, South Africa. Forest Resources Development Service. Forest Management Division, Forestry Department, Food and Agriculture Organization of the United Nations Fire Management Working Paper FM/27/E (FAO, Rome, Italy).

Anderson MK (2005) *Tending the wild: Native American knowledge and the management of California's natural resources*. (University of California Press: Berkeley, CA)

Anderson M (2014) Carrying the torch from Arkansas to Zambia. *Nature Conservancy Magazine* October/November, 24-25.

Anonymous (2004) *Prescribed Burning – Methodologies for Safe and Effective Burning: Participant Resource*, Third Edition, Fire and Emergency Services Authority of Western Australia and Department of Conservation and Land Management of Western Australia, Perth.

Archibald S, Staver AC, Levin SA (2012) Evolution of human driven fire regimes in Africa. *Proceedings of the National Academy of Sciences* 109, 847-852.

Association for Fire Ecology, International Association of Wildland Fire, Tall Timbers Research Station, and The Nature Conservancy (2013) The merits of prescribed fire outweigh potential carbon emission effects.

Altangerel K, Kull CA (2013) The prescribed burning debate in Australia: conflicts and compatibilities. *Journal of Environmental Planning And Management*. 56 (1), 103-120.

Bartram W (1791) *The Travels of William Bartram.* Naturalist Edition. (Ed. F Harper) Reprinted 1998 (University of Georgia Press, Athens, GA)

Belcher CM, Collinson ME, Scott AC (2013) A 450 million year record of fire. pp229-249. In *Fire phenomena and the Earth System: an interdisciplinary guide to fire science (Ed CM Belcher) 1st edition.* (J. Wiley & Sons Ltd, Chichester, England)

Berna F, Goldberg P, Horwitz J, Holt S, Bamford M, Chazan M (2012) Microstratigraphic evidence of *in situ* fire in Archeulean strata of Wondererk Cave, Northern Cape province, South Africa. *Proceedings of the National Academy of Sciences, US* 109 (20), E1215-E1220.

Bigges W (1586) A Summarie And True Discourse Of Sir Francis Drakes West-Indian Voyage. Accompanied With Christopher Carleill, Martin Frobusher, Francis Knollis, With Many Other Captains And Gentlemen. Wherein Were Taken, The Townes Of Saint Jago, Sancto Domingo, Cartagena And Saint Augustine.' (London: Printed for Nicholas Bourne)

Biswell HH (1989) *Prescribed Burning in California Wildlands Vegetation Management.* pp 79-95. (University of California Press, Berkeley, CA)

Blank GB (2006) Two pines and the native tradition of burning North American landscapes. Pps. 299-304 In 'Proceedings of Cultural Heritage And Sustainable Forest Management--the Role Of Traditional Knowledge, 8-11 June 2006, Florence, Italy: Volume 2. Warszawa: Ministerial Conference On The Protection Of Forests In Europe.

Blocker S (1875) Burning of woods—when and how to know. 1996 reprint. Fire Ecology Field Office, US Fish and Wildlife Service, Tall Timbers Research Station, Tallahassee, Florida, *Fire Flame Tips* 4(1): 6.

Bond WJ, Keeley JE (2005) Fire as a global 'herbivore': the ecology and evolution of flammable ecosystems. *Trends in Ecology and Evolution* 20, 387-394.

Bond WJ, Midgley GF (2000) A proposed CO_2-controlled mechanism of woody plant invasion in grasslands and savannas. *Global Change Biology* 6(8), 865–869.

Bond WJ, Midgley JJ (2001) Ecology Of Sprouting In Woody Plants: The Persistence Niche. *Trends In Ecology & Evolution* 16(1), 45-51.

Bond WJ, Parr CH (2010) Beyond the forest edge: ecology, diversity and conservation of the grassy biomes. *Biological Conservation* 143(10), 2395-2404.

Botkin DB (1995) *Our Natural History: The Lessons of Lewis and Clark.* (G.P. Putnam's Sons. NY)

Bowman DMJS, Zhang Y, Walsh A, Williams RJ (2003) Experimental comparison of four remote sensing techniques to map tropical savanna fire-scars using Landsat-TM imagery. *International Journal of Wildland Fire* 12(3-4), 341-348.

Brando PM, Balchd JK Nepstade DC, Mortonf DC, Putzg FE, Coeb MT, Silvérioa D, Macedob MN, Davidsonb EA Nóbregaa CC, Alencara A, Soares-Filhoj BS (2014) Abrupt increases in Amazonian tree mortality due to drought–fire interactions. *Proceedings of the National Academy of Sciences of the United States* 111 (17), 6347-6352.

Brain CK, Sillent A (1988) Evidence from the Swartkrans cave for the earliest use of fire. *Nature* 336, 464 – 466.

Brose PH, Dey DC, Waldrop TA (2014) The fire-oak literature of eastern North America: synthesis and guidelines. USDA Forest Service, Northern Research Station. General Technical Report NRS–135 (Newtown Square, PA)

Brown JK (2000) Chapter 1: Introduction. In 'Wildland Fire in Ecosystems: Effects of Fire on Flora.' (Eds. JK Brown, JK Smith) pp 1-8. USDA Forest Service, Rocky Mountain Research Station. General Technical Report RMRS-GTR-42-vol. 2 (Ogden, UT).

Brown JK, Smith JK (eds.) (2000) Wildland fire in ecosystems: effects of fire on flora. USDA Forest Service, Rocky Mountain Research Station. General Technical Report RMRS-GTR-42-vol. 2 (Ogden, UT).

Buckley M, Beck N, Bowden P, Miller ME, Hill B, Luce C, Elliot WJ, Enstice N, Podolak K, Winford E, Smith SL, Bokach M, Reichert M, Edelson D, Gaither J (2014) 'Mokelumne watershed avoided cost analysis: Why Sierra fuel treatments make economic sense'. A report prepared for the Sierra Nevada Conservancy, The Nature Conservancy, and USDA Forest Service. (Sierra Nevada Conservancy, Auburn, California). Available at: http://www.sierranevadaconservancy.ca.gov/mokelumne.

Buma B, Brown C, Fontaine J, Donato D, Johnstone J (2014) Exploring the limits of resilience: impacts of changing disturbance regimes on serotinous plant populations and communities. *Bioscience* 63(11), 866-876.

Butz R (2009) Traditional fire management: historical fire regimes and land use change in pastoral East Africa. *International Journal of Wildland Fire* 18, 442-250.

Byram GM (1958) Some basic thermal processes controlling the effects of fire on living vegetation. USDA, Forest Service, Southeastern Forest Experiment Station Research Note 114 (Asheville, NC)

Campbell F (2011) *Tempered by Fire,* Bushfire Front , Perth, WA.

Catesby M (1731 *The Natural History of Carolina, Florida and the Bahama Islands: Containing the Figures of Birds, Beasts, Fishes, Serpents, Insects, and Plants*. Two volumes. (London, for the author)

Christensen NL, Burchell RB, Liggett A, Simms EI (1981) The structure and development of pocosin vegetation. Pages 43-61 In 'Proceedings of conference. pocosin wetlands: an integrated analysis of coastal plain freshwater bogs in North Carolina', 3-4 January 1980, Beaufort NC (ED. CJ Richardson)

Cochrane MA (2002) Spreading Like Wildfire—Tropical Forest Fires in Latin America and the Caribbean: Prevention, Assessment and Early Warning (United Nations Environment Programme, Panama)

Cochrane MA (2003) Fire science for rainforests. *Nature* 421, 913-919.

Cochrane MA, Moran CJ, Wimberly MC, Baer AD, Finney MA, Beckendorf KL, Eidenshink J, Zhu Z (2012) Estimation of wildfire size and risk changes due to fuels treatments. *International Journal of Wildland Fire* 21, 357-367. http://dx.doi.org/10.1071/WF11079.

Collingwood GH (1940) Pond pine. *American Forests* 46, 462 463.

Cooper RW (1973) Fire and sand pine. *In* Proceedings, Sand Pine Symposium. p. 207-212. USDA Forest Service, Southeastern Forest Experiment Station, General Technical Report SE-2. (Asheville, NC)

Covington W (2014) accessed on line at: http://news.yahoo.com/two-wildfires-everyone-talking-op-ed-154416902.html?soc_src=mediacontentstory.

Cram D; Baker T; Boren J (2006) Wildland fire effects in silviculturally treated vs. untreated stands in New Mexico and Arizona. USDA Forest Service, Rocky Mountain Research Station, Research Paper RMRS-RP-55 (Ogden, UT)

Cronon W (1983) '*Changes in the Land: Indians, Colonists, and the Ecology of New England.*' (Hill and Wang, New York)

Cumming JA (1964) Effectiveness of prescribed burning in reducing wildfire damage during periods of abnormally high fire danger. *Journal of Forestry* 62, 535-537.

Dale L, Aplet G, Wilmer B (2005) Wildland fire use and cost containment: a Colorado case study. *Journal of Forestry* 103, 314-318.

DeBano LF, Neary DG, Ffolliott PF (1998) *Fire's Effects on Ecosystems*. (John Wiley & Sons Inc., NY)

de Laudonniere R (1587) A notable hiitorie containing foure voyages made by certain French Captains into Florida.. In 'The Principal Navigations Voyages Traffiques and Discoveries of the English Nation' 1904. (Ed. & Translator R Hakluyt)), Volumes 8 and 9. (McMillan Company, NY)

Denevan WM (1992) 'The Pristine Myth: The Landscape of the Americas in 1492.' *Annals of the American Geographers* 82(3), 369-385.

Devlin V (2014) CSKT official says forests managed better with less on reservation. (MISSOULIAN [MT] April 12 edition). Available at http://missoulian.com/news/local/cskt-official-says-forests-managed-better-with-less-on-reservation/article_9d2b7bd6-c1d4-11e3-8147-001a4bcf887a.html

Diekmann L, Panich L, Stiplen C (2007) Native American Management and The Legacy of Working Landscapes In California. *Rangelands* 29(3), 46-50.

DeVivo M.S. 1991. Indian use of fire and land clearance in the southern Appalachians. pps 306-310 In USDA, Forest Service, Southeastern Forest Experiment Station General Technical Report SE-GTR-69. (Asheville, NC)

Douglas J, Mills TJ, Artly D, Ashe D, Bartuska A, Black RL, Coloff S, Cruz J, Edrington M, Edwardson J, Gale RT, Goodman SW, Hamilton L, Landis R, Powell B, Robinson S, Schuster RJ, Stahlschmidt PK, Stires J, Van Wagtendonk J (2001) Review and update of the 1995 federal wildland fire management policy. (US Department of the Interior and US Department of Agriculture, Washington, D.C.)

Eldredge IF (1911) Fire problems on the Florida National Forest. pp 166-170. In 'Proceedings, Society of American Foresters annual meeting (Society of American Foresters, Washington DC)

Emery MR, Wrobel A, Hansen MH, Dockry M, Moser WK, Stark J, Gilbert JH (2014) Using traditional ecological knowledge as a basis for targeted forest inventories: Paper birch (*Betula papyrifera*) in the US Great Lakes Region. *Journal of Forestry* 112(2), 207-214.

Filho J, Horridge M (2011) Ethanol expansion and indirect land use change in Brazil. General Paper No G-218. Center of Policy Studies Morash (University Clayton, Victoria, Australia)

Flannigan M, Cantin A, de Groot W, Wotton M, Newbery A, Gowman L. 2013. Global wildland fire severity in the 21st century. *Forest Ecology and Management* 294, 54-61.

Flatley WT, Lafon CW, Grissino-Mayer HD, LaForest LB (2013) Fire history, related to climate and land use in three southern Appalachian landscapes in the eastern United States. *Ecological Applications* 23(6), 1250–1266.

Frost CC (1998) Presettlement fire frequency regimes of the United States: a first approximation. pp 70-81. In 'Fire in ecosystem management: shifting the paradigm from suppression to prescription'; proceedings of the 20th Tall Timbers fire ecology conference. (Tall Timbers Research Station, Tallahassee, FL)

Gammage B (2011) *The Biggest Estate on Earth: How Aborigines Made Australia*. (Allen & Unwin, Sydney)

Garren KH (1943) Effects of fire on vegetation of the southwestern United States. *Botanical Review* 9, 617-654.

Giglio L; Csiszar I; Justice CO (2006) Global distribution and seasonality of active fires as observed with the Terra and Aqua Moderate Resolution Imaging Spectroradiometer (MODIS) sensors *Journal of Geophysical Research* 111, G02016 doi:10.1029/2005JG000142.

Giglio L, Randerson JT, van der Werf GR, Kasibhatla, PS, Collatz, GJ, Morton DC, DeFries RS (2010) Assessing variability and long-term trends in burned area by merging multiple satellite fire products. *Biogeosciences* 7, 1171–1186.

Giglio L, Randerson JT, van der Werf GR. (2013) Analysis of daily, monthly, and annual burned area using the fourth-generation global fire emissions database (FED4). *Journal of Geophysical Research*: *Biogeosciences* 118 (1), 317-328.

Glitzenstein JS, Streng DR, Wade DD (2003) Fire frequency effects on longleaf pine (*Pinus palustris* P. Miller) vegetation in South Carolina and Northeast Florida, US. *Natural Areas Journal* 23(1), 22-37.

Gramley M (2005) Fire in the South. Southern Group of State Foresters. Available at www.southernwildfirerisk.com/reports/FireInTheSouth1.pdf

Gruell GE (1985) 'Fire on the early western landscape: an annotated record of wildland fires 1776-1900.' *Northwest Science* 59(2), 97-107.

Guyette RP, Dey DC, Stambaugh MC, Muzika RM (2006) Fire scars reveal variety and dynamics of eastern fire regimes. Pp. 20-39. In 'Proceedings Fire in eastern oak forests: delivering science to land managers conference,' 15-17 November 2005, Columbus, OH. USDA Forest Service, Northern Research Station, General Technical Report NRS-P-1 (Newtown Square, PA)

Hallam, SJ (2014): *Fire and Hearth. A study of Aboriginal use and European usurpation in south-western Australia.* Revised Edition. (UWA Publishing, Perth, Western Australia)

Hardesty J, Myers RL, Fulks W (2005) Fire, ecosystems, and people: a preliminary assessment of fire as a global conservation issue. *The George Wright Forum* 22:78-87.

Hardy CC, Menakis JP, Long DG, Brown JK, Bunnell DL (1998). Mapping historic fire regimes for the western United States: integrating remote sensing and biophysical data. pp. 288-300. In 'Proceedings of the seventh biennial forest service remote sensing applications conference', 6-9 April, 1998, Nassau Bay, TX. (American Society for Photogrammetry and Remote Sensing, Bethesda, MD)

Hare RC (1961) Relation of bark thickness and species to cambial fire resistance USDA Forest Service, Southern Forest Experiment Station, Office Report, 16 Pages. Available at http://www.fireresearchinstitute.org/

Hakluyt R (1582) '*Divers Voyages touching the Discovery of America and the Islands adjacent.*' Republished 2010 (Ed. JW Jones) (Cambridge Library Collection, Hakluyt First Series, Cambridge, UK).

He T, Pausas JG, Belcher CM, Schwilk DW, Lamont BB (2012) Fire-adapted traits of *Pinus* arose in the fiery Cretaceous *New Pytologist* 194(3), 599–883.

Hermann SM, Kush JS (2010) Initial Effects From Re-introducing Fire in Alabama Montane Longleaf Stands: Fifty Years Since Last Burn. Pp.263-266. In 'Proceedings 14th Biennial Southern Silvicultural Research Conference', USDA, Forest Service, Southern Research Station, General Technical Report SRS-GTR 121 (Asheville, NC)

Higgins KF, Kruse AD, Piehl JL (1989) Effects of fire in the Northern Great Plains. US Fish and Wildlife Service and Cooperative Extension Service, South Dakota State University, Brookings, South Dakota. Extension Circular 761. Jamestown, ND: Northern Prairie Wildlife Research Center Online at http://www.npwrc.usgs.gov/resource/habitat/fire/index.htm (Version 16MAY2000)

Hood SM (2010) Mitigating old tree mortality in long-unburned, fire dependent forests: A synthesis. USDA Forest Service, Rocky Mountain Research Station, General Technical Report. RMRS-GTR 238 (Fort Collins, CO)

Hough FB (1877) Report upon Forestry. [Excerpt reproduced in 1934 *Journal of Forestry* 32(7), 703]

Huffman JM (2006) Historical fire regimes in southeastern pine savannas. PhD thesis, Louisiana State University, Baton Rouge, LA.

Hutchinson TF, Long RP, Rebbeck J, Sutherland EK, Yaussy DA (2012) Repeat prescribed fires alter gap-phase regeneration in mixed-oak forests, *Canadian Journal of Forest Research* 42, 303-314.

Ito A (2011) Mega fire emissions in Siberia: potential supply of bioavailable iron from forests to the ocean *Biogeosciences* 8, 1679–1697.

Jurskis V, Underwood R. (2013) Human fires and wildfires on Sydney sandstones: History informs fire management. *Fire Ecology* 9, 8-24.

Kallander HR (1958) Prescribed burning of ponderosa pine Fort Apache Indian Reservation, Arizona. pp. 46-48. In 'Proceedings of the Annual Arizona Watershed Symposium,' Number 2.

Kauhanen TA, Mikkola K, Demidova N, Ogibin B (2014) Episodic, patchy disturbances characterize an old-growth Picea abies dominated forest landscape in northeastern Europe. *Forest Ecology and Management* 320, 96–103.

Keeley JE (2009) Fire intensity, fire severity and burn severity: a brief review and suggested USge. *International journal of wildland fire* 18, 116-126.

Keeley JE, Aplet GH, Christensen NL, Conard SC, Johnson EA, Omi PN, Peterson DL, Swetnam TW (2009) Ecological foundations for fire management in North American forest and shrubland ecosystems. USDA Forest Service, Pacific Northwest Research Station General Technical Report PNW-GTR-779 (Portland, OR).

Keeley JE, Bond WJ, Bradstock RA, Pausas JG, Rundel PW (2011) *Fire in Mediterranean climate ecosystems: ecology, evolution and management*. (Cambridge University Press, New York).

Knoepp JD, Tieszen LL, Fredlund GG (1998) Assessing the vegetation history of three Southern Appalachian balds through soil organic matter analysis. USDA Forest Service, Southern Research Station Research Paper SRS-13 (Asheville, NC)

Komarek EV (1965) 'Fire Ecology–Grasslands and Man'. pp. 169-220. In 'Proceedings: Tall Timbers Fire Ecology Conference,' 18-19 March 1965 Tallahassee. Number 4. (Tall Timbers Research Station. Tallahassee, FL)

Komarek EV (1981) History of prescribed fire and controlled burning in wildlife management in the South. In 'Proceedings of a Symposium Prescribed Fire and Wildlife in Southern Forests', 6-8 April 1981; Myrtle Beach, SC. (Ed. GW Wood). (Belle W. Baruch Forest Science Institute of Clemson University: Georgetown, SC)

Kodas M (2014) The measure of a mega-fire. *Wildfire* 23(1), 12-15.

Kull CA (2002) Madagascar's burning issue: the persistent conflict over fire forest. Environment 44 (3): 8-19.

Kush JS (2009) Obituary for Flomaton: One less virgin longleaf stand (Alabama). *Ecological Restoration*. 27(3), 247-248.

Landers L, Wade D (1994) Disturbance, persistence and diversity of the longleaf pine-bunchgrass ecosystem. pp.182-188. In 'Proceedings of the 1993 SAF National Convention',7-10 November 1993, Indianapolis, IN. SAF publication 94-01. (Society of American Foresters, Bethesda, MD)

Lavdas L (1986) An atmospheric dispersion index for prescribed burning. USDA Forest Service, Southeastern Forest Experiment Station Research Paper SE-256 (Asheville, NC)

Lewis HT (1973) 'Patterns of Indian Burning in California: Ecology and Ethnohistory.' Anthropological Papers No. 1. (Ballena Press, Ramona, CA)

Littell J, Mckenzie D, Peterson DL, Westerling AL (2009) Climate and wildfire area burned in western U.S. ecoprovinces, 1916–2003. *Ecological Applications* 19(4), 2009, pp. 1003–1021.

Little EL Jr, Dorman KW (1952) Geographic differences in cone -opening in sand pine. *Journal of Forestry* 50, 204-205.

Little EL, Dorman KW (1954) Slash pine (*Pinus elliottii* including South Florida slash pine normenclature and description. USDA Forest Service, Southeastern Forest Experiment Station, Station Paper 36 (Asheville, NC)

Lotan JE (1975) The role of cone serotiny in lodgepole pine forests. Pps. 471-495, In 'Proceedings: Management of lodgepole pine ecosystem,' October, 1973, Pullman, WA, (Washington State University, Pullman, WA)

Mann C (2005) '*1491: New Revelations of the Americas Before Columbus.* With a New Afterword'. (Vintage Books, New York, NY)

Martin RE, Sapsis DB (1992) Fires as agents of biodiversity: pyrodiversity promotes biodiversity. In 'Proceedings Symposium on Biodiversity of Northern California'. 28-30 October 1991, Santa Rosa, CA. Report. 29., pp 150-157. (Wildland Resources Center. Berkeley, CA)

Mason CL, Lippke BR, Zobrist KW, Bloxton TD, Ceder KR, Comnick JF, McCarter JB, Rogers HK (2006) Investments in fuel removals to avoid forest fires result in substantial benefits. *Journal of Forestry* 104, 27-31.

Mason SLR (2000) Fire and Mesolithic subsistence – managing oaks for acrons in northwest Europe? *Paleogegraphy, Paleoclimatology, Paleoscology.*164, 139-150.

Matthews BE (1992) The suburbanization of America and the future of hunting and fishing: how to get people back in touch with the land, or, Rurbanites, the problem of keeping two roosters in the same henhouse. Annual Conference of the Outdoor Writers Association of America 30 June 1992 (Bismarck, ND).

Mattoon WR (1915). Life history of shortleaf pine. Agricultural Bulletin 244. (U.S. Department of Agriculture, Washington, DC).

Maxwell P, Stephen Z, Mahamba SM (2011) Community based fire management; experiences from the FAO funded Project in the provinces of Manicaland and Matabeleland north, Zimbabwe. FAO at the Vth International Wildland Fire Conference, 9-13 May, 201,1Sun City, South Africa. Forest Resources Development Service. Forest Management Division, Forestry Department, Food and Agriculture Organization of the United Nations Fire Management Working Paper FM/27/E (FAO, Rome, Italy)

Meier AR, Saunders MR, Michler CH (2012) Epicormic buds in trees: A review of bud establishment, development and dormancy release. *Tree Physiology* 32, 565–584.

Melvin MA (2012) National prescribed fire use survey report. Coalition of Prescribed Fire Councils Technical Report 01-12 (Joseph W Jones Ecological Research Station, Newton, GA)

Midgley JJ (2000) What are the relative costs, limits and correlates of increased degree of serotiny? *Australian Ecology* 25(1), 65-68.

Miller C (2010) Exploring the roots: A critical review of a 1912 article that appeared in *Forest Quarterly* (10, 184-194) 'Light burning versus forest management in northern California' by R Boeker. *Journal of Forestry* 108, 46.

Miller M (2000) Chapter 2: Fire Autecology in (Brown J K, Smith JK Eds.) Wildland fire in ecosystems: effects of fire on flora. USDA Forest Service, Rocky Mountain Research Station, General Technical Report RMRS-GTR-42-vol. 2 (Ogden, UT)

Mitchell RJ, Liu Y, O'Brien JJ, Elliott KJ, Starr G, Miniat CF, Hiers JK (2014) Future climate and fire interactions in the southeastern region of the United States. *Forest Ecology and Management* available online 7 January 2014.

Morgan B (2013) Logging, palm oil plantations, and Indonesia's summer of smoke. *Earth Island Journal*. Accessed on line 11/27/2013 at http://www.earthisland.org/ journal/index.php/elist/eListRead/logging_palm_oil_plantations_and_indonesia's_summ er_of_smoke/

Moritz MA, Batllori E, Bradstock RA, Gill M, Handmer J, Hessburg PF, Leonard J, McCaffrey S, Odion D, Schoennagel T (2014) Learning to coexist with wildfire. Nature 515, 58-66.

Mouillot F, Field CB (2005) Fire history and the global carbon budget: a 1° x 1° fire history reconstruction for the 20th century. *Global Change Biology* 11, 398-420.

Mutch RW (2014) Fire behavior science and experience: Framing successful megafire solutions for policy makers, politicians, and the public. In 'Proceedings of 4th Fire Behavior and Fuels Conference,' 18-22 February 2013, Raleigh, NC and 1-4 July, 2013 St. Petersburg, Russia. (Eds. DD Wade and RI Fox) (CD ROM) (International Association of Wildland Fire: Missoula, MT).

Myers R (2000) Chapter 7: Fire in tropical and subtropical ecosystems. Pp 161-174. In 'Wildland Fire in Ecosystems: Effects of Fire on Flora.' (Eds. JK Brown, JK Smith). USDA Forest Service, Rocky Mountain Research Station. General Technical Report RMRS-GTR-42-vol. 2 (Ogden, UT)

Myers R (2006) Living with Fire—Sustaining Ecosystems and Livelihoods Through Integrated Fire Management. (The Nature Conservancy Global Fire Initiative: Tallahassee, FL) http://tncfire.org/documents/Integrated_Fire_ Management_ Myers_2006.pdf.

National Commission on Wildfire Disasters (1995) Report of the National Commission on Wildfire Disasters. Washington DC: American Forests.

Ne'eman G, Goubitz S, Nathan R (2004) reproductive traits of *Pinus halepensis* in the light of fire- a critical review. *Plant Ecology* 171(1-2), 53-68.

Nowacki GJ, Abrams MD (2008) The demise of fire and "mesophication" of the Eastern United States. *BioScience*. 58, 123–138.

Noss R (2013) *Forgotten Grasslands of the South: Natural History and Conservation*. (Island Press, Washington, DC).

Ogle CE (1920) Light burning. *The Timberman* 21(9), 106-108.

Omi P, Martinson E (2004) Effectiveness of thinning and prescribed fire in reducing wildfire severity. USDA Forest Service, Pacific Southwest Research Station, General Technical Report PSW- GTR-193.

Putz FE (2003) Are rednecks the unsung heroes of ecosystem management? *WildEarth* 13(2/3), 10-14.

Pyne SJ (1982) *Fire in America: A Cultural History of Wildland and Rural Fire.* (Princeton University Press, Princeton, NJ)

Pyne SJ (1995) *World Fire: The Culture of Fire on Earth.* (University of Washington Press: Seattle, WA)

Pyne SJ (2004) *Tending Fire: Coping With America's Wildfires.* (Island Press: Washington, DC)

Pyne SJ (2007) Problems, paradoxes, paradigms: triangulating fire research. *International Journal of Wildland fire* 16, 272-276.

Pyne SJ (2010) *America's Fires: A Historical Context for Policy and Practice.* (Island Press, Washington, DC)

Pyne SJ (In Press) *To the last smoke.* Refer to: http://firehistory.asu.edu/florida-a-fire-mosaic/

Raymond C, Peterson D (2005) Fuel treatments alter the effects of wildfire in a mixed-evergreen forest, Oregon. *Canadian Journal of Forest Research* 35, 2981-2995.

Reifsnyder WE, Herrington LP, Spalt KW (1976) thermophysical properties of bark of shortleaf, longleaf, and red pine. Yale University, School of Forestry Bulletin 70.

Reynolds RT, Sánchez Meador AJ, Youtz JA, Nicolet T, Matonis MS, Jackson PL, DeLorenzo DG, Graves AD (2013) Restoring composition and structure in Southwestern frequent-fire forests: A science-based framework for improving ecosystem resiliency. USDA Forest Service, Rocky Mountain Research Station, General Technical Report RMRS-GTR-310 (Fort Collins, CO)

Robinson J (1987) The role of fire on earth: A review of the state of knowledge and a systems framework for satellite and ground-based observations. PhD Thesis University of California at Santa Barbara. Cooperative Thesis No. 112.

USFS Rocky Mountain Research Station (2000-2012). Wildland fire in ecosystems: effects of fire on fauna, flora, cultural resources and archeology, soils and water, air, and nonnative invasive species. USDA Forest Service, Rocky Mountain Research Station General Technical Report RMRS-GTR 42, volumes 1-6. (Fort Collins, CO)

Rogers J (2014) Climate control. *Nature Conservancy Magazine* April/May, 46-49.

Roos CI, Bowman DMJS, Balch JK, Artaxo P, Bond WJ, Cochrane M, D'Antonio CM, DeFries R, Mack M, Johnston FH, Krawchuk MA, Kull CA, MoritzMA, Pyne S, Scott AC, Swetnam TW (2014) Pyrogeography, historical ecology, and the human dimensions of fire regimes *Journal of Biogeography* 41(4),833–836.

Roos CI, Swetnam TW (2012) A 1414-year reconstruction of annual, multidecadal, and centennial variability in area burned for ponderosa pine forests of southern Colorado Plateau region, Southwest US. *The Holocene* 22, 281-290.

San-Miguel-Ayanz J, Moreno JM, Camia A 2013. Analysis of large fires in European Mediterranean landscapes: Lessons learned and perspectives. *Forest Ecology and Management* 294, 11-22.

Sargent CS (1884). *Report on the forests of North America (exclusive of Mexico).* Volume 9 of the Tenth Census of the United States 1880. (Government Printing Office: Washington, DC)

Saveland JM (1987) Using prescribed fire to reduce the risk of large wildfires: A break-even analysis. pp 119-122. In 'Proceedings of Ninth Conference on Fire and Meteorology,' 21-24 April, 1987, San Diego, CA. (American Meteorological Society, Boston, MA)

Scheiter S, Higgins SI, Osborne CP, Bradshaw C, Lunt D, Ripley BS, Taylor LL, Beerling DJ (2012) Fire and fire-adapted vegetation promoted C4 expansion in the late Miocene. *New Phytologist* 195, 653-666.

Schiff AL (1962) *Fire And Water: Scientific Heresy In The Forest Service.* (Harvard University Press, Cambridge, MA)

Scott AC, Bowman DMJS, Bond WJ, Pyne SJ, Alexander ME (2014) *Fire on Earth: An Introduction.* (Wiley-Blackwell, Chichester, England))

Scott AC, Jones TJ (1994) The nature and influence of fires in Carboniferous ecosystems. *Palaeogeography, Paleoclimatology, Paleoecology* 106, 91-112.

Shlisky A, Waugh J, Gonzalez P, Gonzalez M, Manta M, Santoso H, Alvarado E, Ainuddin Nuruddin A Rodriguez-Trejo DA, Swaty R, Schmidt D, Kaufmann M, Myers R, Alencar A, Kearns F, Johnson D, Smith J, Zollner D, Fulks W (2007) Fire, ecosystems and people: threats and strategies for global biodiversity conservation. GFI Technical Report 2007-2. (The Nature Conservancy: Arlington, VA)

Silva JS, Rego F, Fernandes P, Rigolot E (Eds) (2010) Towards Integrated Fire Management – Outcomes of the European Project Fire Paradox. European Forest Institute Research Report 23. (European Forest Institute, Joensuu, Finland)

Stambaugh ML, Marschall JM, Guyrttr RP (2014) Linking fire history to successional changes of xeric oak woodlands *Forest Ecology and Management* 320, 83–95.

Stanturf JL, Goodrick SA (2013) Chapter 17. Fire. in 'The southern forest futures project: technical report'. (Eds D Wear, D N Greis J) USDA Forest Service, Southern Research Station, General Technical Report SRS-GTR-178. (Asheville, NC)

Steelman TA, McCaffrey SM (2011) What is limiting more flexible fire management—public or agency pressure? *Journal of Forestry* 109 (8), 454-461.

Stephens SL, Agee JK, Fulé PZ, North MP, Romme WH, Swetnam TW, Turner MG (2013) Managing forests and fire in changing climates. *Science* 342, 41-42.

Stephens SL, Burrows N, Buyantuyev A, Gray RW, Keane RE, Kubian R, Liu S, Seijo F, Shu L, Tolhurst KG., van Wagtendonk JW (2014) Temperate and boreal forest mega-fires: characteristics and challenges. *Frontiers in Ecology and the Environment* 12, 115–122.

Stoddard H (1961) The use of fire on southeastern game lands. [1935]. In 'The Cooperative Quail Study Association: May 1, 1931 - May 1, 1943', (Eds. HL Stoddard, HL Beadel, EV Komarek) pp. 47-63. Tall Timbers Miscellaneous Publication No. 1. (Tall Timbers Research Station, Tallahassee, FL)

Sugihara NG, Van Wagendonk JW, Shaffer KE, Fites-Kaufman J, Thode AF (Eds). (2006) *Fire in California's ecosystems.* (University of California Press, Berkley, CA)

Sun C, Tolver B (2012) Assessing administrative laws for forestry prescribed burning in the southern United States. *International Forestry Review* 14(3), 337-348.

Underwood R (2007) 'The bushfire disaster in Greece was predictable' URL: http://jennifermarohasy.com/2007/09/the-bushfire-disaster-in-greece-was-predictable-a-note-from-roger-underwood/. Accessed October 6, 2014.

Underwood R (2010) Bushfire politics and management in Australia: where are the gatekeepers? In'Proceedings of 3rd Fire Behavior and Fuels Conference', 25-29 October 2010, Spokane, WA. (Eds. DD Wade, M Robinson) (CD-ROM) (International Association of Wildland Fire: Birmingham, AL)

Underwood R, Packham D, Cheney PO (2008) Bushfires, prescribed burning and global warming. Bushfire Front Occasional paper 1.

USDA Forest Service (2000) Policy Implications of Large Fire Management: A Strategic Assessment of Factors Influencing Costs. Available online at: http://www.fs.fed.us/fire/management/Large_Fire_Mgt.pdf.

Varner JM, Kush JS, Meldahl RS (2000) Ecological Restoration of an Old-growth Longleaf Pine Stand Utilizing Prescribed Fire. In 'Proceedings 21st Tall Timbers Fire Ecology Conference'. 14-16 April 1998, Tallahassee, FL. (Tall Timbers Research Station, Tallahassee, FL)

Varner JM, Hiers JK, Ottmar RD, Gordon DR, Putz FE, Wade DD (2007) Overstory tree mortality resulting from reintroducing fire to long-unburned longleaf pine forests: the importance of duff moisture. *Canadian Journal of Forest Research* 37, 1349-1358.

Vogl RJ (1976) '*A primer of ecological principles.* book One'. (Pyro Unlimited, Cypress, CA)

Wade DD (1981) Some M*elaleuca* - fire relationships including recommendations for homesite protection. pp.29-35. In 'Proceedings Melaleuca Symposium', 23-24 September 1981, Ft. Myers, FL. (Tallahassee, FL)

Wade DD (1987) Linking fire behavior to its effects on living plant tissue. pp.112-116. In 'Proceedings of the 1986 Society of American Foresters Annual Convention, 5-8 October 1986, Birmingham, AL. (Society of American Foresters; Bethesda, MD)

Wade DD (1991) High intensity prescribed fire to maintain *spartina* marsh at the urban/wildland interface. Pp. 211-216. In 'Proceedings 17th Tall Timbers Fire Ecology Conference,' 18-21 May 199, Tallahassee, FL (Tall Timbers Research Station, Tallahassee, FL)

Wade D (1996) Longleaf pine and fire - a marriage of necessity. Pp. 9-11. In 'Proceedings Longleaf Alliance Conference, 17-19 September, 1996, Mobile, AL. Auburn Univ. School of Forestry Report No. 1. (Auburn, AL)

Wade DD (2011) Your fire management career—make it count! *Fire Ecology* 7(1), 107-122. doi: 10.4996/fireecology.0701107.

Wade DD (2014) Letter to the editor; A need for accurate national prescribed fire statistics. *The Forestry Source* 19(1), 7. (Society of American Foresters: Bethesda, MD.

Wade DD, Brock BL, Brose PH, Grace JB, Hoch GA, Patterson WA III (2000) Chapter 4: Fire in eastern ecosystems. Pp 53-96. In 'Wildland Fire in Ecosystems: Effects of Fire on Flora.' (Eds. JK Brown, JK Smith). USDA Forest Service, Rocky Mountain Research Station. General Technical Report RMRS-GTR-42-vol. 2 (Ogden, UT).

Wade D, Custer G, Thorsen J, Kaskey P, Kush J, Twomey B, Voltolina D, (1998) Reintroduction of fire into fire-dependent ecosystems: some southern examples. pp. 94-98. In 'Proceedings 20th Tall Timbers Fire Ecology Conference', 7-10 May 1996, Boise, ID, (Tall Timbers Research Station: Tallahassee, FL).

Wade DD, Ewel J, Hofstetter R (1980) Fire in south Florida ecosystems. USDA Forest Service, Southeastern Forest Experiment Station, General Technical Report SE-17 (Asheville, NC).

Wade DD; Long M. 1979. New legislation aids hazard reduction burning in Florida. *Journal of Forestry* 77(10), 725-726.

Wade D, Miller S, Stowe J, Brenner J (2006) Rx fire laws:_ tools to protect fire: the 'ecological imperative'. pp 233-262. In 'Proceedings of conference Fire in Eastern Oak Forests: Delivering Science to Land Managers', 15-17 November 2005, Columbus, OH. (Ed MB Dickinson). USDA Forest Service, Northern Research Station, General Technical Report NRS-P-1 (Newtown Square, PA)Wade D, Moss S (1999. The economics of prescribed fire: benefits and costs in an urbanizing environment. Presentation at the Sixth Annual Conference of the Wildlife Society; 9 September, 1999. Austin, TX.

Wagle RF, Eakle TW (1979) A controlled burn reduces impact of a subsequent wildfire in a ponderosa pine vegetation type. *Forest Science* 25, 123-129.

Wahlenberg WG (1946) Longleaf Pine - its use, ecology, regeneration, protection, growth and management. Charles Lathrop Pack Forest Foundation (Washington, DC).

Waldrop TA, Goodrick SL (2012) Introduction to prescribed fire in southern ecosystems. USDA Forest Service, Southern Research Station, Science Update SRS-054 (Asheville, NC)

Weaver H (1947) Management problems in the ponderosa pine region. *Northwest Science* 21(4), 160-163.

Weir JR (2009) Conducting Prescribed Fires: A Comprehensive Manual. (Texas A & M University Press, College Station, TX)

Wells BW (1942) Ecological problems of the Southeastern United States Coastal Plain. *Botanical Review* 8, 533-561.

Wenger KF (1958) Silvical characteristics of pond pine. USDA Forest Service, Southeastern Forest Experiment Station, Station Paper 91 (Asheville, NC).

Westerling A, Hidalgo HG, Cayan DR, Swetnam TW (2006) Warming and earlier Spring increase western US forest wildfire activity. *Science* 313, 940- 943.

Whelan RJ (1995) *The ecology of fire*. (Cambridge University Press, Melbourne).

White BJ, Underwood RJ (1974): *Regeneration in the karri forest community*. Forests Department of Western Australia.

Wilent S 2014a. Lawsuits galore. Is there a solution? *The Forestry Source* 19(8), 2.

Wilent S 2014b. Fire funding fix stalled in congress. *The Forestry Source* 19(9), 1, 4.

Williams G (2003) 'References on The American Indian Use Of Fire In Ecosystems' http://www.wildlandfire.com/docs/2003_n_before/Williams_Bibliography_Indian_Use_of_Fire.pdf.

Williams J (2007) The mega-fire reality: re-directing protection strategies in fire prone ecosystems. Keynote address given at the BUSHFIRE CRC Forum 'Are big fires inevitable?' 27 February 2007,Canberra, Australia.

Williams J, Albright D, Hoffmann AA, Eritsov A, Moore PF, Carlos Mendes De Morais J, Leonard M, San Miguel-Ayanz J, Xanthopoulos G, van Lierop P (2011) Findings and implications from a coarse-scale global assessment of recent selected mega-fires. FAO at the Vth International Wildland Fire Conference, 9-13 May, 201,1Sun City, South Africa. Forest Resources Development Service. Forest Management Division, Forestry Department, Food and Agriculture Organization of the United Nations Fire Management Working Paper FM/27/E (FAO, Rome, Italy).

Williams J 2013. Exploring the onset of high-impact mega-fires through a forest land management prism. *Forest Ecology and Management*. 294, 4–10.

Wright HE Jr, Heinselman ML (1973) The ecological role of fire in natural conifer forests of western and northern North America – introduction. *Quaternary Review* 3, 319-328.

In: Current International Perspectives
Editors: Brigitte Leblon and Martin E. Alexander
ISBN: 978-1-63463-682-7

Chapter 2

USE OF REMOTE SENSING IN WILDFIRE MANAGEMENT

***Brigitte Leblon*[1*], *Laura Bourgeau-Chavez*[1,2] *and Jesús San-Miguel-Ayanz*[3]**

[1]Forestry and Environmental Management, University of New Brunswick, Fredericton, New Brunswick, Canada
[2]Michigan Tech Research Institute, Michigan Technological University, Ann Arbor, Michigan, US
[3]Forest Resources and Climate Unit, Institute for Environment and Sustainability, Ispra, Italy

ABSTRACT

Our chapter presents a review of the use of remote sensing technologies in wildfire management, with a particular emphasis on its applicability to fuel moisture monitoring, fire detection, and burn scar mapping. Remote sensing of fuel moisture was first done with NOAA-AVHRR NDVI images, but NDVI is more related to vegetation greenness rather than water stress. NOAA-AVHRR surface temperature images were also used, alone or in association with NDVI images. Both kinds of images have limited availability due to cloud cover. This is not the case for synthetic aperture radar (SAR) images provided by active sensors, but their use in fuel moisture monitoring is still experimental, because of the noisy effects of several factors. The recent availability of polarimetric radar images opens new possibilities of using radar images in fuel moisture monitoring. With respect to active fire detection and burnt area mapping, remote sensing applications can be considered operational. Active fire detection is mainly based on the use of mid-infrared imagery acquired by polar orbital satellite sensors (e.g., AVHRR/ATSR and MODIS) or sensors onboard geostationary satellites (e.g., GOES or Meteosat). Burn scar mapping is based on the use of optical imagery of different spectral and spatial resolutions, but SAR images are also used in areas where persistent cloud coverage prevents the efficient use of optical imagery.

* Corresponding Author address: University of New Brunswick Faculty of Forestry and Environmental Management, Fredericton, E3B 5A3, Canada. Email: bleblon@unb.ca.

Keywords: Fire, fire detection, fire scar, fuel moisture, optical, polarimetric radar, remote sensing, radar, thermal infrared

INTRODUCTION

Wildfire is one of the most prominent disturbance factors in several forest and grassland ecosystems (e.g., Pyne 1996; Leblon and Bourgeau-Chavez 2013; Scott et al. 2014). Wildfires present a challenge for ecosystem management, because they can be both beneficial and harmful. On the one hand, wildfires are a natural part of several ecosystems for maintaining their health and diversity through regulating plant succession and fuel accumulations, controlling age, structure and species composition, affecting insect and disease populations, influencing nutrient cycles and energy flows, regulating biotic productivity, diversity and stability, and determining wildlife habitats (Pyne 1996; Leblon and Bourgeau-Chavez 2013). On the other hand, wildfires can also become a threat to property, human life and economy, particularly in ecosystems where fires are an uncommon or unnatural (e.g., Pyne 1996; Mollicone et al. 2006; Bowman et al. 2009; Scott et al. 2014). Several hundred million hectares of vegetation are estimated to burn annually throughout the world, consuming billion tons of dry matter and releasing emission compounds that affect the composition and functioning of the global atmosphere and human health (e.g., van der Werf et al. 2010; Weise et al. 2014; FAO 2014). Over the last decade in Canada, wildfires have consumed an average of 2 million ha/year and induced fire suppression costs ranging from about $500 million to $1 billion a year (Canadian Forest Service 2014). In Europe, wildfires burn more than half a million ha of forested areas every year. Over 95% of the burnt areas are located in the Mediterranean region, in which critical fire events have taken place in recent years (http://effis.jrc.ec.europa.eu) (San-Miguel-Ayanz et al. 2013a)

Because of the threat that fires represent, operational systems have been developed for fire management that includes fire danger prediction, fire detection and fire control. Given expected increases in fires across the world due to climate changes (Flannigan et al. 2005) better prediction of fire danger and fire detection will have significant benefits both from the economical and the human safety points of views. There is also the need for accurate assessment of burnt areas because they are related to greenhouse gas emissions into the atmosphere that need to be accounted for following *United Nations Framework Convention on Climate Change* (United Nations 1992) and *Kyoto's protocol* (United Nations 1998) requirements as well as for managing post-fire environmental impacts, such as regeneration and erosion. Indeed, wildfires are important climate forcing factors as they release between 25-35% of the world CO_2 net emissions to the atmosphere (FAO 2012). Spaceborne remotely sensed imagery can play an important role in monitoring wildfires worldwide. Indeed, satellite imagery offers the advantages of extensive regional coverage, zero disturbances of the area to be viewed, as well as a method for acquiring data in less accessible areas on a regular and cost effective basis. In the first part of this chapter, we will present the use of remote sensing in pre-fire conditions management. The second part of the chapter will be dedicated to the use of remote sensing for detecting fires and burn scar mapping. The present chapter will primarily focus on satellite data, although the theory may also be applied to airborne sensors, which are currently used during fire suppression/detection activities rather than as pre-fire management or burn scar mapping tools. In the chapter, while both types of wavelengths are recorded by optical sensors, the term "optical" is used to describe wavelengths between 400 and 2500 nm, in contrast to the thermal infrared bands, which range from 3000 to 15000 nm.

MONITORING PRE-FIRE CONDITIONS

Ignition and spread of wildfires depends on fuel moisture, weather conditions, fuel types and topography. These parameters are inputs to fire danger predicting systems that have been developed for fire management, among others for fire suppression. These systems are among others the *National Fire Danger Rating System* (*NFDRS*) in USA (US Forest Service 2012) and the *Canadian Forest Fire Danger Rating System* (*CFFDRS*) in Canada (Forestry Canada Fire Danger Group 1992). The *CFFDRS* is also used in Alaska and in some other parts of the world, including Europe and Asia. Both systems are based primarily on weather parameters that are point source data which are often acquired in a sparse network of weather stations. The availability of satellite images coupled with the development of geostatistics and spatial analyses using geographic information technology allows moving fire danger rating from point-based estimates from weather stations to spatially-explicit estimates. Indeed, satellite images have the advantages of larger sampling areas, lack of destruction of the studied resource, gathering data on less accessible areas and are measuring, in essence, the integrated response of vegetation (including fuel) to environmental influences (including drought).

Several pre-fire conditions can be monitored using remote sensing. The first one is related to the fuel type and fuel load. They can be mapped, such as classical vegetation mapping, from high spatial resolution optical (e.g., Chuvieco and Martin 1994; Burgan et al. 1998; Chuvieco et al. 1999a; Riaño et al. 2002) or radar images (e.g., Saatchi et al. 2007). These maps can then be linked, within a wildfire threat analysis system, to other pre-fire conditions variables, such as topography, proximity to roads and to urban areas, etc. (Burgan et al. 1998; Chuvieco et al. 1999a, 2010, 2012). Another pre-fire condition, which can be estimated by remote sensing, is the fuel moisture. Two types of fuel moisture can be considered. Live fuel moisture conditions have been quantified through the live fuel moisture content (LFMC) or the equivalent water thickness (EWT). LFMC is defined as the ratio between the quantity of water (fresh weight–dry weight) and either the fresh weight or the dry weight (see the review of Ceccato et al. 2001). EWT is the leaf water content per unit leaf area which is defined as the ratio between the quantity of water and the leaf area (see the review of Ceccato et al. 2001). Live fuel moisture conditions have also been quantified indirectly, through the degree of water stress which is expressed in terms of evapotranspiration rates (Vidal et al. 1994). Another type of fuel moisture is the dead fuel moisture that can be more easily computed from weather data and fuel characteristics, because it is in balance with the surrounding atmosphere (Burgan et al. 1998; Pinol et al. 1998; Chuvieco et al. 1999b; Matthews 2013).

Optical Remote Sensing

The first remote sensing studies on fuel moisture monitoring used optical data, mainly Normalized Difference Vegetation Index (NDVI) or Relative Greeness (RG) images that are computed using the red and near-infrared bands from NOAA-AVHRR or other optical sensors such as LANDSAT-TM and SPOT-HRV (e.g., Paltridge and Barber 1988; Burgan et al. 1998; Chuvieco et al. 1999b; Hardy and Burgan 1999). This supposes that timing and extent of drought can be assessed from vegetation greenness, as retrieved from satellite data. NDVI or RG data were correlated to simulated forest evapotranspiration (e.g., Deblonde and Cihlar 1993), to FWI codes and indices (Dominguez et al. 1994; Camia et al. 1999; Leblon et al. 2001, 2007; Oldford et al. 2006), to grassland degree of curing (Burgan et al. 1998; Newnham et al. 2011), and to fire occurrences (e.g., Lopez et al. 1991; Illera et al. 1996; Burgan et al. 1998). NDVI-based operational systems have been proposed to map fire potential index (FPI) (Burgan et al. 1998) and crop droughts or fire dangers (Kogan, 2001). More recently, empirical relationships have been established between LFMC and the Visible Atmospherically Resistant Index (VARI) which also uses the blue LANDSAT-TM band over Southern California chaparral (Stow et al. 2005, 2006; Roberts et al. 2006; Peterson et al. 2008), and fire-prone Australian shrubland, heathland and sclerophyll forests (Caccamo et al. 2012).

These studies listed several problems related to the use of visible/near-infrared images in fuel moisture mapping, namely the saturation of relationships (Paltridge and Barber 1988), the influence of site wetness on relationships (Deblonde and Cihlar 1993) and the difficulty of using these images over forests, due to the spectral mixture of the overstory with the understory, both being different in nature and in moisture content (e.g., Hardy and Burgan 1999; Leblon et al. 2001). In fact, these images and associated vegetation indices are only indirectly related to fuel moisture conditions, because it rather measures the greenness and the chlorophyllous activity of the vegetation (Ceccato et al. 2001; Leblon 2005). In a study on pre-fire conditions using NOAA-AVHRR over Northwest Territories boreal forests, Oldford et al. (2003) showed that high FWI areas correspond to high surface temperature areas on the surface temperature NOAA-AVHRR image, indicating water stress, but also to high NDVI areas over the NOAA-AVHRR NDVI image, indicating no drought conditions.

Greenness and the chlorophyll activity of the vegetation explained the positive correlations between cumulative NDVI and FWI codes and indices found by Leblon et al. (2001, 2007) over Canadian northern boreal forests, since both types of variables increase in parallel throughout the fire season, but for two different reasons: FWI codes and indices, because of drought, and cumulative NDVI because of vegetation growth. In addition, reduction in NDVI could be induced by factors other than drought, like disease or senescence (Leblon 2005) and shadowing or penumbra (Chuvieco et al. 1999b). For all these reasons, a better use of NDVI images over forests will be to map timing of deciduous leaf flushing, which is critical in fire management, because of its relationship to fire occurrence in mixed-deciduous forests.

Fuel moisture is theoretically better related to another optical band, the shortwave infrared (SWIR) (1300-2500 nm) (e.g., Pierce et al. 1990; Pinol et al. 1998; Ceccato et al. 2001). SWIR-related indices, such as the Normalized Difference Infrared Index (NDII), the Normalized Difference Water Index (NDWI), and the Normalized Multi-Band Drought Index (NMDI) were related to LFMC for grasslands, shrublands, woodlands, chaparral and boreal forests (Chuvieco et al. 2002a, 2004a; Dennison et al. 2005; Stow et al. 2006; Roberts et al. 2006; Yebra et al. 2008; Qi et al. 2012) and to fire occurrences in boreal forests and Mediterranean forests (Akther and Hassan 2011; Bisquert et al. 2014). Maki et al. (2004) combined both NDVI and NDWI data into the vegetation dryness index (VDI) and related it to fire occurrences in boreal forests. Verbesselt et al. (2007) proposed the improved VDI (IVDI) to estimate LFMC of herbaceous savannas. Multispectral MODIS raw data and related indices were used in radiative transfer models to retrieve live fuel moisture content of grasslands (Yebra et al. 2008), shrublands (Yebra et al. 2008; Yebra and Chuvieco 2009a) and woodlands (Hao and Qu 2007; Yebra and Chuvieco 2009b; Jurdao et al. 2013; Wang et al. 2013). Relationships were significant only when the water stress was already well developed (Pierce et al. 1990; Pinol et al. 1998). Reflectance variations associated with water changes were smaller than those associated with leaf structure (Pierce et al. 1990; Ceccato et al. 2001; Bowyer and Danson 2004; Danson and Bowyer 2004; Yebra et al. 2013). In addition, shortwave bands are highly disturbed by atmospheric effects.

Fuel moisture is also probably estimated better using hyperspectral data. Indeed, hyperspectral data allow computing derivative variables which are useful to remove, on reflectance, the effect of leaf structure, of background and of atmosphere as well as to resolve overlapping spectra to better separate components of the global spectrum (see the review of Leblon 2005). Hyperspectral data were related to plant water content through empirical relationships (e.g., Pinol et al. 1998; Gillon et al. 2004; Roberts et al. 2006; Al-Moustafa et al. 2012) or radiative transfer models (e.g., Ustin et al. 1998; Ceccato et al. 2001; Riaño et al. 2005). However, from the operational point of view, hyperspectral data are, up to now, only provided by airborne sensors or the EO-1 HYPERION spaceborne sensor and shortwave infrared data are acquired by only a small number of spaceborne sensors, among others LANDSAT-TM, SPOT-VEGETATION, NOAA-16 and MODIS. While the oldest sensors such as LANDSAT-TM have a long revisit period, the newest sensors, such as SPOT-VEGETATION and the new series of the AVHRR sensors, on board NOAA-16, or MODIS, have the advantage to allow daily image acquisition. This temporal scale may be longer during cloudy periods. The performance of these new sensors is still under evaluation. By contrast, thermal infrared data are collecred more often and mostly at the same time as the optical visible and near-infrared data, by several existing polar-orbiting spaceborne sensors, e.g., NOAA-AVHRR, LANDSAT-TM, ATSR-2, RESURS-01, or MODIS and geostationary spaceborne sensors (e.g., METEOSAT, GOES). Geostationary sensors acquire daily images but at spatially coarser scale than polar-orbiting sensors and it is optimal to use both types of data, such as in Anderson et al. (2011).

Thermal Infrared Remote Sensing

Surface temperature (T_S) was better correlated than NDVI to FWI codes and indices (Dominguez et al. 1994; Camia et al. 1999; Aguado et al. 2003; Oldford et al. 2003, 2006; Leblon et al. 2007), to foliar moisture content (Chuvieco et al. 1999b) and to shrub water potential (Gouyet et al. 1991). It was also useful to detect water-stressed coniferous stands, when extreme differences in canopy water content occurred (Pierce et al. 1990). In fact, the difference between surface and air temperatures is a better spectral index to monitor plant water status than the surface temperature solely, the last being too sensitive to weather conditions (Camia et al. 1999; Duchemin et al. 1999). In addition, according to the energy budget equation, plants respond to water stress by stomata closure, thereby decreasing latent heat transfer from leaf surface to the air and causing an increase in leaf surface temperature (Pierce et al. 1990). Solving the energy budget equation, in which the sensible heat flux (H) is inferred from the difference between surface and air temperatures (T_S-T_a), as a function of the latent heat flux (LE) leads to an analytical relationship between actual evapotranspiration rate (AET) and T_S-T_a. Cumulative T_S-T_a data were well related to monthly fire start numbers throughout the fire season over Mediterranean forests (Prosper-Laget et al. 1995). For the same ecosystem, Vidal et al. (1994) used the energy budget equation to compute the ratio between actual and potential evapotranspirations (AET/PET) from daily NOAA-AVHRR surface temperatures and synoptic air temperatures. The ratio was related to fire occurrences (Vidal et al. 1994) and to two shrub flammability variables (Desbois and Vidal, 1996). The ratio was used to operationally monitor fire danger over Mediterranean forests in 1994 (Desbois and Vidal 1995) and was correlated to FWI codes and indices over Canadian northern boreal forests (Strickland et al. 2001).

However, these studies also showed that estimating AET from T_s-T_a using the energy budget equation is more problematic over forest canopies than over crop canopies (Leblon 2005). First, canopy height makes forests different from a thin leaf surface, as supposed by the energy budget equation, because of an additional level of radiation absorption and convective heat exchange between the ground and the superior stratum. Second, the measured surface radiative surface temperature is different from the aerodynamic surface temperature (T_{rad}) required by the equation, because of an additional excess resistance (known as the kB^{-1} factor) to heat transfer from leaves, which increases with the canopy height. Third, the aerodynamic resistance (r_a) is lower than the canopy resistance (r_c) and T_s-T_a is thus less sensitive to moisture fluctuations. This lower sensitivity is compensated by the sensitivity of satellite signals to ground vegetation patches which are an important fire danger parameter. Also, the clumped nature of canopy elements in tree crowns reduces wind speed near leaves and allows sunlit leaves to have temperatures elevated well above T_a. Wind can affect temporal fluctuations of T_s-T_a, but these fluctuations on the 1 km pixel basis of NOAA-AVHRR may be very small, because eddies near the surface are on a scale of about 10 m and because of the spatial integration over the pixel.

The energy budget equation requires an estimate for T_a. If synoptic T_a measurements are used, they should be corrected for shelter and tree height effects (Prosper-Laget et al. 1995). They can also be estimated as the radiative surface temperature of nearby well-watered canopies (Duchemin et al. 1999), since for not well-watered canopies, a systematical bias has been observed because the difference between surface and air temperatures is an indicator of water stress. T_a was also estimated as the radiative surface temperature corresponding to the extrapolation of the NDVI/T_s relationship to an NDVI of an infinitely thick vegetation canopy (e.g., Goward et al. 1994). However, such an estimate requires first that the range of variation in NDVI and T_S is large enough to accurately define the slope (Pierce et al. 1990). Second, the images should not be contaminated by clouds, snow or standing water, because the slope can then be positive (e.g., Goward et al. 1994). Third, the canopy should be well-watered because the NDVI/T_s slope can changed as a function of the moisture of the canopy.

Indeed, the slope was related to several moisture-related variables which are listed in Section 2.3.

The energy budget equation also requires knowledge of the aerodynamic and canopy resistances which are difficult to estimate (Vidal et al. 1994; Vidal and Devaux-Ros 1995; Strickland et al. 2001) and whose estimates are valid only for small areas. Thereby, other analytical models for computing AET from T_S have been proposed. The first one is the Surface Energy Balance Algorithm for Land (SEBAL) (Bastiaanssen et al. 1998). It computes LE as a residual quantity of the energy budget equation, but H is derived from the vertical difference in air temperature (δT_a) between the surface roughness length to heat transport (z_{oh}) and the reference height (z_h), δT_a being directly inferred from T_S. SEBAL has been validated on both short and tall vegetation (Bastiaanssen et al. 1998).

The second one does not compute LE as a residual of the energy budget equation. It uses the Penman-Monteith approach, in which the vapour pressure deficit of the air (VPD) is estimated from the saturation vapour pressure at the mean daily surface temperature (VP^*_{Ts}) (Granger 1997). Indeed, according to the feedback theory, feedback links between the surface and the overlying air are such that the observed surface temperature is a good indicator of the air humidity over the surface (Granger 1997). It is applicable to both short and tall canopies, because the VPD- VP^*_{Ts} relationship does not depend on the cover type. However, it does not distinguish between vegetated and non vegetated surfaces having the same surface roughness, temperature and air humidity, unless they have a different albedo leading to a different R_n. Its operational use thereby requires a careful land use mapping. There are other more sophisticated approaches to estimate AET from T_S, like soil-vegetation-atmosphere transfer (SVAT) models (see the review in Olioso et al. 1999). SVAT models usually require a high number of input variables and thereby have little operational potentials in fire management.

Synergisms between Optical and Thermal Infrared Remote Sensing

Several empirical studies already showed that inclusion of thermal infrared data improved correlations between optical indices and drought-related variables (Dominguez et al. 1994; Chuvieco et al. 1999b, 2003, 2004b; Aguado et al. 2003; Oldford et al. 2003, 2006; Leblon et al. 2007; Garcia et al. 2008). Oldford et al. (2006) showed that, compared to weather station data interpolation, improved spatial resolution mapping of slow-drying fuel moisture, such as the drought code (DC), can be achieved at the pixel level when DC is computed using a regression model which has surface temperature and NDVI NOAA-AVHRR images sensing data as independent variables (Figure 1). The fire shown in the center of the 15x15 pixel area was classified by the Sustainable Resource Development Department of Alberta as a surface fire, caused by lightning. It is interesting to observe that the fire burned in a closed coniferous forest cover type, which was classified as having a high DC danger rating when the NOAA-AVHRR image was used, but as having a moderate DC danger rating in the weather station-based map.

Combining optical vegetation indices with surface temperature data helps account for the influence of the ground cover rate over the composite surface temperature measured by the sensor. The inverse relationship between NDVI and T_S was related to fire occurrences in

Mediterranean forests (Prosper-Laget et al. 1994) and to moisture-related variables, such as canopy resistance (Nemani and Running 1989), sensible and latent heat flux (Nemani and Running 1989; Olioso et al. 1999), leaf water potential (Goward et al. 1994), accumulated rainfall (Duchemin et al. 1999), FWI codes and indices (Dominguez et al. 1994; Aguado et al. 2003, Oldford et al. 2003, 2006; Leblon et al. 2007), and foliar moisture content (Chuvieco et al. 1999b; 2003). NDVI and T_S images were used to define several drought indices, like the Vegetation and Temperature Condition Index (VT) (Kogan 2001), an empirical index (Chuvieco et al. 2003), the Water Deficit Index (WDI) (Vidal and Devaux-Ros 1995), and the Temperature-Vegetation Wetness Index (TVWI) (Akther and Hassan 2011). WDI was related to the number of fires and the area burned in the case of Mediterranean forests (Vidal and Devaux-Ros, 1995). TVWI maps that have been used together with surface temperature and normalized multiband drought index maps were related to fire occurrence in the case of boreal forests (Akther and Hassan, 2011). Nieto et al. (2010) used NDVI and T_S images to retrieve air temperatures and relative humidity that in turn are used to estimate the equilibrium moisture content of dead fuel in Mediteranean areas. Chowdhury and Hassan (2013) related MODIS 8-day composites of surface temperature, NMDI, and NDVI data to discrete fire danger ratings computed over northern boreal forests.

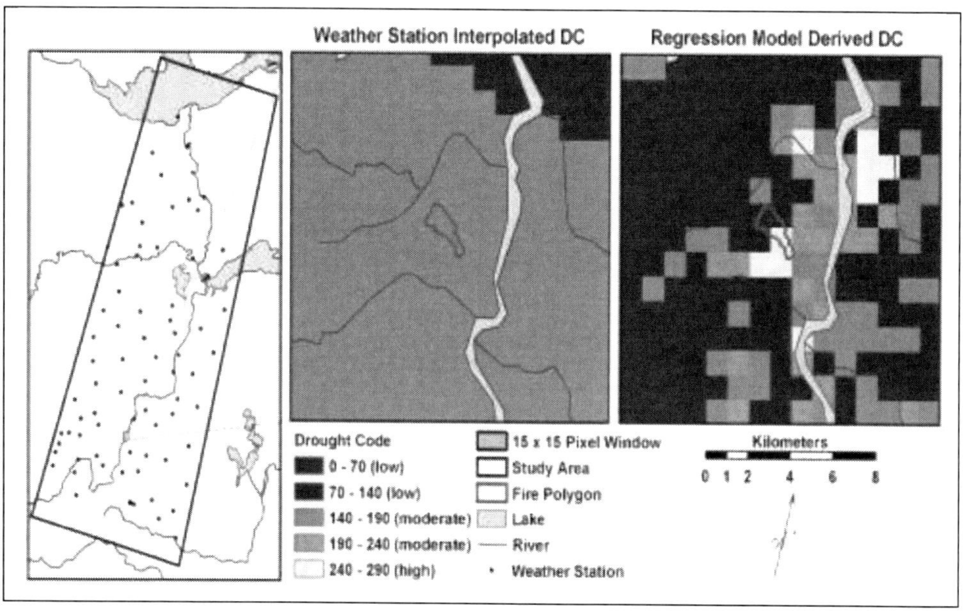

Figure 1. Comparison between DC maps computed from weather station interpolation and from NOAA-AVHRR NDVI and T_s images for an area, where the fire "vq0332" starts. The image was acquired 3 days before fire ignition (after Oldford et al. 2006).

Other synergisms between optical and thermal infrared data can also be considered when estimating AET using the energy budget equation. Indeed, the required net radiation flux (R_n) can be computed from the solar irradiance at the surface or from the surface albedo, both variables being inferred from optical data (e.g., Granger 1997; Bastiaanssen et al. 1998). Also, the ratio between the soil heat flux (G) and R_n can be analytically derived from optical vegetation indices (e.g., Bastiaanssen et al. 1998; Leblon 2005).

Using both thermal infrared and optical images improve the correlation with fuel moisture variables, but these images have the following operational limitations: i) cloud cover and other sun illumination conditions which limit image acquisition (Leblon 2005), ii) decoupling the impact of dry mater content from equivalent water thickness (EWT) on reflectance values (Yebra et al. 2013), and iii) low spatial resolution of optical sensors over large areas (Leblon 2005). As reviewed in Leblon (2005), many strategies can be applied to overcome the problem of cloudy days, like the interpolation of evaporation fractions for the cloudy days, or the use of images acquired by passive or active microwave (radar) sensors, which are able to penetrate cloud cover. Currently, the *SSM/I, AMSRE* and *SMOS* sensors provide images acquired in passive microwave frequencies, but at a coarser spatial resolution than *NOAA-AVHRR* images. These sensors are still under evaluation. By contrast, active microwave sensors acquire images at a finer spatial resolution than most of the optical sensors. These sensors are on board several recent satellites, i.e., *ERS-1/2, ENVISAT, RADARSAT-1/2, ALOS-PALSAR, Sentinel-1,* and *PALSAR-2.*

Radar Remote Sensing

Studies reviewed in Leblon et al. (2002) and in Abbott et al. (2007) have shown that radar backscatter ($\sigma°$) measurements over forested areas depend on (i) vegetation type, species, and structure, (ii) vegetation biomass, (iii) topography and surface roughness and canopy height; (iv) flooding and the presence/absence of standing water, and (iv) moisture. Three sources of moisture variation may contribute to the forest radar backscatter: the forest floor, the canopy (including its woody elements) and the environmental conditions (rain events). Over boreal forests, positive relationships between radar backscatter and cumulative rainfall amounts were found with *ERS-1* C-VV SAR images (Bourgeau-Chavez et al. 1999; Leblon et al. 2002) and with *RADARSAT-1* C-HH SAR images (Abbott et al. 2007). The good correlation between $\sigma°$and weather variables, which are used to compute the various FWI codes and indices, leads to the expectation that these indices and codes are also well related to $\sigma°$. FWI codes and indices were correlated to $\sigma°$ derived from ERS-1 C-VV and *RADARSAT-1* C-HH SAR images acquired over burned and unburned boreal forests located in Alaska (Bourgeau-Chavez et al. 1999, 2007) and in the Northwest Territories, Canada (Leblon et al. 2002; Abbott et al. 2007) (Figure 2). While these studies produced encouraging results, they also showed that single channel C-band SAR images are restricted in their applicability across the landscape primarily due to variations in surface roughness and biomass which act as confounding factors. More recently, fully polarimetric X-, C-, and L-band SAR sensors were launched into orbit (*ALOS-PALSAR* in 2006 and *TerraSAR-X* and *RADARSAT-2* in 2007) that allow for decomposition of the backscattered energy into dominant scattering mechanisms which may prove useful for reducing the confounding factors and allowing improved estimation of the fuel moisture in the absence of ancillary information. Table 1 compares *RADARSAT-2* polarimetric SAR variables calculated from images acquired under the same incident angle and during an extreme dry date and a wet date over two types of savannahs (shrub/tree and herbaceous) (Kong et al. 2014) and a chronosequence of Alaskan boreal black spruce ecosystems (recent burns, regenerating forests dominated by shrubs, open canopied and moderately dense forest cover) (Bourgeau-Chavez et al. 2013a).

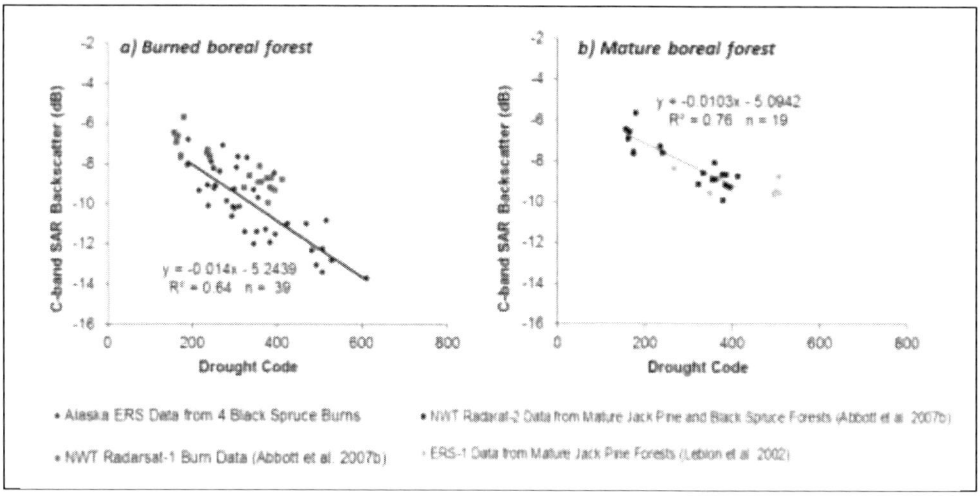

Figure 2. Relationship between the Fire Weather Index Drought Code (DC) and *ERS-1* C-VV and *RADARSAT-1* C-HH SAR images acquired over burned and unburned boreal forests located in Alaska and in the Northwest Territories, Canada (data from Leblon et al. 2002; Abbott et al. 2007; and Bourgeau-Chavez et al. 2007) (adapted from Bourgeau-Chavez 2013).

In both cases, there was a significant difference between the wet and the dry dates for all backscatter polarizations. Among the linear singular polarized backscatter, HH polarization was amongst the most sensitive. For the circular singular polarized backscatter, the best polarization depends on the vegetation type. For the polarimetric decomposition parameters, the Freeman-Durden (Freeman and Durden 1998) and van Zyl decomposition (van Zyl et al. 2011) parameters were more different between the wet and dry dates than the Cloude-Pottier decomposition (Cloude and Pottier 1997) parameters, particularly for the parameters corresponding to the double-bounce scattering in the case of the savannahs and to the surface scattering in the case of the Alaskan boreal forests (Table 1). The lack of significance of the Cloude-Pottier decomposition can be explained by the fact that these parameters work with the polarimetric state only, and do not consider the span information (i.e., radar intensity). In particular, two Cloude-Pottier decomposition parameters, H and A, describe the randomness of scattering and the importance of one scattering mechanism to the others and do not represent intensity information. In contrast, the two other decompositions (Freeman-Durden and Van Zyl) provide estimates of scattering intensity from known physical mechanisms (surface, double-bounce, and volume scatter). Among all the polarimetric discriminators defined in(Touzi et. al. (1992), the minimum scattering intensity (S_{min}), the maximum scattering intensity (S_{max}), the minimum received power ($P_{r,min}$), and the maximum received power ($P_{r,max}$) produced significant positive differences on both savannahs and boreal forest sites (except for the dense mature boreal forest). The minimum of the degree of polarization (d_{min}) produced stronger differences than d_{max}. In both cases, the differences are negative for both the savannah sites and the sparse or dense spruce forests. Both parameters were found to be related to the biomass structure and surface roughness variations. Indeed, they are a measure of the polarization state of the received signal which is highly related to the vegetation structure.

Table 1. Normalized difference (%) between the median values of various radar and polarimetric radar variables corresponding to RADARSAT-2 images acquired during a wet and dry date over South Africa savannahs and Alaskan boreal forests

Variable	Savannahs(1)		Boreal forest(2)					
	Shrub/Tree	Herbaceous	Low Burn Severity	Moderate Burn Severity 1	Moderate Burn Severity 2	Shrubby Regrowth	Sparse Spruce Forest	Dense Spruce Forest
C-HH	40.53	64.12	54.93	62.71	62.27	41.11	43.72	17.34
C-HV	45.45	72.55	45.16	51.09	60.24	30.42	31.42	21.02
C-VV	17.12	41.46	53.64	58.92	61.92	34.68	38.71	15.83
C-RR	48.04	75.15	55.78	53.01	60.15	21.70	39.33	15.64
C-LR	34.42	58.39	55.33	62.49	61.20	43.04	39.90	11.55
C-LL	54.54	75.13	50.07	53.59	62.29	29.12	34.64	15.21
CP *H*	8.08	8.83	-1.86	-11.40	-0.67	-9.68	-1.74	1.42
CP *A*	0.49	1.91	11.44	3.89	-1.41	-6.75	5.42	7.95
CP α (°)	16.09	21.73	0.09	-13.03	1.65	-15.74	-1.55	3.40
vZ vol	44.33	71.39	49.95	53.53	62.10	29.01	34.61	15.24
vZ surf	36.07	62.38	56.28	63.71	61.12	46.96	41.68	10.19
vZ dbl	62.48	79.12	58.03	51.41	58.53	13.62	41.73	17.65
FD vol	46.24	73.02	50.07	53.59	62.43	28.82	34.62	15.19
FD surf	36.91	56.07	57.35	65.26	60.71	54.08	45.14	7.59
FD dbl	77.28	79.00	54.06	29.21	43.71	-18.35	30.47	-9.42
d_{max}	-2.35	-2.59	5.34	0.76	1.78	8.76	0.26	-2.98
d_{min}	-44.01	-86.02	12.45	6.32	4.02	40.66	-8.97	-15.38
Δd	3.11	16.42	-6.25	-10.87	-3.38	-27.13	6.39	8.09
Pr_{max}	42.08	65.29	54.41	59.90	55.86	40.23	23.97	8.26
Pr_{min}	48.00	72.77	33.33	57.69	51.67	16.67	21.05	-2.30
FP	-2.39	-2.25	4.84	0.81	1.35	8.19	0.13	20.93
S_{max}	42.14	66.49	53.13	59.61	55.27	38.19	24.56	11.46
S_{min}	36.75	62.42	55.68	60.28	54.90	40.74	25.56	-3.03
ND_S	7.84	19.96	66.27	81.87	75.37	58.38	63.30	76.07

(1) from Kong et al. 2014; (2) from Bourgeau-Chavez et al. 2013a.

Except for the dense spruce forest, the fractional polarization (FP) produced a smaller difference than its individual components (Pr_{min} and Pr_{max}).These polarimetric parameters from a multi-date dataset (across a range of soil moisture conditions) were used to develop empirical algorithms to estimate the volumetric soil moisture over the Alaskan boreal test area (Bourgeau-Chavez et al. 2013b). The lowest error was produced through a multiple regression model that uses C-VH and the maximum of degree of polarization (d_{max}) with or without Pr_{min} for the burned and unburned sites respectively. d_{max} appears to provide information on the structural complexity of the landscape (surface roughness and/or high biomass) to aid in extraction of surface soil moisture from the SAR data. The resulting maps derived from two polarimetric *RADARSAT-2* images acquired during a) a dry date (23 August 2010) and b) a moderately dry date (11 July 2009) over sites within the regenerating black spruce boreal forests of Alaska, near Delta Junction are presented in Figure 3.

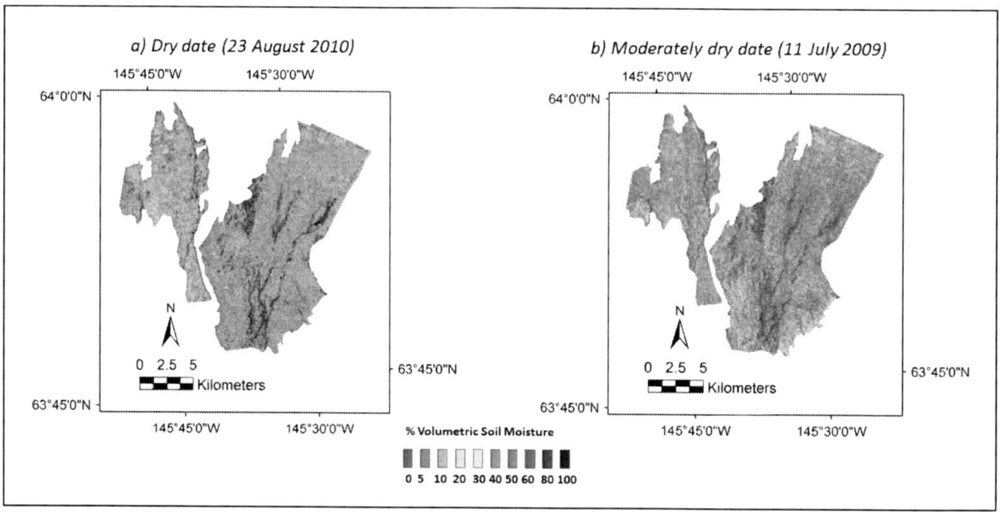

Figure 3. Maps of % volumetric soil moisture content (VMC) derived from two polarimetric RADARSAT-2 data acquired during a) a dry date (23 August 2010) and b) a moderately dry date (11 July 2009) over sites within the regenerating black spruce boreal forests of Alaska, near Delta Junction (biomass < 1.7 kg/m^2). Algorithm used: %VMC = 319.31 D_{max} -612.50 $Unpol_{max}$ + 5190.4 C-VH - 266.37 (R^2 = 0.77). These maps show improvement in moisture retrieval over single channel SAR by 27-33%VMC accuracy. RMS Error (based on *in situ* data from both dates) = 6.7 %VMC.

The algorithm used to derive these maps uses one single channel SAR intensity C-VH and two polarimetric variables, d_{max} and $Unpol_{max}$. These maps show improvement in moisture retrieval over single channel SAR by 27-33%VMC accuracy. The root mean square error (based on *in situ* data from both dates) was 6.7 %VMC.

Although radar images are theoretically available independently of the weather conditions, the availability of images acquired with the old radar sensors was limited because of the long revisit cycle of the satellites, such as ERS-1/2 (35 days). Fortunately the revisit period is shorter for the new radar satellites such as RADARSAT-1/2 (3-6 days). With *RADARSAT-1* images, Abbott et al. (2007) was able to get two images per day because of the pointing capability of the sensor. Such revisiting time will be improved in the near-future with the planned *RADARSAT-3* constellation and the new *SENTINEL* missions. In addition, often the radar images have a finer spatial resolution than optical or thermal infrared images.

Thus, by contrast to the conclusions of Chowdhury and Hassan (2014), we can conclude that radar remote sensing is a viable solution in fire danger monitoring as it represents a data source that is complementary to optical or thermal infrared data. Consequently, synergisms between optical or thermal infrared bands and radar bands should be investigated.

Fire Detection and Burnt Area Mapping

Fire detection is one critical stage of wildfire management, which is aimed at either fighting or monitoring the fire. For fire fighting the early detection is essential. So far, fire detection for fire fighting is based on human observation, the use of fixed optical cameras to monitor the surrounding environment, or aerial survey. The revisit time provided by current satellite sensors is not considered sufficient for fire fighting operations by forest fire mangers. However, the monitoring of wildfires and wildfire effects for large territories is mainly based on satellite remote sensing. Mapping of burnt areas and assessment of wildfire effects is one of the most successful applications of satellite remote sensing. Satellite remote sensing provides the means for acquiring comprehensive and harmonized information on wildfire effects for large territories at low cost. For this purpose, burnt area mapping is performed with a wide variety of remote sensors and techniques. A wide variety of optical and radar sensors have been used for fire detection and burnt area mapping, from local to global scales. This section reviews the application of remote sensing in active fire detection and the assessment of fire damages through the mapping of the extent of burnt areas.

Fire Detection

Fires produce anomalies that are detectable in many different parts of the electromagnetic spectrum and are therefore suitable for detection with the use of remote sensing techniques. Although fire detection is possible in the microwave range of the spectrum, these techniques are not used operationally because of the high cost of the sensors and the low nominal achievable spatial resolution of the detection (Kempka et al. 2006). Therefore the focus of this section is on the detection of fires from optical remote sensors.

Firstly, active fires can be detected from the light they emit in the visible part of the spectrum; however, the discrimination of the fire-emitted light is only possible at night (Cahoon et al. 1992; Elvidge 2001). Since most of the fires occur and have their highest intensity during the day, the detection of them solely at night is not of high interest for operational fire management. Secondly, fires can be detected by the smoke plume they produce (Figure 4). This detection method is widely used at local scale, as an alternative to visual detection by human operators (e.g., Bugarić et al. 2014). Image processing algorithms can be used to single out this smoke plume in contrast to its background, and associate it to a fire. Although these systems eliminate false alarms produced by overheating of ground areas, they also present some limitations. The limitations arise from two facts: first, the smoke plume can only be detectable some time after the fire has started; and second, smoke is often conducted along the surface and emerges in an area different from that where the fire started. Ground automatic detection systems can make use of cameras mounted in towers, buildings or masts with good visibility of the surveyed terrain. The cameras can be fixed (attached to the structure) or mounted on a positioning system to vary the azimuth and elevation angles. A positioning system can be used to survey the entire environment by varying automatically the scanning angles. The detection delay depends on the scan velocity given by the motors and the optical system in the camera, noting that the image processing requirements for automatic detection are higher when using mobile sensors. The sensor technologies used in today's automatic ground detection systems are mainly infrared and visual cameras (San-Miguel-Ayanz et al. 2005). Often these devices are coupled with intelligent algorithms that speed up the processing and recognition of signals from either smoke or distinct hot temperatures produced by wildfires (Rosas-Romero 2014).

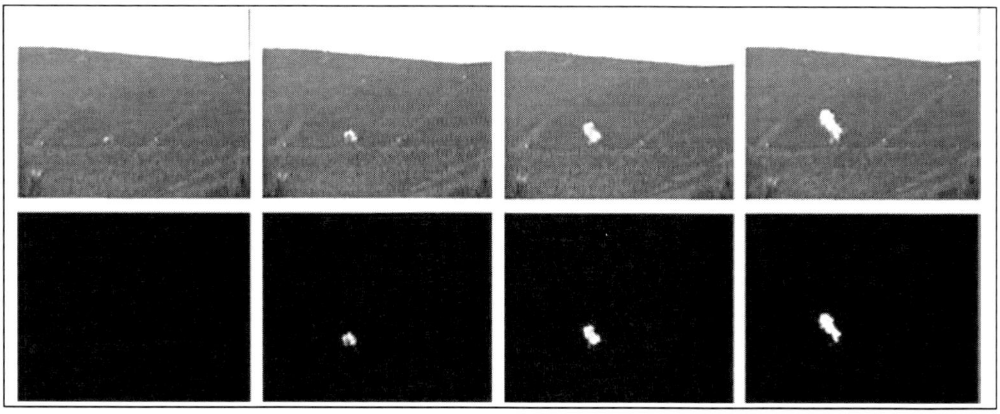

Figure 4. Fire smoke plume identification with optical cameras. The sequence shows smoke detection from fire ignition (left of sequence) to the full development of the smoke plume (far right of the sequence) (from San-Miguel-Ayanz et al. 2005).

Lastly, and most commonly, fires are detected due to the distinct high temperature they produce, which results in a high reflection signal in the mid-infrared and thermal electromagnetic spectra. Active fires produce temperatures ranging between 800 K and 1200 K, although they can reach up to 1800 K. These temperatures are easily detectable in the mid-infrared part of the spectrum (Matson and Dozier 1981). The principle of wildfire detection using near and middle infrared cameras can be considered fully developed and is applied in many areas of the world (Wu et al. 2014; Johnston et al. 2014). These devices become highly efficient in monitoring large areas of fairly flat terrain or if installed in high look out points that provide a wide view of the surrounding forest areas. A recent review of these methods is provided in Alkhatib (2014).

In addition to the detection of wildfires from ground-installed devices, the most widely used method for fire monitoring is the use of remote sensors on board aerial vehicles or satellites. The mid-infrared spectral window is suitable for fire detection from aerial and space borne sensors because it is far from the peak of the Earth and Solar radiations at 0.5 μm and 9.7 μm, respectively (Figure 5). Fires also radiate in the thermal part of the spectrum, i.e., between 8 μm and 12 μm. However, the peak radiation at these wavelengths corresponds to a normal environmental temperature of 300 K. Fires can be detected as local or absolute maximum in the mid-infrared and thermal spectra. An absolute (or regional) maximum is used in the so-called thresholding algorithms. Any area above a given threshold temperature is considered a fire. However, differences in fire characteristics among regions in the world lead to problems of false alarms and/or missed fires using this method. Although fixed thresholding algorithms were used in the past (Arino et al. 2012; Mouillot et al. 2014), most current techniques fore fire detection make use of the so-called contextual algorithms. Contextual algorithms detect local maxima. Multispectral criteria are aimed at detecting the difference between a fire pixel (active fire) and the background temperature (environmental temperature in the proximity of the fire pixel (Flasse and Ceccato 1996, Giglio et al. 2003, 2006). These algorithms are used for fire detection in operational geostationary (e.g., *METEOSAT*, *GOES*) and geosyncronous satellite sensors (e.g., *MODIS*, *VIIRS*).

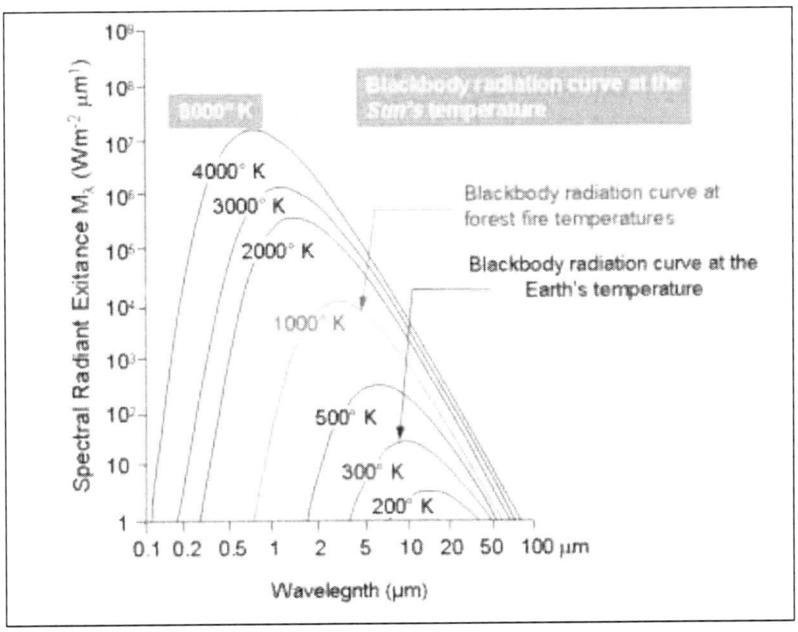

Figure 5. Spectral radiant exitance as a function of the temperature of the black body. The figure shows that forest fires being hotter than the Earth's surface exhibit a peak in their spectral exitance at a shorter wavelength than the Earth's surface (adapted from Kennedy et al. 1994).

Operational active wildfire monitoring is performed through the use of geo-stationary or geo-syncronous satellite sensors. Geo-station sensors are the GOES Imager (Geostationary Operational Environmental Satellite) (e.g., Zhang et al. 2014; Hao and Larkin 2014), the SEVIRI on board the *Meteosat Second Generation* (*MSG*) satellite (e.g., Calle et al. 2006; Roberts et al. 2006; Freeborn et al. 2014) and *Moderate-resolution optical imager* on board the *Himawari Multifunciton Transport Satellite* (*MTSAT-2*) (Hyer et al. 2013). The most used geo-synchronous satellite sensors are the *AVHRR* on board the *NOAA* meteorological satellite, the *ATSR* (*Along Track Scanning Radiometer*) on board the *ERS-1* and *2* and *Envisat*, and the *MODIS* (*Moderate Resolution Imaging Spectroradiometer*) on board the *Terra* and *Aqua* satellites. The new suit of data provided by the *Visible Infrared Imaging Radiometer* (*VIIRS*) on board the *Suomi National Polar-Orbiting Partnership* (*S-NPP*) satellite the launced in 2012 will provide continuity to the active fire products provided by *MODIS*. In addition to operational satellites, fire detection from space has been successfully implemented using other scientific satellite missions such as *BIRD* (*Bispectral and Infrared Remote Detection*) of the *German Aerospace Centre* (*DLR*) (Briess et al. 2003; Zhukov et al. 2006) and *ASTER* (*Advanced Spaceborne Thermal Emissions and Reflection Radiometer*) (Giglio et al. 2008).

GOES and *SEVIRI* provide high frequency coverage in the order of 30 min and 15 min, respectively. They are thus suitable for monitoring most wildfire processes (Prins and Menzel 1992; Prins et al. 1998). Compared to ATSR, this is a relative advantage for monitoring fire activities with *AVHRR* that provide a maximum of 1 daily pass, or *MODIS*, which provides 2 daily passes. However, due to its high spatial resolution, good fire detection capabilities and global coverage, *MODIS* has become the standard sensor for active fire monitoring at regional to global scales (Hao and Larkin 2014). The *Aqua MODIS* instrument acquires data twice daily (1:30 PM and AM), as does the *Terra MODIS* (10:30 AM and PM). These four daily *MODIS* fire observations serve to advance global monitoring of fire processes and their effects on ecosystems, the atmosphere, and climate (Giglio et al. 2006). Active fire data from modis are distributed through the *FIRMS* (*Fire Information for Resource Management System*) of *NASA* and are widely used in regional and global fire information systems. Several authors analyzied the reliabitliy of the *MODIS* active detection algorithms and the resulting product through the comparision with other satellite sensor active fire detection (Morisette et al. 2005; Csiszar et al. 2006; Schroeder et al. 2008). Some operational fire monitoring systems using *MODIS* active fire detection include the *Canadian Wildland Fire Information System* (*CWFIS*) (http://cwfis.cfs.nrcan.gc.ca), the *USA Active Fire Mapping Service*, or the *European Forest Fire Information System (EFFIS)* (http://effis.jrc.ec.europa.eu). In the case of *EFFIS*, post-processing filters based on landcover ancillary data are applied to the *MODIS* product to reduce the number of false alarms produced by non-fire hot surfaces (e.g., industrial areas, hot ground soils) and therefore increase the reliability of the active fire detection (San-Miguel-Ayanz et al. 2012). Figure 6 shows operational fire detection monitoring in the European region within *EFFIS*. Newly available satellite-based fire products that provide global coverage currently span a decade, as is the case of *MODIS*, or longer as in the case of the *ATSR* and *AATSR* series. Satellite derived information will continue to be available as fire detection products are generated from new and future missions, such as *VIIRS* (Justice *at al.* 2012) and *Sentinel-3 SLSTR* (Wooster et al. 2012).

Burnt Area Mapping

Remotely sensed data have been extensively used for burnt area mapping. Fires produce a significant change in the structure and the reflectance of vegetation and the soil properties within the burnt area that are noticeable in the microwave, visible and especially the infra-red part of the electromagnetic spectrum. Since a wide variety of satellite sensors provide data on the above-mentioned spectral ranges, the choice of the sensor is driven by the required spatial coverage of the area of interests, i.e., global, regional, or national/local, and the persistence of cloud coverage over this area. While low and medium spatial resolution sensors (*AVHRR, ATSR, VEGETATION,* and *MODIS*) can provide daily (or more frequent) coverage of any region, higher spatial resolution sensors have imited time coverage. Generally, the higher the spatial resolution of the sensor, the lower the revisit time over a given area. Sensors like *WiFS, LISS, LANDSAT,* and *SPOT HRV* are used for burnt area mapping at national and local scales. Lastly, actvie sensors like the Synthetic Aperture Radar are used in areas where the persistent cloud coverage prevents the effient use of optical sensors.

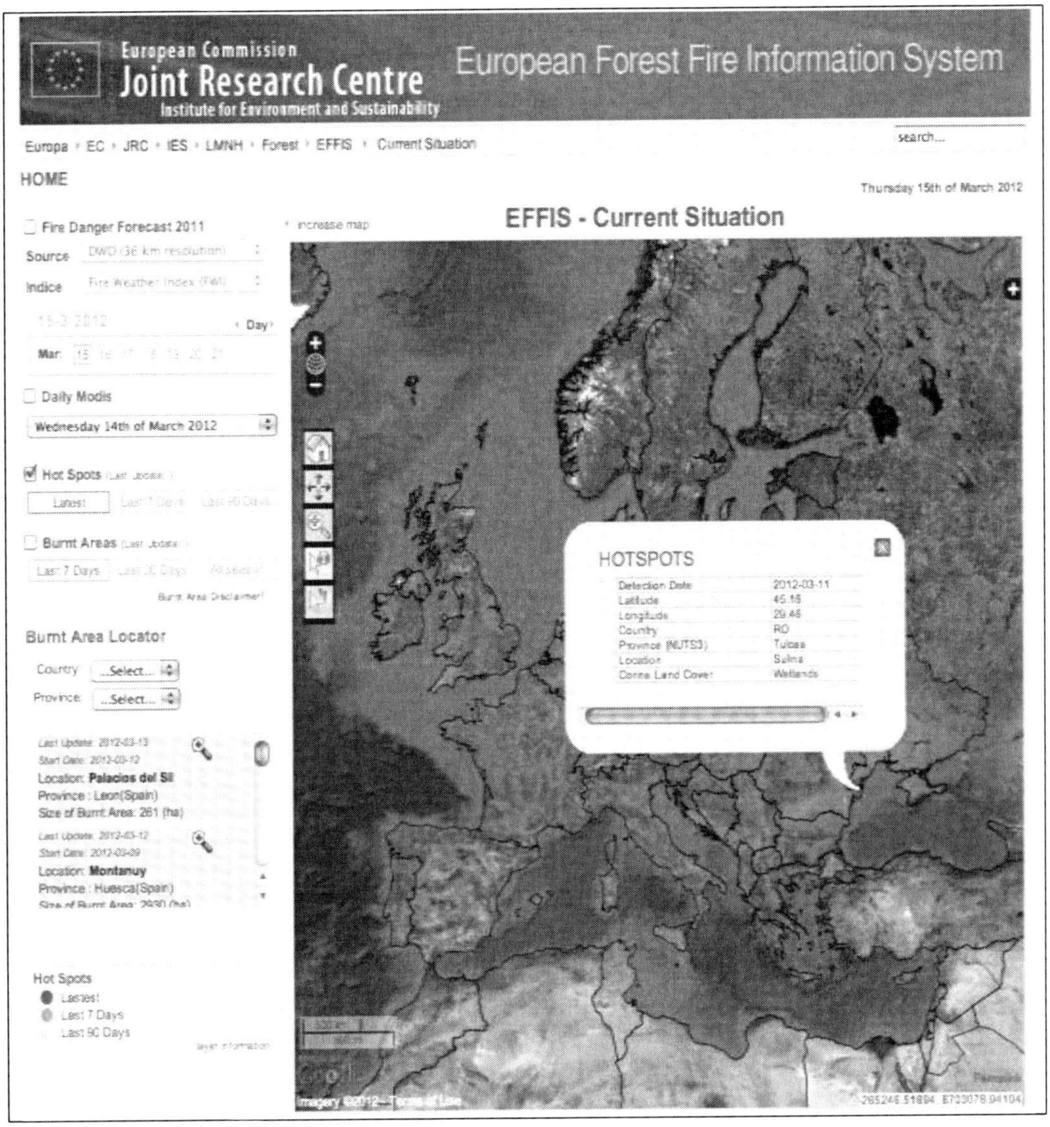

Figure 6. MODIS-based active fire detection in the European Forest Fire Information System (EFFIS) (http://effis.jrc.ec.europa.eu).

At the global scale, *NOAA-AVHRR* data were extensively tested in the 1990s. Studies differed mainly on the use of diverse spectral indices, although most commonly, burn scar areas were discriminated from a multi-temporal comparison of NDVI (Kasischke and French 1993; Martin and Chuvieco 1995; Pereira 1999). More recently other global burnt area datasets were derived from *SPOT Vegetation* and the *ATSR-2* on board *Envisat* (Grégoire et al. 2003; Tansey et al. 2004; Simon et al. 2004). Although these data provide gross estimates of burnt areas at the global level, the lack of extensive validation and agreement between them limit their use at regional or national levels. Nevertheless, partial validations of the *MODIS* global burnt area products were performed by Roy et al. (2005) and Boschetti et al. (2007, 2010). Pereira (1999) showed that the accuracy of the results for mapping burnt areas with *AVHRR* data in the Mediterranean region of Europe was about 80% for large fires. The methods were considered suitable only for fires larger than 1000 ha, and reliable for fires larger than 2000 ha. However, the mapping of those fires would correspond only to approximately 30% and 21%, respectively, of the total yearly burnt area in the European Mediterranean region.

With the launch of the *MODIS* sensor on board of the *TERRA* and *AQUA* satellites, a new capability for regional mapping of burnt areas was put in place and use for fire monitoring at regional to global scales (Giglio et al. 2006, 2010; Chuvieco et al. 2008). The availability of free data of medium spatial resolution from the *MODIS* sensors since 2000 provided a definite impulse for the use of remote sensing at the regional and global scales (Justice et al. 2002). Better radiometry and higher spectral information of the *MODIS* sensor provided the right data for the discrimination of burnt areas at these scales (Boschetti et al. 2010). The simultaneity in the operation of both satellites provides higher frequency in data acquisition and enough revisit time for accurate mapping of burnt areas. At the global scale, the *MODIS* program has released a standard product on burned areas that is based on a multitemporal change detection approach to analyze differences between modeled and actual reflectance, and to take into account Bidirectional Reflectance Distribution Function (BRDF) corrections (Roy et al. 2002, 2005). The resulting products were compared to the active fire product of *MODIS* (Roy et al. 2008) reaching a high level of agreement. Other European sensors of similar spatial and spectral resolutions to those of *MODIS*, such as the *MERIS* on board the *ENVISAT* satellite, were not used extensively in fire applications due the lack of availabitly of data in near real time.

At the regional scale, *MODIS* is operationally used in systems such as the Canadian *CWFIS* (http://cwfis.cfs.nrcan.gc.ca/home) and the European *EFFIS* (San-Miguel-Ayanz et al. 2012), mentioned above. Two full mosaics of *MODIS* data are received and processed daily in *EFFIS* to provide near-real time monitoring of wildfires and map burnt areas. The systems is thus updated up to two times daily, providing accurate information of fire impacts in Europe (San-Miguel-Ayanz et al. 2009) and support to decision and policy making at the European level (San-Miguel-Ayanz et al. 2013b). The use of higher spatial resolution imagery from *Advanced Wide Field Sensor* (*AWiFS*) for regional coverage in Europe was recently tested in Europe. However, the results of this exercise showed that the benefits derived from the use of high spatial imagery in term of detailed mapping of fire perimeters are obscured by the limitations in the revisit time of the sensor. These results did not enhance those of the standard *Rapid Damage Assessment* module of *EFFIS* based on *MODIS* imagery (Sedano et al. 2012, 2013). Figure 7 shows the extent of burnt area as they were mapped from MODIS and AWIFS imagery. However, the recent near-real time availability of high-spatial resolution images from the Landsat program provides new means for the potential use of these images for detailed burnt scar mapping at regional to global scales. The required computational requirements for this exercise may be provided by the *Google Earth Engine* (https://earthengine.google.org , which recently published the first 30 m resolution maps of global forest change from 2000-2012 (Hansen et al. 2013).

Given the multiplicity of data sources and resulting burnt area products, a recent joint initiative of the *Group on Earth Observations* (*GEO*) and the *Global Observation of Forest Cover Fire Implementation Team* (*GOFC Fire IT*) aims at the establishment of a *Global Wildfire Information System (GWIS)* where information provided by different sources at regional to global scales would be accessible, facilitating the inter-comparison of products and the accessibility of those to a network of potential users across the globe (ftp://ftp.earthobservations.org/GEO-X/GEO-X_10_GEO%202012-2015%20Work%20Plan %20Update.pdf).

At national to local scales, the wide variety of remotely sensed products at medium to high resolution (10 m to 30 m ground spatial resolution) makes it possible the accurate mapping of burnt areas. However, as mentioned above, the increase in spatial resolution is often accompanied by a decrease in revisit time of the sensor, which prevents the acquisition of this imagery for large areas. High-spatial burnt area mapping has been performed with Landsat Thematic Mapper imagery (Pereira and Setzer 1993; Michalek et al. 2000; Koutsias and Karteris 2000; Mitri and Gitas 2004) complemented in some cases by the *SPOT* and *ASTER* sensors. Some analyses made use of the *LISS-3* sensor of the *IRS* Indian satellite and the *RESURS MSU-K* (San-Miguel-Ayanz et al. 1998). A variety of indices computed from the original spectral bands and many different techniques were used to enhance the mapping of burnt areas (e.g., Pereira 1999; Li et al. 2000; Chuvieco et al. 2002b; Mitri and Gitas 2004; Bastarriak et al. 2010; Libonati et al. 2011; Gitas et al. 2012). However, this exercise was, in most cases, limited to the mapping of burnt areas at local and sub-national scale. An exception to this is the case of Portugal, where an operational system capable of processing Landsat TM scenes for mapping burnt areas was set up (Oliveira et al. 2012). Recently, mapping of burnt areas with Landsat TM and ETM imagery was done in Australia (Goodwin et al. 2014), the Mediterranean region (Katagis et al. 2014), and the Eurasian boreal region (Chen et al. 2014).

Figure 7. Perimeter of burnt areas obtained from *MODIS* (yellow) and *AWIFS* (blue) in one of the large fires in the Peloponnese province, Greece in 2007 `(from San-Miguel-Ayanz et al. 2012).

The use of high resolution remote sensing in the management of critical wildfires has improved dramatically in the last decade. The variety of remote sensing imagery of high and very-high spatial resolutions such as those provided by *QUICKBIRD, IKONOS, FORMOSAT, EARLYBIRD,* and *RAPIDEYE* has permitted the rapid coverage of critical fire events. The processing of this imagery provides a great level of spatial detailed that is needed for the accurate analysis of fire damages and the sound planning of restoration measures. Data provision for critical fire events has been supported by the agreement of the space agencies in the so-called International Space Charter, which allows rapid provision of available remotely sensed data from a series of satellites, including *RADARSAT, ERS, ENVISAT, SPOT, IRS, SAC-C, NOAA, LANDSAT, ALOS, DMC* supporting crisis management. Furthermore, programs like the European Copernicus, provide the means to the civil protection services in the countries to activate the request of data acquisition and processing (http://emergency. copernicus.eu/mapping/).

Although most of the studies on burnt area mapping were based on the use of optical imagery, there are a series of examples in which data from active sensors such as the Synthetic Aperture Radar (SAR) were used. Most of the studies were carried out in boreal forest (Kasischke et al. 1994, French et al. 1996; Bourgeau-Chavez et al. 1997, 2002), but some examples exist elsewhere, such as for the south-east Asian tropical rainforest (Siegert and Ruecker 2000), the savannas (Menges et al. 2004), and the Mediterranean area (Gimeno and San-Miguel-Ayanz 2004; Gimeno et al. 2005). Rather than the changes in vegetation condition and structure, the detection of burnt areas from SAR is based on the changes on moisture content in the burnt surface with respect to the unburned areas. Burnt areas tend to have higher moisture content than unburned areas, which increases the backscatter. Thus, burnt areas appear as bright objects in relation to the surrounding non-affected areas when they are wet. More recently, *RADARSAT-2* polarimetric SAR images have been tested to map fire scars. For mapping 8 year old fire scars over a flat area in Northeast China, Goodenough et al. (2011) applied the unsupervised Freeman-Durden-Wishart classification on a *RADARSAT-2* polarimetric SAR image having about 39° incidence angle and a west-looking direction. They were able to retrieve 85% of the fire scar areas of a SPOT5 classified image. They also applied an unsupervised K-Nearest Neighbors Graph Clustering classification method that is based on the Cloude-Pottier decomposition to map 9 year old fire scars over a rough area in Alberta (Canada) on a *RADARSAT-2* polarimetric SAR image having about 30° incidence angle and an east-looking direction. In this case, 84% of the *RADARSAT-2* fire scars pixels fell inside the perimeters of the GIS polygon. This unsupervised classifier was shown to be more tolerant of topographic variations. However, for mapping recent burns of less than two months over a rough terrain in the Calabria peninsula, Southern Italy, with images acquired at 32.4°-34° incidence angle and an east-looking direction, Leblon (2011) showed that the Freeman-Durden decomposition images were better than the Cloude-Pottier decomposition images. They explained this difference by the fact that the Freeman-Durden decomposition uses more information from the imaged area because it implicitly considers the intensity information, which is not the case of the Cloude-Pottier which only relies on the polarization state.

CONCLUSION

In this chapter, we reviewed studies using optical, thermal infrared and radar images for pre-fire and post-fire conditions monitoring. For the pre-fire conditions, our review had a particular emphasis on the studies using satellite data to monitor fuel moisture. Remote sensing of fuel moisture was first done with *NDVI* images (mainly from *NOAA-AVHRR*, based on the assumption that the greenness of the scene is a good indicator of fuel moisture and fire danger. *NDVI* images are used operationally to map fire potentials, but the reviewed studies have shown that *NDVI* is sensitive to the chlorophyll activity of the vegetation rather than to its actual changes in moisture content. By contrast, thermal infrared images allow the computation of surface temperature which is analytically related to surface moisture-related variables, like evapotranspiration, through the energy budget equation. This analytical approach was used in an operational fire danger monitoring system over Mediterranean forests. More recent studies on the use of remote sensing in fuel moisture monitoring use both *NDVI* and surface temperature images. This is done at no additional cost of data acquisition, since both kinds of images are provided at the same time by numerous existing satellites, like *NOAA-AVHRR, LANDSAT-TM, ATSR-2, RESURS-01, METEOSAT, GOES, EOS-ASTER,* and *EOS-MODIS*.

An important operational limitation of using optical and thermal infrared data is image availability, which still depends on weather and illumination conditions. These images also have the advantage of having a finer spatial resolution. Good relationships were found between *ERS-1* and *RADARSAT-1* radar backscatters and *FWI* codes and indices over northern boreal forests, but these studies also showed that radar backscatter is affected by several confounding factors other than those related to moisture, such as surface roughness and biomass. More recently, the availability of polarimetric SAR images allows for decomposition of the backscattered energy into dominant scattering mechanisms which may prove useful for reducing the confounding factors. Statistically significant differences between wet and dry dates were observed in the case of several polarimetric variables extracted from *RADARSAT-2* C-band polarimetric SAR images, such as the Freeman-Durden and van Zyl decomposition parameters. Models that can use such *RADARSAT-2* polarimetric variables for estimating fuel moisture were also established. One limitation of the operational use of SAR images in fire danger monitoring is timely image availability, since there are often long revisit periods (~35 days) for many existing radar satellites and by the commercial operating mode of some new radar satellites, like *RADARSAT-2*. However, the availability in the near future of SAR satellite constellations such as the planned *RADARSAT-3* mission, will decrease the revisit period. Eventually, further studies are needed to assess the combination of optical and thermal infrared images to radar images for monitoring pre fire conditions.

Much of the reviewed remote sensing studies for pre-fire condition management are based on the estimation of dead fuel moisture or FWI codes and indices through empirical relationships. Further research is needed to develop analytical methods. On the other hand, all required input variables of an operational fire danger system, like wind parameters, will surely not be derived from remote sensing data and additional ground-based and weather information will always be required to effectively monitor fire danger. In this regard, *in situ* sensing systems, such as the one described in Teillet et al. (2001), will be useful.

Regarding fire detection, remote sensing techniques can be considered fully operational. At the local scale, active fire detection is mainly based on the use of visible and infra-red cameras for the detection of thermal anomalies or smoke plumes. Fire detection devices are often mounted on watchtowers providing an ample view to monitor potential fire outbreaks in the sorunding areas. Fire detection at this scale is focused on improving the timing in the detection of the fire outbreaks to support to forest fire fighting operations. At regional to global scales, information on active fires is provided by geo-stationary satellite sensors (*GOES, SEVIRI,* and *MTSAT-2*) or geo-syncronous sensors (*AVHRR, ATSR,* and *MODIS*). Fires are detected using fixed threshold or contextual algorithms able to detect local temperature maxima in the mid-infrared spectrum. The high revisit time of the geostationary satellites provides frequent information (15 to 30 min) that is needed for monitoring fire processes and fire effects. By contrast, non-geostationary satellites provide a lower revisit time (1 to 2 daily passes), but they provide global fire information that is essential for the monitoring of wildfire processes and their effects on ecosystems, the atmosphere, and climate. Global active fire detection products were used to determine fire regimes and seasonality in different regions of the world and at the global scale. However, differences in accuracy among the existing products still persist.

Burnt area mapping from remote sensing has been on-going for nearly 30 years. Most of these applications are based on passive optical remote sensing imagery at global and regional scales. Global burn area datasets were derived from *AVHRR, ATRS, SPOT-VEGETATION,* and recently from *MODIS*. National or local burnt area maps were obtained from higher spatial resolution imagery (e.g., *LANDSAT, LISS-3,* and *SPOT-HRV*). At the local scale, and also in some cases at the regional scale, active sensors such as the *ERS* and *RADARSAT* SARs have proven their capacity for monitoring fires under all-weather conditions. Currently, the data acquired by the *MODIS* sensor has become the standard for fire monitoring at regional to global scales and is used for environmental policy and decision-making. At the local level, numerous examples of the use of high-spatial resolution imagery exist. However, the lack of operational routines for the processing of satellite imagery and the difficulties in acquiring cloud free imagery due to the low revisit time of the sensors has prevented the full operationalization of remote sensing. Agreements have been recently established among the Space Agencies in the International Space Charter for the provision of remote sensing data for wildfire crisis management, which permits the rapid monitoring of critical fire events.

ACKNOWLEDGMENTS

The pre-fire conditions study is a compendium of the works from the following students: L. Gallant, S. White, M. Doyle, M. Abbott, G. Strickland, K. Abbott, S. Oldford, P.A Fernandez-Garcia, M. Kong and L. Bourgeau-Chavez. *ERS-1* and *RADARSAT-1* SAR image processing were assisted by N. French and G. Staples Field support for the *RADARSAT-2* study was provided by A. Shenoy, E. Kasischke, K. Riordan, K. Manies, and N. French. The whole study was funded by *CIFFC, MacDonald, Dettwiler and Associates Ltd. (MDA),* and by several *NSERC* Discovery grants awarded to B. Leblon. Additional funding for this research was provided by *NASA* grants NNX09AM15G and NNG04GR24G. *NOAA-AVHRR* images were provided under the *EODS* program, thanks to J. Cihlar and J. Chen, Canada

Centre of Remote Sensing, Ottawa. *ERS-1* SAR images were provided by E.S. Kasischke. RADARSAT-1 images were acquired under the *ADRO-2* program of the Canadian Space Agency. *RADARSAT-2* polarimetric images were provided by two Canadian Space Agency *SOAR* grants (*SOAR#445* and *SOAR-E#009*). Weather data were obtained from Marty Alexander, Canadian Forest Service, Edmonton and Rick Lanoville, Government of Northwest Territories, Fort Smith. Examples for the sections on active wildfire detection and burnt area mapping were extracted from the European Forest Fire Information System (EFFIS). This is the result of a team of scientists working in different fire-related disciplines at the at the European Commission Joint Research Centre (http://effis.jrc.ec.europa.eu)

REFERENCES

Abbott K, Leblon B, Staples G, Alexander ME, MacLean D (2007) Fire danger monitoring in a northern boreal forest region using RADARSAT-1 imagery. *International Journal of Remote Sensing* 28(5-6), 1317–1338.

Aguado I, Chuvieco E, Salas J (2003) Assessment of forest fire danger conditions in southern Spain from NOAA images and meteorological indices. *International Journal of Remote Sensing* 24(8), 1653–1668.

Akther MS, Hassan QK (2011) Remote sensing-based assessment of fire danger conditions over boreal forest. *IEEE Journal of Selected Topics in Applied Earth Observations and Remote Sensing* 4(4), 992–999.

Alkhatib AA (2014) A review on forest fire detection techniques. *International Journal of Distributed Sensor Networks (in press)*

Al-Moustafa T, Armitage RP, Danson FM (2012). Mapping fuel moisture content in upland vegetation using airborne hyperspectral imagery. *Remote Sensing of Environment* 127, 74–83.

Amraoui M, Leberato MLR, Calado TJ, DaCamara CC, Pinto Coelho LP, Trigo RM, Gouvela CM (2013) Fire activity over Mediterranean Europe based on information from Meteosat-8, *Forest Ecology and Management* 294, 62-75.

Anderson MC, Hain C, Wardlow B, Pimstein A, Mecikalski JR, Kustas WP (2011) Evaluation of drought indices based on thermal remote sensing of evapotranspiration over the continental United States. *Journal of Climate* 24, 2025–2044.

Arino O, Casadio S, Serpe D (2012) Global night-time fire season timing and fire count trends using the ATSR instrument series. *Remote Sensing of Environment* 116, 226–238.

Bastarrika A, Chuvieco E, Martin MP (2010) Mapping burned areas from Landsat TM/ETM+ data with a two-phase algorithm: Balancing omission and commission errors. *Remote Sensing of Environment* 115, 1003-1012

Bastiaanssen WGM, Menenti M, Fedded RA, Holtslag AAM (1998) A remote sensing surface energy balance algorithm for land (SEBAL). 1. Formulation. *Journal of Hydrology* 212–213, 198–212.

Bisquert M, Sánchez JM, Caselles V (2014) Modeling fire danger in Galicia and Asturias (Spain) from MODIS images. *Remote Sensing* 6, 540–554.

Boschetti L, Roy D. Barbosa P, Boca R (2007) A MODIS assessment of summer 2007 extent burned in Greece. *International Journal of Remote Sensing* 8, 2433-2436.

Boschetti L, Roy D, Justice C, Giglio L (2010) Global assessment of the temporal reporting accuracy and precision of the MODIS burned area product. *International Journal of Wildland Fire* 19, 705–709.

Bourgeau-Chavez LL (2013) *Improving estimates of organic layer fuel moisture in boreal ecosystems with polarimetric sar for fire danger assessment*, Ph.D. thesis, University of New Brunswick, Faculty of Forestry and Environmental Management, 254 pages.

Bourgeau-Chavez LL, Harrell PA, Kasischke ES, French NHF (1997) The detection and mapping of wildfires. *International Journal of Remote Sensing* 18, 355-373.

Bourgeau-Chavez LL, Kasischke ES, Rutherford MD (1999) Evaluation of ERS SAR data for prediction of fire danger in a boreal region. *International Journal of Wildland Fire* 9(3), 183–194.

Bourgeau-Chavez LL, Kasischke ES, Brunzell S, Mudd JP (2002) Mapping fire scars in global boreal forests using imaging radar data. *International Journal of Remote Sensing* 23, 4211-4234.

Bourgeau-Chavez LL, Garwood G, Riordann K, Cella B, Alden S, Kwart M, Murphy K (2007). Improving the prediction of wildfire potential in boreal Alaska with satellite imaging radar. *Polar Record* 43(4), 321–330.

Bourgeau-Chavez LL, Leblon B, Buckley J, Charbonneau F (2013a) Assessment of polarimetric SAR data for fuel moisture estimation: analysis of wet versus dry conditions. *International Journal of Remote Sensing* 34(16), 5709–5730.

Bourgeau-Chavez LL, Leblon B, Buckley J, Charbonneau F (2013b) Estimating fuel moisture using RADARSAT-2 polarimetric SAR images, *Remote Sensing of Environment* 132, 71–85.

Bowman DMJS, Balch JK, Artaxo P, Bond WJ, Carlson JM, Cochrane MA, D'Antonio CM, DeFries RS, Doyle JC, Harrison SP, Johnston FH, Keeley JE, Krawchuk MA, Kull CA, Marston JB,. Moritz MA, Prentice IC, Roos CI, Scott AC, Swetnam TW, van der Werf GR, Pyne SJ (2009) Fire in the Earth System. *Science* 324, 481-484.

Bowyer P, Danson FM (2004). Sensitivity of spectral reflectance to variation in live fuel moisture content at leaf and canopy level. *Remote Sensing of Environment* 92, 297–308.

Briess K, Jahn H, Lorenz E., Oertel D, Skrbek, W, Zhukov, B (2003) Fire recognition potential of the Bi-spectral InfraRed Detection (BIRD) satellite. *International Journal of Remote Sensing* 24(4), 865–872.

Bugarić M, Jakovčević T, Stipaničev, D (2014) Adaptive estimation of visual smoke detection parameters based on spatial data and fire risk index. *Computer Vision and Image Understanding* 118, 184–196.

Burgan RE, Klaver RW, Klaver JM (1998) Fuel models and fire potential from satellite and surface observations. *International Journal of Wildland Fire* 8(3), 159–170.

Caccamo G, Chisholm LA, Bradstock RA, Puotinen ML, Pippen BG (2012) Monitoring live fuel moisture content of heathland, shrubland and sclerophyll forest in south eastern Australia using MODIS data. *International Journal of Wildland Fire* 21, 257–269.

Cahoon DR, Stocks BJ, Levine JS, Cofer WR, Chung CC (1992) Evaluation of a technique for satellite-derived area estimation of forest fires, *Journal of Geophysical Research* 97, 0148-0227.

Calle A, Casanova JL, Romo, A (2006) Fire detection and monitoring using MSG Spinning Enhanced Visible and Infrared Imager (SEVIRI) data. *Journal of Geophysical Research - Biogeosciences* 111(G4), 1–13.

Camia A, Bovio G, Aguado I, Stach N (1999) Meteorological fire danger indices and remote sensing. In 'Remote Sensing of Large Wildfires in the European Mediterranean Basin'. (Ed. E. Chuvieco) pp. 39–59. (Springer-Verlag, Berlin)

Forestry Canada Fire Danger Group (1992) Development and structure of the Canadian forest fire behaviour prediction system. Forestry Canada, Information Report ST-X-3 (Ottawa, ONT, Canada).

Canadian Forest Service (2014) Facts about wildland fires in Canada. Available from http://www.nrcan.gc.ca/forests/fire/13143.

Ceccato P, Flasse S, Tarntola S, Jacquemoud S, Grégoire JM (2001) Detecting vegetation leaf water content using reflectance in the optical domain. *Remote Sensing of Environment* 77(1), 22–33.

Chen D, Loboda T, Channan S, Hoffman-Hall A (2014) Long-term record of sampled disturbances in northern Eurasian boreal forest from pre-2000 Landsat data. *Remote Sensing* 6(7), 6020–6038.

Chowdhury EH, Hassan QK (2013) Use of remote sensing-derived variables in developing a forest fire danger forecasting system. *Natural Hazards* 67(2), 321–334.

Chowdhury EH, Hassan QK (2014) Operational perspective of remote sensing-based forest fire danger forecasting systems. *ISPRS Journal of Photogrammetry and Remote Sensing (in press).*

Chuvieco E (2009) Earth observation of wildland fires in Mediterranean ecosystems, 257 pages (Springer-Verlag, Berlin).

Chuvieco E, Martin MP (1994) Global fire mapping and fire danger estimation using AVHRR images. *Photogrammetry Engineering and Remote Sensing* 60(5), 563–570.

Chuvieco E, Salas FJ, Carvacho L, Rodirguez-Silva F (1999a) Integrated fire risk mapping. In 'Remote Sensing of Large Wildfires in the European Mediterranean Basin'. (Ed. E. Chuvieco) pp. 61–84. (Springer-Verlag, Berlin).

Chuvieco E, Deshayes M, Stach N, Cocero D, Riaño D (1999b) Short-term fire risk foliage moisture content estimation from satellite data. In 'Remote Sensing of Large Wildfires in the European Mediterranean Basin'. (Ed. E. Chuvieco) pp. 17–38. (Springer-Verlag, Berlin).

Chuvieco E, Riaño D, Aguado I, Cocero D (2002a) Estimation of fuel moisture content from multitemporal analysis of Landsat Thematic Mapper reflectance data: Applications in fire danger assessment. *International Journal of Remote Sensing* 23, 2145–2162.

Chuvieco E, Martin MP, Palacios A (2002b) Assessment of different spectral indices in the red-near-infrared spectral domain for burned land discrimination. *International Journal of Remote Sensing* 23, 5103-5110.

Chuvieco E, Aguado I, Cocero D, Riaño D (2003) Design of an empirical index to estimate fuel moisture content from NOAA-AVHRR analysis in forest fire danger studies. *International Journal of Remote Sensing* 24(8), 1621–1637.

Chuvieco E, Cocero D, Aguado I, Palacios-Orueta A, Prado E (2004a). Improving burning efficiency estimates through satellite assessment of fuel moisture content. *Journal of Geophysical Research-Atmospheres* 109(D14S07), 1–8.

Chuvieco E, Cocero D, Riaño D, Martín MP, Martínez-Vega J, de la Riva J, Pérez F (2004b). Combining NDVI and surface temperature for the estimation of live fuel moisture content in forest fire danger rating. *Remote Sensing of Environment* 92, 322–331.

Chuvieco E, Giglio L, Justice C (2008) Global characterization of fire activity: Toward defining fire regimes from Earth observation data. *Global Change Biology* 14, 1488–1502.

Chuvieco E, Aguado I, Yebra M, Nieto H, Salas J, Pilar M, Vilar L, Martínez J, Martín S, Ibarra P, de la Riva J, Baeza J, Rodríguez F, Molina JM, Herrera MA, Zamora R. (2010). Development of a framework for fire risk assessment using remote sensing and geographic information system technologies. *Ecological Modelling* 221(1), 46–58.

Cloude SR, Pottier E (1997) An entropy based classification scheme for land applications of polarimetric SAR. *IEEE Transactions Geoscience and Remote Sensing* 35(1), 68–78.

Csiszar IA, Morisette JT, Giglio L (2006) Validation of active fire detection from moderate-resolution satellite sensors: The MODIS example in northern Eurasia. *IEEE Transactions on Geoscience and Remote Sensing* 44, 1757–1764.

Danson FM, Bowyer P (2004) Estimating live fuel moisture content from remotely sensed reflectance. *Remote Sensing of Environment* 92, 309–321.

Deblonde G, Cihlar J (1993) A multiyear analysis of the relationship between surface environmental variables and NDVI over the Canadian landmass, *Remote Sensing of Environment* 7, 151–177.

Dennison PE, Roberts DA, Peterson SH, Rechel J (2005) Use of Normalized Difference Water Index for monitoring live fuel moisture. *International Journal of Remote Sensing* 26, 1035–1042.

Desbois N, Vidal A (1995) La télédétection dans la prévision des incendies de forêt, *Ingénieries- EAT* 1, 21–29.

Desbois N, Vidal A (1996) Real-time monitoring of vegetation flammability using NOAA-AVHRR thermal infrared data, *EARSeL Advanced in Remote Sensing* 4(4), 25–32.

Dominguez L, Lee BS, Chuvieco E, Cihlar J (1994) Fire danger estimation using AVHRR images in the Prairie provinces of Canada. *Proceedings 2nd International Conference on Forest Fire Research*, 2(17), pp. 679–690 (Coimbra, Portugal).

Duchemin B, Guyon D, Lagouarde JP (1999) Potential and limits of NOAA-AVHRR temporal composite data for phenology and water stress monitoring of temperate forest ecosystems. *International Journal of Remote Sensing* 20(5), 895–917.

Elvidge CD, Imhoff ML, Kimberly EB, Hobson VR, Nelson I, Safran J, Dietz JB, Tuttle TB (2001) Night-time lights of the world: 1994-1995, *ISPRS Journal of Photogrammetry and Remote Sensing*, 56, 81-89.

Flannigan, MD, Amiro, BD, Logan, KA, Stocks, BJ, and Wotton BM (2005). Forest fires and climate change in the 21st century. *Mitigation and Adaptation Strategies for Global Change* 11 (4), 847–859.

Flasse SP, Ceccato P (1996) A contextual algorithm for AVHRR fire detection. *International Journal of Remote Sensing* 17, 419-424.

Food and Agriculture Organization (FAO) (2012). About the Global Fire Information Management System (GFIMS). Available from http://www.fao.org/nr/gfims/about/en/

Food and Agriculture Organization (FAO) (2014) *Forest and other Vegetation Fires,* http://www.fao.org/forestry/firemanagement/en/

Freeborn PH, Wooster MJ, Roberts G, Xu W (2014) Evaluating the SEVIRI fire thermal anomaly detection algorithm across the Central African Republic using the MODIS active fire product. *Remote Sensing* 6(3), 1890–1917.

Freeman A, Durden SL (1998) A three-component scattering model for polarimetric SAR data. *IEEE Transactions Geoscience and Remote Sensing* 36(3), 963– 973.

French NH, Kasischke ES, Bourgeau-Chavez LL, Harell, PA (1996) Monitoring variations in soil moisture on fire disturbed sites in Alaska using ERS-1 SAR imagery. *International Journal of Remote Sensing* 17, 3037-3053.

García M, Chuvieco E, Nieto H, Aguado I (2008) Combining AVHRR and meteorological data for estimating live fuel moisture content. *Remote Sensing of Environment* 112, 3618–3627.

Giglio L, Descloitres, J, Justice C, Kaufman, Y (2003) An enhanced contextual fire detection algorithm for MODIS. *Remote Sensing of Environment* 87, 273–282.

Giglio L, Csiszar I, Justice C (2006) Global distribution and seasonality of active fires as observed with the Terra and Aqua Moderate Resolution Imaging Spectroradiometer (MODIS) sensors. *Journal of Geophysical Research – Biogeosciences* 111(G2), 1–12.

Giglio L, Csiszar I, Restás Á, Morisette JT, Schroeder W, Morton D, Justice CO (2008). Active fire detection and characterization with the advanced spaceborne thermal emission and reflection radiometer (ASTER). *Remote Sensing of Environment* 112(6), 3055–3063.

Giglio L, Randerson JT, van der Werf GR, Kasibhatia PS, Collatz GJ, Morton DC, DeFries RS (2010) Assessing variability and long-term trends in burned area by merging multiple satellite fire products. *Biogeosciences* 7, 1171-1186.

Gillon D, Dauriac F, Deshayes M, Valette JC, Moro C (2004) Estimation of foliage moisture content using near infrared reflectance spectroscopy. *Agricultural and Forest Meteorology* 124, 51–62.

Gimeno M, San-Miguel-Ayanz J (2004) Evaluation of RADARSAT-1 data for identification of burnt areas in Southern Europe. *Remote Sensing of Environment* 92, 370-375.

Gimeno M., San-Miguel-Ayanz J, Schmuck G (2005) Identification of burnt areas in Mediterranean forest environments from ERS-2 SAR time series. *International Journal of Remote Sensing* 25, 4873-4888.

Gitas IZ, Mitri G, Veraverbeke S, Polychronaki A (2012). Advances in remote sensing of post-fire monitoring: a review. In 'Remote Sensing of Biomass: Principles and Applications' (Ed. T. E. Fatoyinbo), pp. 143-176 (InTech Pub., Rijeka, Croatia).

Goodenough DG, Chen H, Richardson A, Cloude S, Hong W, Li Y (2011). Mapping fire scars using Radarsat-2 polarimetric SAR data, *Canadian Journal of Remote Sensing* 37(5), 500–509.

Goodwin, NR, Collett, LJ (2014) Development of an automated method for mapping fire history captured in Landsat TM and ETM+ time series across Queensland, Australia. *Remote Sensing of Environment* 148, 206–221.

Gouyet JF, King C, Le Gleau H, Malon JF, Phulpin T, Valette JC (1991) Apport des données satellitaires NOAA-AVHRR dans le suivi de la végétation forestière. *Proceedings 5th International Colloquium on Physical Measurements and Signatures in Remote Sensing*, pp. 625–629 (Courchevel, France).

Goward SN, Waring RH, Dye DG, Yang J (1994) Ecological remote sensing of OTTER satellite macroscale observations. *Ecological Applications* 4(2), 322–343.

Granger RJ (1997) Comparison of surface and satellite-derived estimates of evapotranspiration using a feedback algorithm. *Proceeding. 3rd International Workshop on Application of Remote Sensing in Hydrology*, pp. 71–81, (Goddard Space Flight Center, Washington, DC, USA).

Grégoire JM, Tansey K, Silva JMN (2003) The GBA2000 initiative: Developing a global burnt area database from SPOT-VEGETATION imagery. *International Journal of Remote Sensing* 24, 1369–1376.

Hansen MC, Potapov PV, Moore R, Hancher M, Turubanova SA, Tyukavina A, Thau D, Stehman SV, Goetz SJ, Loveland TR, Kommareddy A, Egorov A, Chini L., Justice CO, Townshend JRG(2013) High-resolution global maps of 21st century forest cover change. *Science* 342(6160), 850–853.

Hao XJ, Qu JJ (2007) Retrieval of real-time live fuel moisture content using MODIS measurements. *Remote Sensing of Environment* 108, 130–137.

Hao WM, Larkin NK (2014) Wildland fire emissions, carbon, and climate: wildland fire detection and burned area in the United States. *Forest Ecology and Management* 317, 20-25.

Hardy CC, Burgan RE (1999) Evaluation of NDVI for monitoring live moisture in three vegetation types of the Western U.S. *Photogrammetry Engineering and Remote Sensing* 65, 603–610.

Hyer EJ, Reid JS, Prins EM, Hoffman JP. Schmidt CC, Miettinen JI, Giglio L (2013) Patterns of fire activity over Indonesia and Malaysia from polar and geostationary satellite observations. *Atmospheric Research* 122, 504–519.

Illera P, Fernandez A, Delgado JA (1996) Temporal evolution of the NDVI as an indicator of forest fire danger. *International Journal of Remote Sensing* 17(6), 1093–1105.

Johnston F, Henderson S, Chenc Y, Randerson J, Marlier M, DeFries R, Kinney P, Bowman D, Frauer M (2012) Estimated global mortality attributable to smoke form landscape fire. *Environmental Heath Perspectives* 120, 695-701.

Jurdao S, Yebra M, Guerschman JP, Chuvieco E. (2013). Regional estimation of woodland moisture content by inverting radiative transfer models. *Remote Sensing of Environment* 132, 59–70.

Justice CO, Townshend JRG, Vermote EF, Masuoka E, Wolfe RE, Saleous M, Roy DP, Morisette JT (2002) An overview of MODIS land data processing and product status. *Remote Sensing of Environment* 83, 3-15.

Justice CO, Roman MO, Csiszar I, Vermote EF, Wolfe RE, Hook SJ, Friedl M, Wang Z, Schaaf CB, Miura T, Tschudi M, Riggs G, Hall DK, Lyapustin AL, Devadiga S, Davidson C, Masuoka EJ (2013) Land and cryosphere products from Suomi NPP VIIRS: overview and status. *Journal of Geophysical Research-Atmospheres* 118, 9753–9765.

Kasischke ES, French NHF (1993) Locating and estimating the areal extent of wildfires in Alaskan boreal forests using multiple-season AVHRR NDVI composite data. *Remote Sensing of Environment* 51 263-275

Kasischke ES, Bourgeau-Chavez LL, French NHF (1994) Observations of variations in ERS-1 SAR image intensity associated with forest fires in Alaska. *IEEE Transacions on Remote Sensing* 32, 206-210.

Katagis T, Gitas IZ, Toukiloglou P, Veraverbeke S, Goossens R (2014) Trend analysis of medium-and coarse-resolution time series image data for burned area mapping in a Mediterranean ecosystem, *International Journal of Wildland Fire* (*in press*).

Kennedy, PJ, Belward, AS, Grégoire, JM (1994) An improved approach to fire monitoring in West Africa using AVHRR data. *International Journal of Remote Sensing,* 15(11), 2235–2255.

Kempka T, Kaiser T, Solbach K (2006) Microwaves in fire detection. *Fire Safety Journal* 41, 327–333.

Kogan FN (2001) Operational space technology for global vegetation assessment. *Bulletin of the American Meteorological Society* 82(9), 1949–1964.

Leblon B (2005). Using remote sensing for fire danger monitoring. *Natural Hazards* 35(3), 343–359.

Leblon B (2011) Research and training on the use of RADARSAT-2 polarimetric SAR images in several environmental applications. In Proceeding 8th Advanced SAR workshop, October 2011 (Canadian Space Agency, Montréal, Québec).

Leblon B, Chen J, Alexander ME, White S (2001) Fire danger monitoring using NOAA-AVHRR NDVI images in the case of northern boreal forests. *International Journal of Remote Sensing* 22(14), 2839–2846.

Leblon B, Kaschike ES, Alexander ME, Doyle M, Abbott M (2002) Fire danger monitoring using ERS-1 SAR images over northern boreal forests. *Natural Hazards* 27(3), 231–255.

Leblon B, Fernandez-Garcia, PA, Oldford S, MacLean D, Flannigan M (2007) Using NOAA-AVHRR cumulative indices for estimating fire danger codes in northern boreal forests. *International Journal of Applied Earth Observations and Geoinformation* 9(3), 335–342.

Leblon B., Bourgeau-Chavez LL 2013. Wildfire. In Encyclopedia of Natural Hazards (Ed. P. Bobrowsky), p. 1102 (Springer-Verlag, Berlin)

Libonati R, DaCamara CC, Pereira JMC, Peres JF (2011) On a new coordinate system for improved discrimination of vegetation and burned areas using MIR/NIR information. *Remote Sensing of Environment* 115, 1464–1477.

Lopez, S., Gonzalez, F., Llop, R., & Cuevas, J.M. (1991). An evaluation of the utility of NOAA-AVHRR images for monitoring forest fire risk in Spain, *International Journal of Remote Sensing* 12(9), 1841–1851.

Maki M, Ishiahra M, Tamura M (2004) Estimation of leaf water status to monitor the risk of forest fires by using remotely sensed data. *Remote Sensing of Environment* 90, 441-450.

Matthews S (2013) Dead fuel moisture research: 1991–2012. *International Journal of Wildland Fire* 23(1), 78–92.

Matson M, Dozier J (1981) Identification of sub resolution high temperature sources using a thermal IR sensor. *Photogrammetric Engineering and Remote Sensing* 47, 1311-1318.

Menges CH, Bartolo RE, Bell D, Hill GJE (2004) The effect of savanna fires on SAR backscatter in northern Australia. *International Journal of Remote Sensing* 25(22), 4857-4871.

Michalek JL, French NHF, Kasischke ES, Johnson RD, Colwell JE (2000) Using Landsat TM data to estimate carbon release from burned biomass in an Alaskan spruce forest complex. *International Journal of Remote Sensing* 2, 323-338.

Mitri GH, Gitas IZ (2004) A semi-automated object-oriented model for burned area mapping in the Mediterranean region using Landsat TM imagery. *International Journal of Wildland Fire* 13, 367-376.

Mollicone D, Eva HD, Achard F (2006) Human role in Russian wild fires. *Nature* 440, 436-437.

Morisette J, Giglio L, Csiszar I, Justice C (2005) Validation of the MODIS active fire product over Southern Africa with ASTER data. *International Journal of Remote Sensing* 26, 4239–4264.

Mouillot F, Schultz MH, Yue C, Cadule P, Tansey K, Ciais, P. Chuvieco E (2014) Ten years of global burned area products from spaceborne remote sensing – A review: Analysis of needs and recommendations for future developments. *International Journal of Applied Earth Observation and Geoinformation* 26, 64-79.

Nemani RR, Running SW (1989) Estimation of regional surface resistance to evapotranspiration from NDVI and thermal-IR AVHRR data. *Journal of Applied Meteorology* 28(4), 276–284.

Newnham GJ, Verbesselt J, Grant IF, Anderson SAJ. (2011) Relative greenness index for assessing curing of grassland fuel. *Remote Sensing of Environment* 115, 1456–1463.

Nieto H, Aguado I, Chuvieco E, Sandholt I (2010) Dead fuel moisture estimation with MSG-SEVIRI data. Retrieval of meteorological, data for the calculation of equilibrium moisture content. *Agricultural and Forest Meteorology* 150, 861–870

Oldford S, Leblon B, Gallant L, Alexander M. (2003). Mapping pre-fire conditions in the Northwest Territories, Canada, using NOAA-AVHRR images. *Geocarto International* 18(4), 21–32.

Oldford S, Leblon B, MacLean D, Flannigan M. (2006). Predicting slow drying fuel moisture codes using NOAA-AVHRR images. *International Journal of Remote Sensing* 27(18), 3881–3902.

Olioso A, Chauki H, Courault D, Wigneron JP (1999). Estimation of evapotranspiration and photosynthesis by assimilation of remote sensing data into SVAT models. *Remote Sensing of Environment* 68, 341–356.

Oliveira S, Pereira JMC, Carreiras J (2012) Fire frequency analysis in Portugal (1975–2005), using Landsat-based burnt area maps. *International Journal of Wildland Fire* 21, 48–60.

Paltridge GW, Barber J (1988) Monitoring grassland dryness and fire potential in Australia with NOAA-AVHRR data. *Remote Sensing of Environment* 25(3), 381–394.

Pereira JMC (1999) A comparative evaluation of NOAA/AVHRR vegetation indices for burned surface detection and mapping. *IEEE Transactions on Geoscience and Remote Sensing* 37, 217–226.

Pereira JMC, Setzer AW (1993) Spectral characteristics of fire scars in Landsat-5 TM images of Amazonia. *International Journal of Remote Sensing* 14, 2061-2078.

Peterson S, Roberts DA, Dennison PE (2008) Mapping live fuel moisture with MODIS data: A multiple regression approach. *Remote Sensing of Environment* 112, 4272–4284.

Pierce LL, Running SW, Riggs GA (1990). Remote detection of canopy water stress in coniferous forests using the NS001 Thematic Mapper Simulator and the Thermal Infrared Multispectral Scanner. *Photogrammetry Engineering and Remote Sensing,* 56(5), 579–586.

Pinol J, Filella I, Ogaya R, Penuelas J. (1998). Ground-based spectroradiometric estimation of live fine fuel moisture of Mediterranean plants. *Agricultural and Forest Meteorology* 90(3), 173–186.

Prosper-Laget V, Douguedroit A, Guinot JP (1994) A satellite index of forest fire occurence risk in summer in the Mediterranean area. In *Proceedings 2nd International Conference on Forest Fire Research*, 2, pp. 637–646 (Coimbra, Portugal).

Prosper-Laget V, Wigneron JP, Guinot JP, Seguin B (1995) Utilisation du satellite NOAA pour la détection des risques d'incendies de forêts. *La Météorologie* 8(10), 28–38.

Pyne SJ (1996) Intrduction to Wildland Fire (J. Wiley and Sons, New York)

Qi Y, Dennison PE, Spencer J, Riaño D (2012) Monitoring live fuel moisture using soil moisture and remote sensing proxies. *Journal of the Association for Fire Ecology* 8, 71–87.

Riaño D, Chuvieco E, Salas J, Palacios-Orueta A, Bastarrika A (2002) Generation of fuel type maps from Landsat TM images and ancillary data in Mediterranean ecosystems. *Canadian Journal of Forest Research* 32(8), 1301–1315

Riaño D, Vaughan P, Chuvieco E, Zarco-Tejada P, Ustin SL (2005). Estimation of fuel moisture content by inversion of radiative transfer models to simulate equivalent water thickness and dry matter content: Analysis at leaf and canopy level. *IEEE Transactions on Geoscience and Remote Sensing* 43, 819–826.

Roberts DA, Peterson S, Dennison PE, Sweeney S, Rechel J (2006) Evaluation of Airbone Visible/Infrared Imaging Spectrometer (AVIRIS) and Moderate Resolution Imaging Spectrometer (MODIS) measures of live fuel moisture and fuel condition in a shrubland ecosystem in southern California. *Journal of Geophysical Research-Biogeosciences*, 111 (GO4S02), 1–16.

Roberts GJ, Wooster MJ (2008) Fire detection and fire characterization over Africa using Meteosat SEVIRI. *IEEE Transactions on Geoscience and Remote Sensing* 46(4), 1200–1218.

Rosas-Romero R (2014) Remote detection of forest fires from video signals with classifiers based on K-SVD learned dictionaries. *Engineering Applications of Artificial Intelligence* 33, 1–11.

Roy D, Boschetti L, Justice C, Ju J (2008) The collection 5 MODIS burned area product—Global evaluation by comparison with the MODIS active fire product. *Remote Sensensing of Environment* 112, 3690–3707.

Saatchi SS, Halligan K, Despain DG, Crabtree RL (2007) Estimation of forest fuel load from radar remote sensing. *IEEE Transactions on Geoscience and Remote Sensing* 45, 1726–1739.

San-Miguel-Ayanz J, Schulte E, Schmuck G, Camia A, Strobl P, Libertà G, Giovando C, Boca R, Sedano F, Kempeneers P, McInerney D, Withmore C, Santos de Oliveira S, Rodrigues M, Durrant T, Corti P, Oehler F, Vilar L, Amatulli G (2012) Comprehensive monitoring of wildfires in Europe: the European Forest Fire Information System (EFFIS). In 'Approaches to Managing Disaster and Assessing Hazards, Emergencies and Disaster Impacts' (Ed. J. Tiefenbacher), pp. 87–105 (InTech Pub., Rijeka, Croatia).

San-Miguel-Ayanz J, Moreno JM, Camia A (2013a) Analysis of large fires in European Mediterranean landscapes: Lessons learned and perspectives. *Forest Ecology and Management* 294, 11-22

San-Miguel-Ayanz J, Schulte, E, Schmuck G. Camia, A (2013b) The European Forest Fire Information System in the context of environmental policies of the European Union. *Forest Policy and Economics* 29, 19-25.

Schroeder W, Prins E, Giglio L, Csiszar I, Schmidt C, Morisette J, Morton D (2008) Validation of GOES and MODIS active fire detection products using ASTER and ETM plus data. *Remote Sensing of Environment* 112, 2711–2726.

Scott AC, Bowman DMJS, Bond WJ, Pyne SJ, Alexander ME (2014) Fire on Earth: an Introduction, 434 pages (Wiley-Blackwell, Oxford, UK).

Sedano F, Kempeneers P, Strobl P, McInerney D, San-Miguel-Ayanz J (2012) Increasing spatial detail of burned scar maps using IRS-AWiFS data for Mediterranean Europe. *Remote Sensing* 4, 726–744.

Sedano F, Kempeneers P, San-Miguel-Ayanz J, Strobl P, Vogt P (2013) Towards a pan-European burnt scar mapping methodology based on single date medium resolution optical remote sensing data. *International Journal of Applied Earth Observation and Geoinformation* 20, 52–59.

Siegert F, Ruecker G (2000) Use of multitemporal ERS-2 SAR images for identification of burnt scars in south-east Asian tropical rainforest. *International Journal of Remote Sensing* 21, 831–837.

Simon M, Plummer S, Flerens F, Hoelzemann JJ, Arino O (2004) Burnt area detection at global scale using ATSR-2: The GLOBSCAR products and their qualification. *Journal of Geophysical Research* 109(D14S02), 1-16.

Stow D, Niphadkar M, Kaiser J (2005) MODIS-derived visible atmospherically resistant index for monitoring chaparral moisture content. *International Journal of Remote Sensing* 26, 3867–3873.

Stow D, Madhura N, Kaiser J (2006) Time series of chaparral live fuel moisture maps derived from MODIS satellite data. *International Journal of Wildland Fire* 15, 347–360.

Strickland G, Leblon B, Gallant L, Alexander ME (2001) Monitoring fire danger of northern boreal forests from optical and thermal infrared NOAA-AVHRR images. In Proceedings of the 23th Canadian Remote Sensing Symposium, pp. 667–676, (Ste-Foy, Canada).

Tansey K, Grégoire JM, Defourny P, Leigh R, Pekel JP, van Bogaert E, Bartholomé E (2008) A new, global, multi/annual (2000-2007) burnt area product at 1 km resolution. *Geophysical Research Letters* 35(L01401), 1–6.

Teillet PM, Dudelzak AE, Pultz TJ, McNairn H, Chichagov A (2001) A framework for *in-situ* sensor measurement assimilation in remote sensing applications. In Proceedings of the 23th Canadian Remote Sensing Symposium, pp. 111–118 (Ste-Foy, Canada).

Touzi R, Goze S, LeToan T, Lopes A, Mougin E (1992) Polarimetric discriminators forSAR images. *IEEE Transactions on Geoscience and Remote Sensing* 30(5), 973–980

United Nations (1992) United Nations Framework Convention on Climate Change, 33 pages, http://unfccc.int/key_documents/the_convention/items/2853.php.

United Nations (1998) Kyoto Protocol to The United Nations Framework Convention on Climate Change, 21 pages, http://unfccc.int/kyoto_protocol/items/2830.php.

US Forest Service 2012. The Wildland Fire Assessment System. Available from http://www.wfas.net/

Ustin S, Roberts DA, Pinzon J, Jacquemoud S, Gardner M, Scheer G, Castaneda CM, Palacios-Orueta A (1998) Estimating canopy water content of chaparral shrubs using optical methods. *Remote Sensing of Environment* 65(3), 280–291.

Van der Werf GR, Randerson JR, Giblio L, Collatz, GJ, Mu M, Kasibhatia PS, Morton DC, DeFries RS, Jin Y, van Leeuwen TT (2010) Global fire emissions and the contribution of deforestation, savanna, forest, agricultura, and peat fires (1997-2009). *Atmospheric Chemistry and Physics* 10, 11707-11735.

Van Zyl JJ, Arii M, Kim Y (2011) Model-based decomposition of polarimetric sar covariance matrices constrained for nonnegative eigenvalues. *IEEE Transactions Geoscience and Remote Sensing* 49(9), 3452–3459.

Verbesselt J, Somers B, Lhermitte S, Jonckheere I, van Aardt J, Coppin P (2007) Monitoring herbaceous fuel moisture content with SPOT VEGETATION time-series for fire risk prediction in savanna ecosystems. *Remote Sensing of Environment* 108, 357–368.

Vidal A, Devaux-Ros C (1995) Evaluating forest fire hazard with a Landsat TM derived water stress index. *Agricultural and Forest Meteorology* 77(3-4), 207–224.

Vidal A, Pinglo F, Durand H, Devaux-Ros C, Maillet A (1994) Evaluation of a temporal fire risk index in Mediterranean forests from NOAA thermal IR. *Remote Sensing of Environment* 49(3), 296–303.

Wang L, Hunt ER Jr, Qu JJ, Hao X, Daughtry CST (2013) Remote sensing of fuel moisture content from ratios of narrow-band vegetation water and dry-matter indices. *Remote Sensing of Environment* 129, 103–110.

Weise DR, Wright CS (2014) Wildland fire emissions, carbon and climate: characterizing wildland fuels. *Forest Ecology and Management* 317, 26-40.

Wooster MJ, Xu W, Nightingale T (2012) Sentinel-3 SLSTR active fire detection and FRP product: Pre-launch algorithm development and performance evaluation using MODIS and ASTER datasets. Remote Sensing of Environment 120, 236–254.

Wu Z, He HS, Yan J, Liu Z, Liang Y (2014) Relative effects of climatic and local factors on fire occurrence in boreal forest landscapes of northeastern China. *Science of the Total Environment* 493, 472-480.

Yebra M, Chuvieco E (2009a) Linking ecological information and radiative transfer models to estimate fuel moisture content in the Mediterranean region of Spain: Solving the ill-posed inverse problem. *Remote Sensing of Environment* 113, 2403–2411.

Yebra M, Chuvieco E (2009b) Generation of a species-specific look-up table for fuel moisture content assessment. *IEEE Journal of Selected Topics in Applied Earth Observations and Remote Sensing* 2, 21–26.

Yebra M, Chuvieco E, Riaño D (2008) Estimation of live fuel moisture content from MODIS images for fire risk assessment. *Agricultural and Forest Meteorology* 148, 523–536.

Yebra M, Dennison PE, Chuvieco E, Riaño D, Zylstra P Jr, Danson FM, Qi Y, Jurdao S (2013) A global review of remote sensing of live fuel moisture content for fire danger assessment: Moving towards operational products. *Remote Sensing of Environment* 136, 455-468.

Zhang X, Kondragunta S, Roy DP (2014) Interannual variation in biomass burning and fire seasonality derived from geostationary satellite data across the contiguous United States from 1995 to 2011. *Journal of Geophysical Research-Biogeosciences* 119(6), 1147–1162.

Zhukov B, Lorenz E, Oertel D, Wooster M, Roberts G (2006) Spaceborne detection and characterization of fires during the bi-spectral infrared detection (BIRD) experimental small satellite mission (2001–2004). *Remote Sensing of Environment* 100(1), 29-51.

In: Current International Perspectives
Editors: Brigitte Leblon and Martin E. Alexander
ISBN: 978-1-63463-682-7

Chapter 3

FIRE DANGER RATING AS A TOOL FOR FIRE MANAGEMENT. IMPLEMENTATION OF THE CFFDRS: THE ARGENTINE EXPERIENCE

***M. C. Dentoni*[1,*]*, M. M. Muñoz*[2] *and D. S. Marek*[3]**
[1]Servicio Nacional de Manejo del Fuego and Universidad Nacional de la Patagonia San Juan Bosco Sede Esquel, Chubut, Argentina
[2]Consejo Agrario Provincial de Santa CruzVivero Forestal, Santa Cruz, Argentina
[3]Provincial Wildfire Coordination Centre. Ministry of Forest, Lands and Natural Resource Operations,Kamloops, BC, Canada

ABSTRACT

Since the early 1920s different approaches have been used to assess the complex environmental conditions that determine the variation in fire danger, from simple indices to complex systems composed of different indicators related to specific fire behaviour variables. Complex systems provide much more useful information that facilitate and improve the accuracy of management decisions, but their development requires longer term research, and their implementation poses more difficulties. When deciding how to evaluate danger in a certain fire environment, a deep understanding of the relative importance of the weather, fuel and topography is necessary. Availability of meteorological data and fire records often restricts the adjustment of indicators. The socio-cultural context in which the system will be implemented has to be considered, as well as the human and economic resources available. This was done during the development and implementation of a fire danger rating system (FDRS) for Argentina. The country has a wide diversity of fire environments and the characteristics of fire management agencies vary widely from one region to another. In 2000, a decision was made to adapt the Fire Weather Index (FWI), a component of the Canadian Forest Fire Danger Rating System (CFFDRS), for use in Argentina. The decision was based on the capacity of the index to be adjusted to different climates, the availability of the basic

[*]CorrespondingAuthoraddress: Servicio Nacional de Manejo del Fuego and Universidad Nacional de la Patagonia San Juan Bosco Sede Esquel, C.C. 14 – (9200) Esquel, Chubut, Argentina. Email: mcdentoni@gmail.com

meteorological data to calculate it and the wide range of operational applications. Also, FWI would serve as the necessary first step in developing a Fire Behaviour Prediction System (FBP) based on local fuels. The implementation has been gradual in the country and has faced great challenges because of the cultural and social constraints of introducing a foreign system and the long term economic and institutional support required. The strategy has been to train local personnel in its use and interpretation, and to involve them in the collection of data for adjustments. Case studies were done for the different regions and a training program for firefighters has been developed. The process has been very positive in the sense that it has induced a better understanding of the fire environment, has showed the benefits of making decisions based on quantified indicators, and has promoted a more proactive, safe attitude.

Keywords: Fire danger rating, fire management, fire weather indices, fire behaviour

INTRODUCTION

The importance of understanding the influence of the environment on the way fires burn, has interested foresters since the beginning of the last century. They noticed that fire behavior was closely related to the moisture content of dead material on the forest floor. These wildland fire research pioneers in fire behavior learned to predict how intense fires might be, simply by feeling moisture of the upper litter layer (Chandler 1991). In the early 1920s, researchers in Canada and the United Statesbegan systematically studying the relationship between moisture content of dead fuels and fire behavior, and the complex effects of weather on both. This research led to the development of fire danger meters to assist forest rangers in planning for anticipated fires and their control.These fire danger meters together with fire weather forecasts, significantly improved the task of determining staffing needs and equipment placement as fire danger changed; they became the '*first line of defense against fire during danger season*' (Zon 1941). Early workers in the USSR and Australia also developed equations to mathematically predict fire danger that led to the Nesterov Ignition Index (1949) and the McArthur Index (1966) respectively.These were followed by many other indices developed in various countries.

With the increasing diversity of fire danger indices, the terminology related to the subject became confusing and up to the present there is not full agreement in the meaning of terms such as `danger` and `risk´. To address terminology problems, the FAO (1986) published a glossary that states `*fire danger is the result of both constant and variable danger factors affecting the ignition, spread and difficulty of control of fires and the damage they cause*´, while fire risk is defined as `*the chance of fire starting, as affected by the nature and incidence of causative agencies; an element of fire danger in any area*´. It follows that a reliable index to evaluate danger should be comprehensive, taking into account all the aspects involved in the concept.

Fire danger indices were built with different mathematical approaches and input variables, to quantify the effects of factors relevant to fire behavior. In most cases, indices accurately represent well the fire environment for which they were developed, but they may be less suitable in other environments. Many fire danger indices give a broad idea of the fire danger condition for an area, but are not directly related to any specific aspect of fire behavior. More complex FDRS's are composed of an array of indicators that address different aspects of fire danger. That is the case for both, the National Fire Danger Rating System (NFDRS) developed in the US and the Fire Weather Index (FWI) System developed in Canada, the two most widely used (Dowdy et al. 2009). Both systems provide relative ratings that can be interpreted in terms of fire occurrence and behavior (Deeming et al. 1977; Van Wagner 1987).The development and implementation of this type of system is more complex, but its outputs are very useful tools for a wide range of operational decisions.

The process of systematically evaluating and integrating the individual and combined factors influencing fire danger is referred to as `fire danger rating´ (FDR) (Merrill and Alexander 1997). Its importance is recognized by fire management organizations of many countries. Alexander (2008), states *`The value of forest and rural fire management is a proven fact: fire control planning and decision making without a system of rating fire danger is like operating a boat without a rudder´.*

The FAO *Voluntary Guidelines for Principles and Fire Management Strategic Actions* (2006), recommend countries establish a FDRS or adapt an existing one to the local environment. To adapt an existing system, the relationships between indicators and fire behavior in the new environment have to be ensured (Fogarty et al. 1998). In both cases, training in the interpretation and use of fire-danger in fire operations is required, as well as a diagnosis of institutional structures and technological capabilities of operational agencies (Taylor and Alexander 2006). Key to the successful implementation of any implementation strategy is consideration of local culture and economic circumstances.

Wildfires occur year-round in Argentina, due to the wide diversity of ecosystems (Figure 1). In northern regions with summer rains, fire season starts in late fall and extends to early spring, while in the South fire season starts in spring and extends to fall.With the creation of the Dirección Forestal in 1943, the need for more effective forest fire control was recognized (Tortorelli 1947). A present, the type and level of organization varies among the different provinces. Large destructive fires in the late 1980's resulted in substantial growth of the northern Patagonia fire services. As a result, Patagonian fire agencies have led the country in the modernization of fire management organizations for decades.With the creation in 1996 of the Plan Nacional de Manejo del Fuego (PNMF), under the Secretaria de Ambiente y Desarrollo Sustentable (SAyDS), which recently gave way to the *Sistema Federal de Manejo del Fuego* (*SMNF*), fire management was organized at the national level. *SMNF* coordinates fire management activities with local fire agencies, standardizes procedures and promotes the development and application of new technologies.

The first fire danger indices used in the country were the Rodriguez y Moretti Index (1988) and the Monte Allegre Index (Sampaio et al. 1998). The Rodriguez y Moretti Index was used for several years in northwestern Patagonia, while the Monte Allegre Index was mainly used in the northeast of the country. These indices are both reliable in the regions where constructed, but they did not perform satisfactory in other fire environments. With increasing interaction between national and local fire agencies, the benefits of a single nation-wide index that would allow comparison of fire dangers throughout Argentina was recognized. Also, some fire managers started to recognize the need for danger indicators with more operational applications.

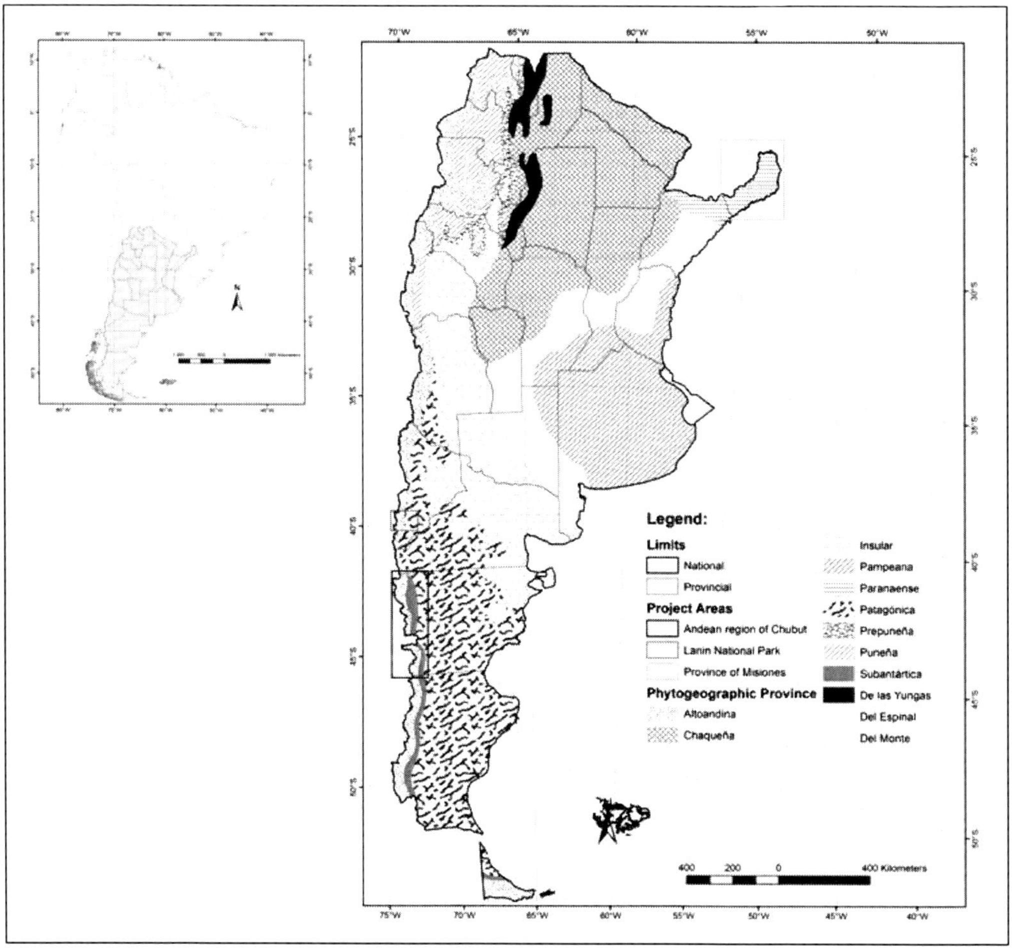

Figure 1. Phytogeographic regions of Argentina, based on Cabrera (1971), and pilot project areas for FWI implementation.

The development of the FDRS began in 2000, with the assistance of the British Columbia Forest Service (BCFS) Protection Branch and of the Canadian International Development Agency (CIDA). As a first step in the development of a FDRS, a review of different existing systems was undertaken (Dentoni and Muñoz 2012). This was followed by a project to test the meteorological subsystem FWI of the CFFDRS in three pilot project areas representing different fire environments, cultures, and levels of organization in fire management: two located in north-western Patagonia and the third in the province of Misiones (see Figure1).

Later, in his report of a visit to Argentina, Taylor (2001) stated `The CFFDRS can provide a suitable framework to organize existing data, scientific and technical expertise and knowledge of fire science and fire management in Argentina. There is a need to adapt technical aspects of the system for fire environment and develop mechanisms to collect, disseminate, and interpret and implement outputs appropriate to the existing agencies and technical situation. While a national fire danger rating system provides a common scientific core, implementation mechanisms will vary by agency and province´. In subsequent years, FDRS implementation was extended to other regions. Performance of FWI varied by region, as well as its acceptance by fire managers as a tool for decisions. The process has promoted a more proactive attitude, has shown the benefits of basing decisions on objective indicators, and has strengthened the desire for safer operations. It is the objective of this chapter, to analyze the different aspects involved in a FRDS development and implementation, through the experience carried on in Argentina.

PRELIMINARY CONSIDERATIONS

The expectations were that the implementation of a FDRS would strengthen the provincial and national fire management organizations by providing a fire management tool upon which to base operational decisions, making them safer and more efficient. During revision of the different systems, it was found that the Canadian Forest Fire Weather Index (FWI) Systemwill perform the best (Figure 2a). It would have advantages as being calculated with simple weather observations that could come from manual weather stations and using hand tables. These aspects were important because many of the existing stations in the country were, and still are, manually operated and because most of the fire agencies at that time did not have access to computers. FWI had been adapted to widely different environments in numerous countries such as New Zealand, China, Mexico, and US states, such as Alaska and Florida. Some Mediterranean countries of the European Union, had also adopted parts of the CFFDRS (Taylor 2001). Considering the different fire environments present in Argentina, it seemed very possitive to use a system that could be adjusted to different weather and fuel conditions. The limitations for certain environments, namely shurblands and grasslands, for which it was not developed, were clear. The challenge was to base fire danger rating on a common nation-wide system with necessary adaptations in each region; and its six components account for the effects of fuel moisture and wind on fire behaviour. The first three components, the fuel moisture codes, are numeric ratings of the moisture content of litter and other fine fuels, the average moisture content of loosely compacted organic layers of moderate depth, and the average moisture content of deep, compact organic layers.The remaining three components are fire behavior indices, which represent the rate of fire spread, the fuel available for combustion, and frontal fire intensity; their values change as the fire danger changes. Each of them relates to a different aspect of fire behaviour and gives useful information for different operational applications. As example, the Fine Fuel Moisture Code (FFMC) indicates ignition in fine fuels, Drought Code (DC) relates to mop up difficulties and, Initial Spread Index (ISI) is a good indicator of initial attack preparedness. As shown in Figure 2a, FWI combines the codes and indices in a single number that relates to fire intensity and is used to define fire danger classes with different approaches, such as scaling FWI and delimiting danger classes by percentiles (Turner 1973) or using a criteria based on Byram´s fire intensity concept (Alexander 2008). Fire agencies in the country with a high level of organization can use the system to its full capability, while others might use only FWI classes for simple applications such as providing fire danger levels for information to the public.

The potential of fire behaviour prediction systems like BEHAVE (Rothermel 1983) was already known and some Argentine fire managers tried to apply them, but the lack of fuel models was a strong limitation. The modular structure of the CFFDRS (Figure 2b) would facilitate adjustments during implementation of FWI and, at the same time, serve as the base for future development of fire behaviour prediction (FBP) systems for local fuels.

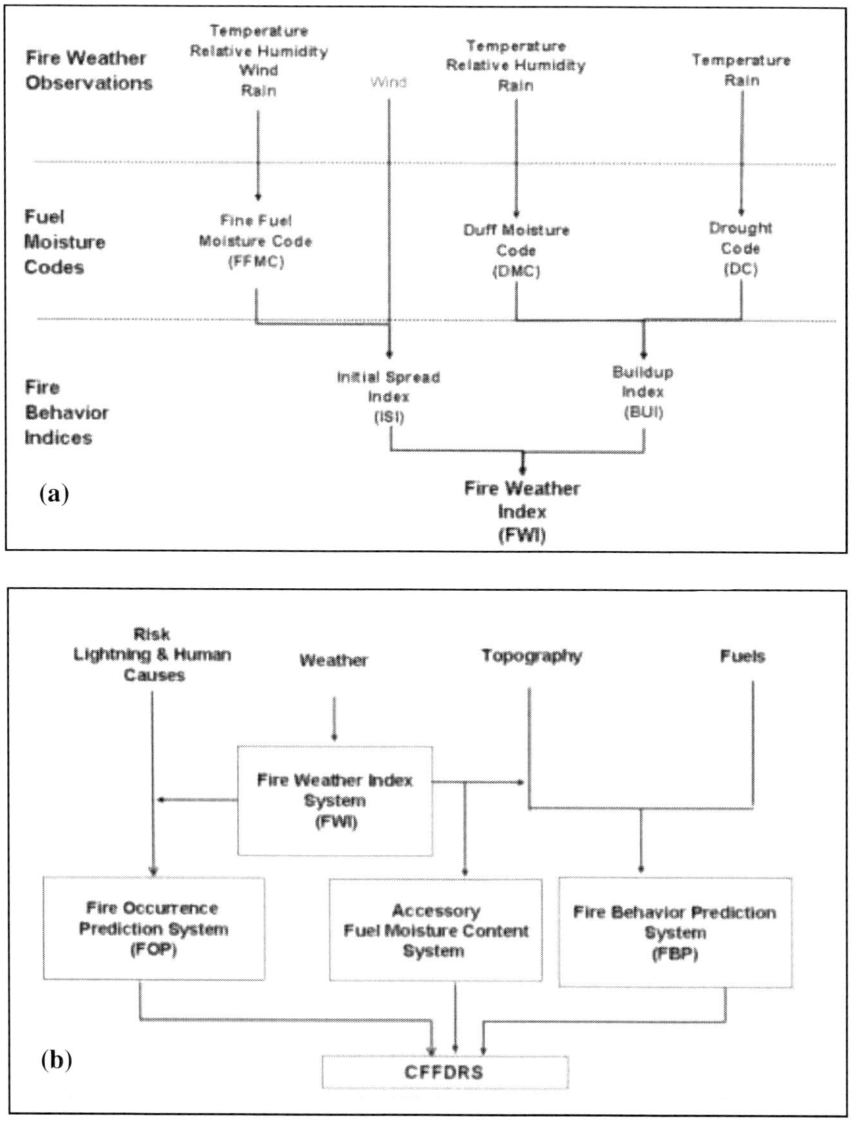

Figure 2. Diagram of the structure of (a) the Canadian Fire Weather Index (FWI) (after Van Wagner 1987) and (b) the Canadian Forest Fire Danger System (CFFDRS) (after Stocks 1989).

METHODS

Adjustments and Implementation in the Pilot Project Areas

A work plan was proposed in *Project 2000: Fire Management and Technology Transfer,* developed by BCFS and PNMF, which provided recommendations with regard to selection of project areas, data collection and analysis, training and meteorological support required. The outlined training program consisted of three courses:

1. *"The Fire Environment"*, to introduce dynamic relationship among the factors of weather, topography and fuels that affect fire behaviour;
2. *"Fire Weather"*, for a basic understanding of meteorological variables and weather phenomena crucial for predicting firebehaviour, and; and
3. *"Fire Danger Indices"*, to explain different ways of rating danger, in particular the structure, calculation, and interpretation of FWI.

In addition, training was implemented to show how to make temporal (hourly) and special (height) adjustments to specific codes and indices, to collect field samples, and to take and record fire behaviour observations. Publications in English relevant for the understanding of the FWI system were translated into Spanish.

A preliminary analysis of distribution of frequency of days in the fire season, number of fire starts, and area burned by FWI danger classes (Taylor 2001) was conducted for each project area. The individual components that make up the FWI were also tested against the features of fire activity they represent: for example, FFMC and Duff Moisture Code (DMC) against anthropogenic and natural-caused starts respectively (Dentoni et al. 2006). Values of codes and indices associated with large fires were compared with thresholds found in other countries (Alexander 1991; Kourtz and Todd 1992; Alexander and Cole 2001; de Groot et al. 2005). This initial analysis was restricted to the limited historic data available at the time, which included meteorological records, fire occurrence, and fire behaviour measurements / estimates.

Fire management organizations differed among project areas. In Patagonia there was a longer experience in fire management and a thus a more complex organizations. In the northeast, the state provincial fire management organizations were more rudimentary, and the fire environment itself was changing with increasing presence of plantations, mainly of pine and eucalyptus. Forest companies in the region were starting forest protection programs and led the process of FDR implementation in the region. In both cases the implementation steps were as follows:

a - A survey of meteorological stations was conducted, to select those that could be used for FWI calculations. The criteria was that the stations should fulfil established standards (Turner and Lawson 1978) and be able to collect and transmit data at 12 am noon LST everyday, the time at which FWI is calculated (often, data is collected only Monday through Friday at manual stations).

b - A communications network was established for daily collection of data and calculation of FWI. The advice was that calculations be done using hand tables for an introductory time period to reinforce understanding of the effect each weather variable has on the codes and indices.

c - Methods for delimiting five danger classes (extreme, very high, high, moderate and low) were developed:

  c.1 The two pilot project areas in northern Patagonia, Parque Nacional Lanin in the Province of Neuquén and the Andean region of Province of Chubut, are located in the rain shadow of the Andes, with a strong west-east precipitation gradient. Vegetation varies from western rain forest to the shrublands of the forest-steppe ecotone in the east. In the ecotone, there is also an increasing presence of pine plantations. Broad fuel type were defined as: forest type A, a closed, wet forest, with less wind penetration, and the presence of an organic layer of varying depth and the presence a significant amount of medium and large fuels on the floor, where the dominant species are lenga (*Nothofaguspumilio*) and ñire (*Nothofagusantarctica*) (Figures 3a and 3b); forest type B, dominant species are cipres de la cordillera (*Austrocedruschilensis*) or Pehuen (*Araucaria araucana*) with a more open canopy, exposed to a stronger wind influence, where the organic soil layer is almost absent and there is little to none dead medium and large material on the forest soil (Figures 3c and 3d); shrublands, which include various lower canopy species, grasslands, and pine plantations. In these fuel types the relative weight of BUI and ISI varies with the presence of dead fuels with different lag times, type of soil, or wind exposure; definition of danger classes based on this concept would give a more realistic evaluation of fire danger. Danger classes in forest type A were determined using combinations of BUI and ISI values, used in an area of BC with a similar climate. Based on the concept of associating danger classes with intensity classes (Alexander 2008) models C-7 and C-6 of the Canadian FBP system (Taylor et al.1996) were suggested respectively for forest type B and plantations (Taylor 2001). Model O1-b was used to evaluate danger in grasslands; the use of this model requires estimates of grass curing determined through sampling during the fire season. For shrublands, a model developed in New Zealand (Pearce 2001) was utilized. At noon, when indices are calculated, fire managers should check danger in the different fuel types present, and decide the danger class for the day.

Figure 3. Views of (a) and (b) forest type A and (c) and (d) forest type B.

c.2 The climatic characteristics in the north-east of the country differ from those of the climatic regions for which FWI has to date been implemented. For this reason, and because of the type of fire organizations, the decision was made to scale FWI and delimit danger classes by percentiles (Turner 1973; Van Wagner 1987). The more sophisticated decision process proposed for Patagonia was not considered appropriate for the small, less complex fire agencies of the region. As these organizations grow and mature, the plan was for them to test, modify as necessary, and then include grass, plantation, and slash models to complement existing danger ratings.

d - Operational prevention and presuppression guidelines were prepared, as decision aids for resources assignments according to danger class. They were built through consultation with local fire managers, considering the respective resources and capabilities.

e - Periodic meetings were scheduled to evaluate the potential danger situation, by addressing several basic questions (Beck 2006): Will a fire start? If so, how many? Where are they likely to occur? What is their expected behaviour and what values will theythreaten? What resources will be required to respond? The meteorological data and the codes and indices should be displayed as each of them explains a different feature of fire behaviour. Based on the defined final danger class, prevention and presuppressiondecissons are taken, according to the measures indicated in the operational guidelines for each class. A table proposed by Alexander (2008), that indicates fire activity expected for each danger class was provided to fire agencies as an aid for decisson making.

Extension to Other Regions of the Country

The fire danger climatology of different environments was accomplished by the analysis of historic FWI records. The database of daily weather records for calculation of FWI was provided by the *Servicio Meteorológico Nacional (SMN)*. The analysis used data from 37 meteorological stations, for the period 1994 to 2007. Prior to 1993 many SMN stations did not record noon data.Fire danger climatology is updated periodically as additional years are added to the data base. As an initial approach, danger classes were delimited by percentiles, assigning the 10 % of days with higher FWI values to the `Extreme´ class, next 15 % to `Very High´ danger class, 25 % to `High´ danger, 25 % to `Moderate´ and 25 to `Low´ danger class (Figure 4). Daily averages and maximums for each code and index were calculated for each series, to allow comparison of on-going seasons with statistical values. FWI associated with previous extreme fires were checked to ensure that the extreme climatic class was not underestimating danger, and a disproportionate number of large fires occurred in lower classes.

The extension was progressive and accompanied the development of fire agencies. The procedure in each region followed the same steps as explained for the pilot project areas. The time from initial introductory talks to system implementation took approximately 3 years. Periodically, technicians from the *BCFS Protection Branch Program* visited the country to supervise the activities. The program did not have a permanent source of funding which resulted in some delays.

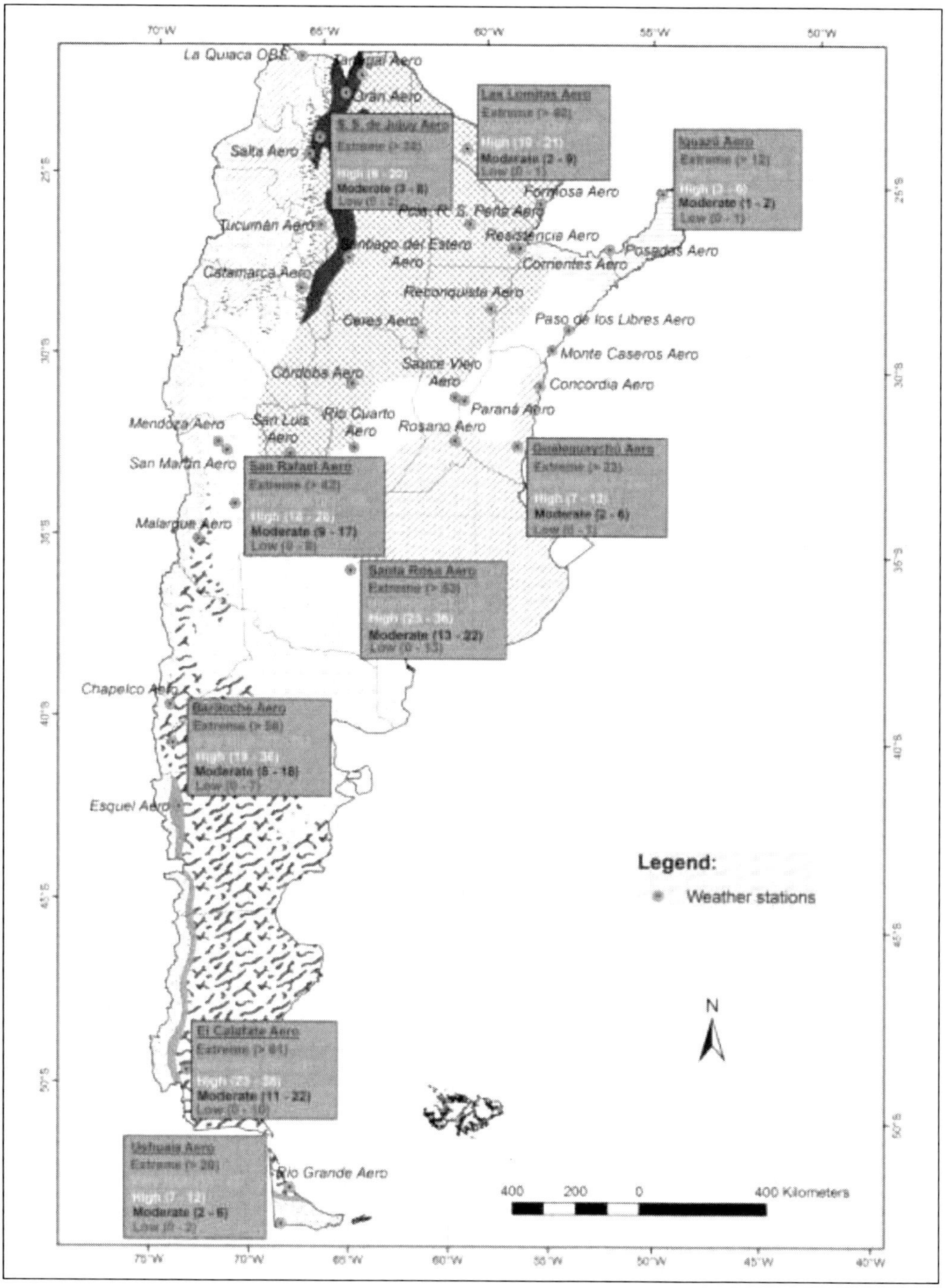

Figure 4. FWI danger classes for different regions of Argentina, based on data provided by the Servicio Meteorológico Nacional (SMN) for the 1994-2010 period.

Available fire records usually contained starting date, location, and vegetation affected, but nota detailed description of fire behaviour. Thus, the analysis was limited to the distribution of fire starts and area affected by danger class, and the fire starts were compared to FFMC to find threshold values for ignition. In the training of local technicians, special emphasis was placed on the observation of fire behaviour, data recording, and building a file with case studies. de Groot et al. (2006) foundthat hot spots increased over threshold FFMC values. This tool is being explored to calibrate FFMC in areas where fire records are extremely scarce or nonexistent.

The level of operational applications varied widely. In some areas, danger classes are only for general public information, but in other areas, the three levels of information provided by the system are used: FWI defines the danger class, ISI and BUI provide information on fire spread and fuel consumption, and FFMC, DMC, and DC indicate dryness of surface dead fuels and soil (Van Wagner 1987).

Meteorological Support

Through an agreement between the *SNMF* and the *SMN*, the necessary meteorological information is provided in a timely, consistent manner to operationally use the FDRS. This included the historic data base for statistical analysis, surface data for daily and forecasted indices calculation, upper air data, radar and satellite information and outputs of weather prediction models. SMNF forecasters work jointly with SMN forecasters in support of fire management.

Forecasts describing meteorological phenomena relevant to fire occurrence and behaviour are issued periodically. Weather events such as the approach of a cold front or lighting can result in a significant change in fire danger. For example, the presence of a subsidence inversion associated with a high pressure system in complex terrains requires adjustments of danger indices with elevation (Shunk et al. 2013). Also, a forecasted wind increase warns of the need to make ISI and FWI hourly adjustments.

RESULTS

Since 2000, the FWI has been partially or fully implemented in almost two thirds of the country (Figure 5). In some cases, the introduction of a foreign system was questioned. In most cases, a clear explanation of the reasons and expected benefits of FWI along with involvement of local people in making adjustments and implementation wassufficient to overcome initial reluctance. Although names of codes and indices were translated into Spanish in manuals and interpretative products, the initials were kept in English and are today common language in the Argentine fire community.

Along these years, over 40 technicians of provincial and national fire management agencies have been trained in the calculation and interpretation of the FWI system, taking and recording weather and fire behaviourobservations, as well as participating in fuel sampling protocols. Also, firefighters and fire managers have been trained to interpret the effects that weather, terrain, and fuels have on fire behaviour. The understanding of codes and indices has enhanced the observation of fire behaviour variables, such as frontal fire intensity, spread, flammability, or smouldering, thereby making their job safer.

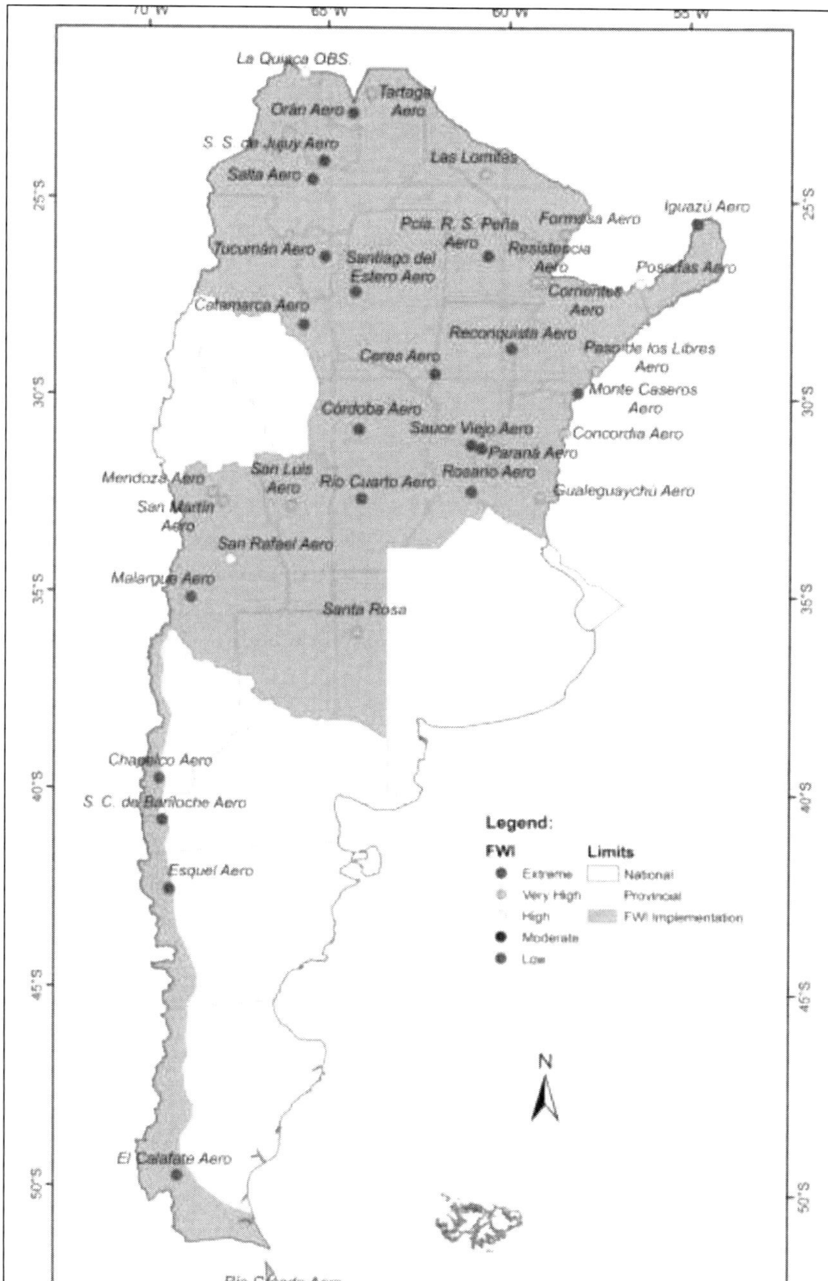

Figure 5. Danger class map published jointly by the *SNMF* and the *SMN*, for those regions of the country where FWI has been implemented. (http://www.ambiente.gob.ar/?idseccion=27; http://www.smn.gov.ar/?mod=dpd&id=4).

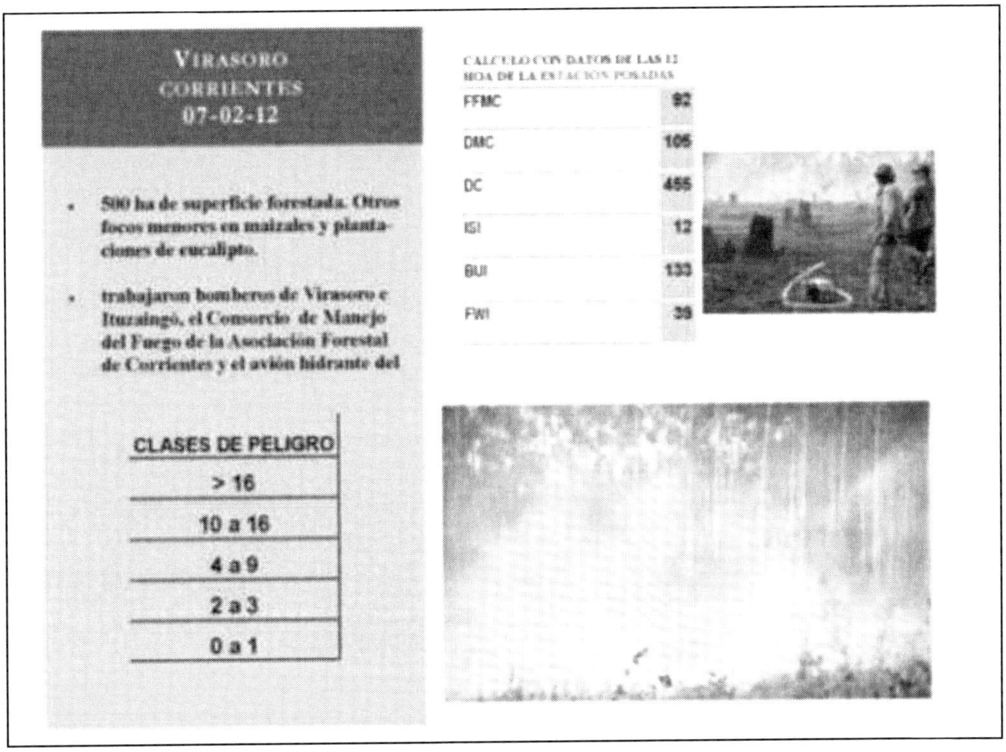

Figure 6. Example of interpretative poster relating codes and indices to fire behaviour.

Preliminary data analysis of number of starts and extent of area affected by danger class and FFMC threshold for ignition has been carried on for different regions with satisfactory results (Bianchi 2008). Promoting data analysis by technical and operational personnel has been extremely positive for acceptance of the system and in fine-tuning adjustments. Firefighters and officers were trained to develop or participate in case studies (Bari and Dinamarca 2013; Romero et al. 2013), and interpretative products are being developed to help associate the type of fire with code and index values (Figure 6).

It is strongly recommended that FWI danger classes defined by percentage of days in each classbe used with care, and operational applications limited to public information in shrubland environments. Adjustments to class limits were done according to FWI values associated with extreme fires. This is the case for danger classes based on data from the *Catamarca Aero* station, located in the Prepuneña phytogeographic region (Figure 4) where limits for danger classes originally defined by FWI percentiles were modified due to large firesthat occurred in the moderate class. Similar adjustments are being considered for locations in the Chaqueña, del Espinal, and del Monte regions. Fernandez (2005) proposed FWI/ISI relationships be used to better relate danger to fire behaviour in shrublands. These relationships should be investigated to improve fire danger predictions in Argentine shrublands. The model proposed to evaluate danger in grasslands has proved to be challenging because of the time and personnel required for sampling. For this, projects to develop photographic catalogues or use satellite data to replace sampling have been initiated.

At present, the index is calculated daily using the 72-hour forecast for nearly 60 meteorological stations. The calculations are done at local level with data collected according to capability (mailing, websites, and phone). Several fire agencies have acquired meteorological stations for the primary purpose of rating fire dangers. The recommendation of using hand tables has proved positive to help understand the impact of meteorological variables in the codes and indices. Also, Excel spreadsheets and a FORTRAN program are available for calculations. All equations have been adapted to the southern hemisphere. In particular, BUI is monitored during the fire season using data from 37 meteorological stations and compared to climatic trends (Figure 7). As BUI values in a region approach or exceed the mean value for the time of the year, large fires, which overcame local resources, involving medium and large fuels occur. As a preliminary test, BUI values associated with the initiation of 43 fires in randomly selected fire environments were compared to mean and extreme BUI values for the same date (Figure 8). BUI of the starting day was above the climatic average in 84% of the cases. In only one case, BUI was below 30, the threshold for increasing fire activity with increasingfuel availability, while in 26 cases, BUI was above 90, which corresponds to the Canadian threshold for severe fire behaviour potentials (Alexander and Cole 2001). Weather forecasts for prevention and presuppression planning are issued twice a week during the fire season and complement the information provided by the indices.

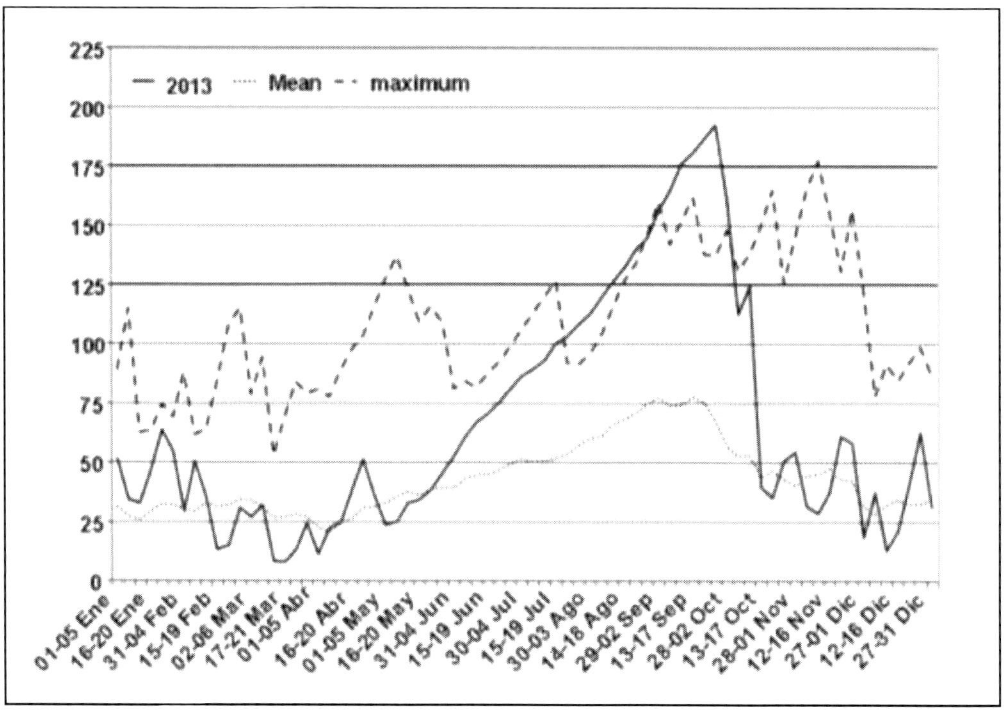

Figure 7. Annual BUI variationin 2013, and mean and maximum BUI over the 1994-2010 period, for the Río Cuarto Aero weather station (see location in Figure 4).

The routine proposed for planning prevention and presuppression was better accepted by young agencies that introduced FDR concepts and practices since their institutional beginnings. Older agencies find it difficult to establish a new process for operational decisions that differs from what they did for years. A special case exists in provinces wherefire management responsibilities are divided among administrative units. Usually prevention is under an environmental Secretary or Ministry, and fire suppression is the responsibility of volunteer or police firefighters. In these cases, operational applications other than roadside letters or press communication of the danger class are very difficult to put into practice. A different strategy is needed for these situations.

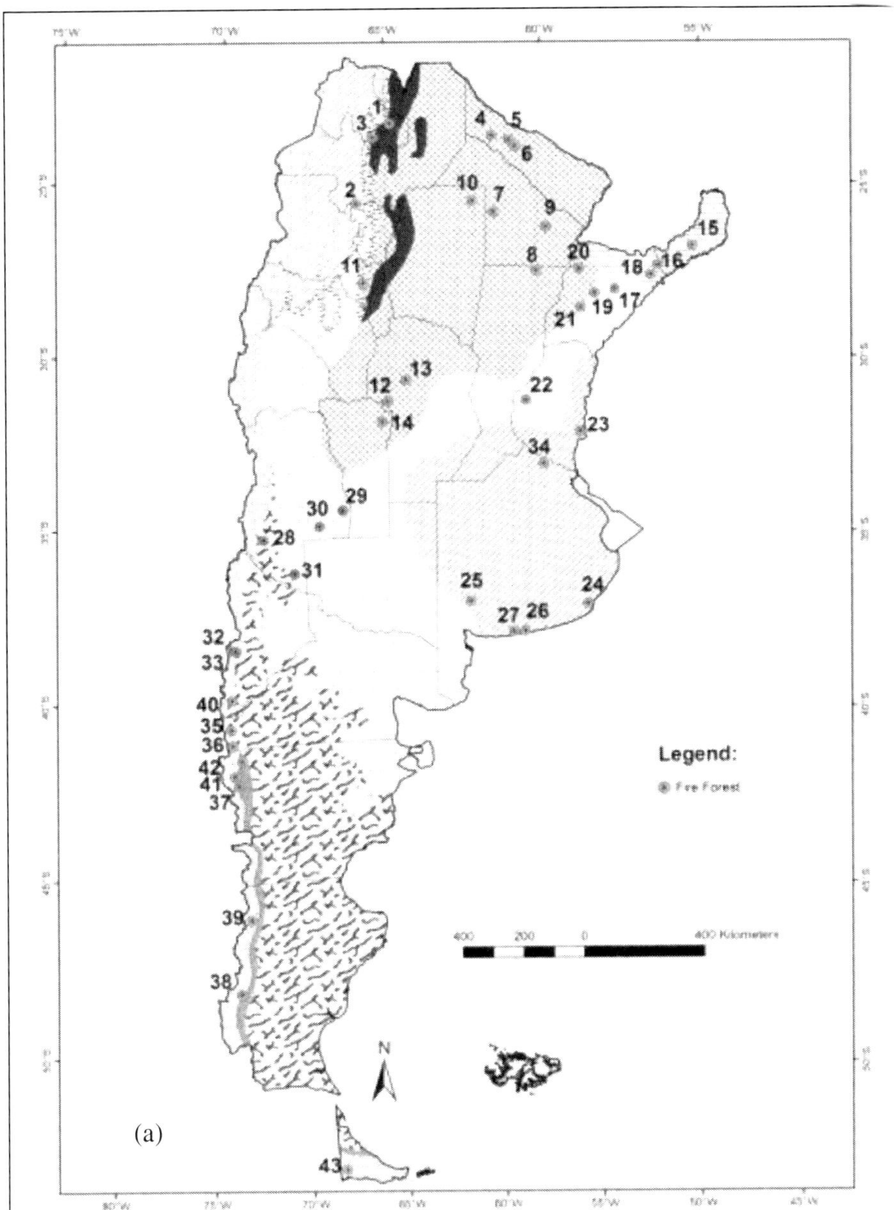

Figure 8. (Continued).

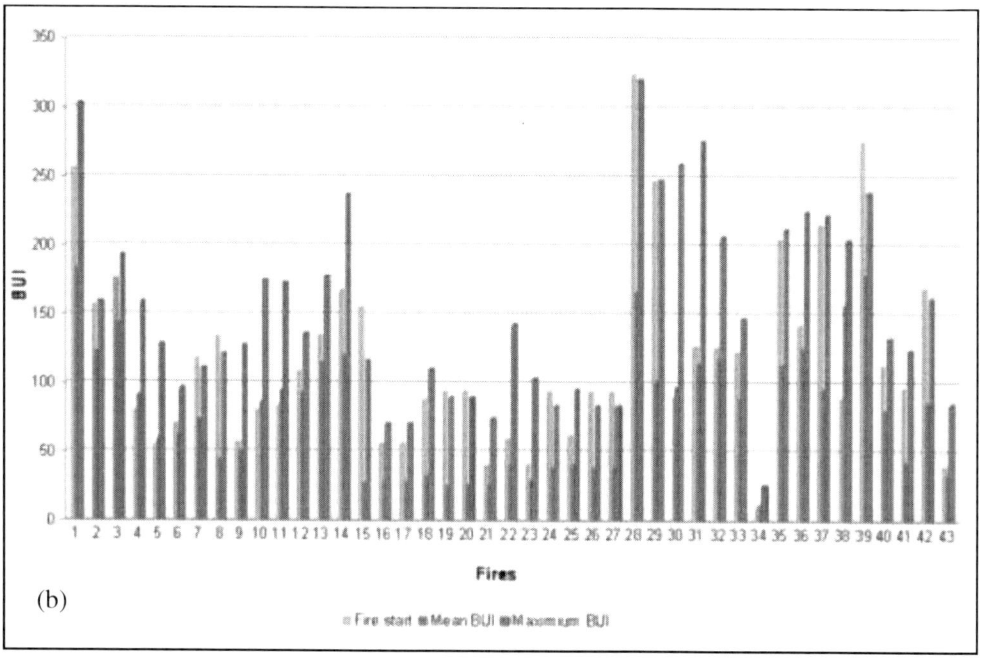

Figure 8. (a) Location of 43 large fires and (b) BUI values associated to the day of the start of each of them, with the mean and maximum BUI over the 1994-2013 period, from the closest SMN meteorological stations.

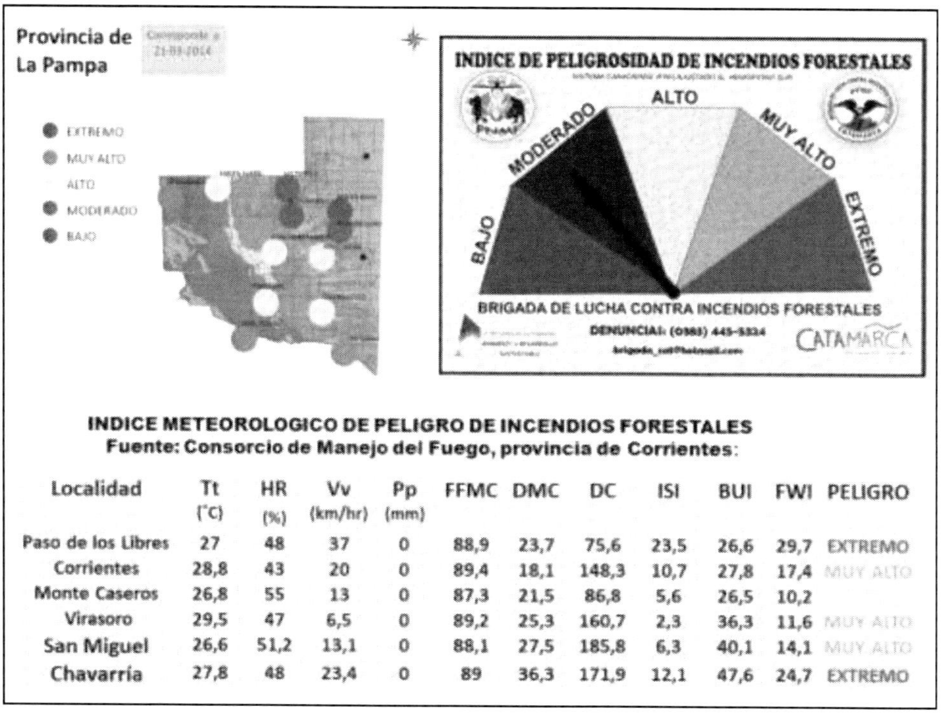

Localidad	Tt (°C)	HR (%)	Vv (km/hr)	Pp (mm)	FFMC	DMC	DC	ISI	BUI	FWI	PELIGRO
Paso de los Libres	27	48	37	0	88,9	23,7	75,6	23,5	26,6	29,7	EXTREMO
Corrientes	28,8	43	20	0	89,4	18,1	148,3	10,7	27,8	17,4	MUY ALTO
Monte Caseros	26,8	55	13	0	87,3	21,5	86,8	5,6	26,5	10,2	
Virasoro	29,5	47	6,5	0	89,2	25,3	160,7	2,3	36,3	11,6	MUY ALTO
San Miguel	26,6	51,2	13,1	0	88,1	27,5	185,8	6,3	40,1	14,1	MUY ALTO
Chavarría	27,8	48	23,4	0	89	36,3	171,9	12,1	47,6	24,7	EXTREMO

Figure 9. Examples of formats used to display and communicate daily danger class in three different provinces of Argentina.

Local agencies communicate fire danger in different formats (Figure 9). At a national level, *SMNF* publishes a daily danger map in cooperation with the *SMN* (Figure 4), It provides revised options for the spatial display of danger in a GIS environment, such as for example the maps displayed by the Indonesia Meteorological and Geophysical Agency (de Groot et al. 2006). The fact that local network stations are not connected to the *SMN* network limits producing a national map by interpolation, as some zones may show a danger class different from local calculations.Although guidelines were provided, interpretation of the meaning of a `fire danger class´ needs to be clarified. There is often the expectation that the class will explain all aspects of dangers. The concept that already stated by Alexander (2008) *`It is difficult to portray all aspects of fire danger in a single number, and only one fire danger class scheme can´t be expected to cover the full range of fire management needs´* has to be strengthened with local examples of past fires, which are still fresh in people's mind.

Success of the Argentine FDRS can be measured by the substantial increase in number of fire managers that recognize the necessity of rating fire dangers since the late 1990's. The development and implementation of FDRS throughout Argentina became mandatory according to the National Fire Management Act of 2013 that specifies that prevention, presuppression, and suppression actions must be taken according to fire danger conditions.

CONCLUSION

In a broad sense, FWI explains the different fire danger climates, and shows fire danger fluctuations through the years. Tests and adjustments have been limited by the restricted number of fire and meteorological records available and the scarce network of meteorological stations. The collection fire behaviour data induced as a result of implementing the FDRS will be extremely helpful for future adjustments. A larger number of reliable meteorological stations and a stronger infrastructure for communicating data in operational time would be of great benefit. Cooperation with research agencies to scientifically organize data, test relationships between weather, fuels, and fire behaviour for local fire environments, and develop new ones when necessary is required.

The concept that a FDRS is needed for fundamental fire management decisions has been established in the country and now has a legal framework. The strategies for implementation have been adapted to the environmental and socio-cultural characteristics of each region, undoubtedly promoting a more proactive attitude. The systematic evaluation of potential danger based on objective indicators and the allocation of resources according to potential fire danger introduces a different work routine that requires institutional and human resource development. Active involvement of local managers, technicians and firefighters, has been key to the acceptance and implementation of the FDRS. Most importantly, the training received especially pertaining to the effects of weather, terrain and fuels on fire behaviour, and the emphasis on the benefits of observing fire behaviour makes their work safer.

ACKNOWLEDGMENTS

We recognize the Canadian International Development Agency (CIDA), the British Columbia Forest Service (BCFS), the Secretaría de Ambiente y Desarrollo Sustemtable (SAyDS) and the Servicio Meteorológico Nacional (SMN), for facilitating the work described in this chapter. We appreciate Dr Judi Beck and Stephen Taylor for their advice and inputs to the project, and Bradley Martin for his assistance in personnel training and revision of this chapter. We thank Gabriel Zacconi for his collaboration in making the figures. Many firefighters, technicians and professionals from different fire agencies along the country, deserve our special recognition for their active participation and enthusiasm. Last and most importantly, we acknowledge the outstanding work and dedication of the SMNF technical staff.

REFERENCES

Alexander ME (1991). The 1985 Butte fire in central Idaho: a Canadian perspective on the associated burning conditions. *USDA Forest Service, Southeastern Forest Experiment* Station General Technical Report GTR-SE-69. (Asheville, NC).

Alexander ME, Cole FV (2001). Rating fire danger in Alaska ecosystems: CFFDRS provides an invaluable guide to systematically evaluating burning conditions. *Fireline* 12(4), 2–3.

Alexander ME (2008). Proposed revision of fire danger class criteria for forest and rural areas in New Zealand. 2nd Edition. National Rural Fire Authority, Wellingtone, in association with Scion, Rural Fire Research Group. (Christchurch, NZ).

Bari M, Dinamarca D (2013). Estudio de caso: incendio Los Rápidos. Plan Nacional de Manejo del Fuego. Informe Técnico 6-ISSN 2313-9420. Available at http://www.ambiente.gob.ar/default.asp?IdArticulo=9964 [downloaded 10 October 2014].

Beck J (2006). Fire danger rating in support of operational fire management. In `*ProceedingsEcofuego 2006.* II. ReunionPatagonica y Tercera Nacional sobre Manejo y Ecología del Fuego´, 25-28 April 2006, (Eds G Defossé, M Dentoni, and O Picco). pp. 35-46. (Esquel, Argentina).

Bianchi L (2008). *Métodos para Definir las Clases de Peligro de Incendios en Regiones de Argentina, basados en un Indicador Meteorológico.*BachelorThesis. Universidad Nacional de la Patagonia San Juan Bosco. (Esquel, Argentina).

Cabrera AL (1971). *Fitogeografía de la República Argentina.* Boletín de la Sociedad Argentina deBotánica.(Buenos Aires, Argentina).

Chandler C, Cheney C, Thomas P, Trabaud L, Williams D (1991). `*Fire in forestry. Forest fire behavior and effects*´. 2nd Edition. (Krigger Publishing Company, Florida).

de Groot, WJ, Wardati, Y Wang (2005). Calibrating the Fine Fuel Moisture Code for grass ignition potential in Sumatra, Indonesia. *International Journal of Wildland Fire* 14(2), 161-168.

de Groot WJ, Field RD, Brady MA., Roswintiarti O, Maznorzian M (2006). Development of the Indonesian and Malayasian Fire Danger Rating System. *Mitigation and Adaptation Strategies for Global Change* 12(1), 165-180.

Deeming, JE, Burgan RE, Cohen JD (1977). *The National Fire Rating System.* USDA Forest Service, Intermountain Forest and Range Experiment Station General Technical Report INT-39. (Ogden, UT).

Dentoni MC, Muñoz MM (2012). *Sistemas de Evaluación de Peligro de Incendios.* Plan Nacional de Manejo del Fuego Informe Técnico 1-ISSN 2313-9420,Available at http://www.ambiente.gob.ar/default.asp?IdArticulo=9964 [Verified 10 October 2014].

Dentoni MC, Marek DS, Bianchi LO, CiampoliHalaman MC, Guglielmin DA, Muñoz MM (2006). Evaluación del peligro de incendios en la región andino patagónica norte. In `ProceedingsEcofuego 2006. II. ReunionPatagonica y Tercera Nacional sobre Manejo y Ecología del Fuego´, 25-28 April 2006, Esquel, Argentina. (Eds G Defossé, M Dentoni, and O Picco). pp. 123. (Esquel, Argentina).

Dowdy AJ, Mills GA, Finkele K, de Groot WJ (2009). *Australian fire weather as represented by the McArthur Forest Fire Danger Index and the Canadian Forest Fire Weather Index.* Australian Government, Bureau of Meteorology, CAWCR Technical Report nº10. (Melbourne, Australia).

FAO (1986) *Wildland Fire Management Terminology.* (Food and Agricultural Organization of the United Nations, Rome).

FAO (2006). Fire Management: Voluntary Guidelines. Principles and strategic actions. *Fire Management Working Paper 17* (Food and Agricultural Organization of the United Nations, Rome).

Fernandez P (2005). *Estudo de Adaptaçäo para Portugal Sistema CanadianodeIndexaçäo do Prigo de Incèndio.* Universidade de Trásos Montes e Alto Duoro. relatorio para Agència para la Prevençäo de Incèndios (Villa Real, Portugal).

Fogarty LG, Pearce HG, Catchpole WR, Alexander ME (1998). Adoption vs. Adaptation: Lessons from Applying the Canadian Forest Fire Danger Rating System in New Zealand. In Proceedings III International Conference on Forest Fire Research. *14th Conference on Fire and Forest Meteorology,* 16-20 November 1998, Luso.Portugal. (Ed. DX Viegas) Vol 1, pp. 1011-1028 (Coimbra, Portugal).

Kourtz P, Todd B (1992). Predicting the daily occurrence of lightning-caused forest fires. Forestry Canada. Petawawa National. *Forestry Institute Information Report PI-X-1 12.* (Chalk River, Ontario).

Lawson BD, Frandsen WH, Hawkes BC, Dalrymple GN (1997). Probability of sustained smoldering ignition for some boreal forest duff types. Natural Resources Canada, Canadian Forest Service, Northern Forestry Centre. *Forest Management Note 63. (Edmonton, Alberta).*

McArthur AG (1966). Weather and grassland fire behaviour. *Department of National Development, Forestry and Timber Bureau Leaflet No. 100.* (Canberra, Australia).

Merril DF, Alexander ME (1997). Glossary of forest fire management terms. Natural Resourses Council Canada Publication NRCC Nº 26516. 4th Edition (Ottawa, Ontario).

Nesterov VG (1949). *Combustibility of the forest and methods for its determination.* (USSR State Industry Press).

Pearce HG (2001). A New Zealand scrubland fire danger model: scientific rigour versus operational need. In *Proceedings Bushfire 2001 Conference,* 3 July 2001. (Eds Pearce G. and L. Lester), pp 53 (Christchurch, New Zealand).

Rodríguez N, Moretti A (1988). Índice de peligro de propagación de Incendios Forestales. In *Proceedings VI Congreso Forestal Argentino,* 16-20 August 1988. Vol 3, pp 5-7. (Santiago del Estero, Argentina).

Romero G, Bonnemezon E, Monasterolo, Rodríguez, Mora L (2013). Estudio de Caso: Incendios La Gama. Plan Nacional Manejo del Fuego. Informe Técnico 5 - ISSN 2313-9420. Available at http://www.ambiente.gob.ar/default.asp?IdArticulo=9964 [Verified 10 October 2014]

Rothermel C (1983). *How to predict the spread and intensity of forest and range fires.*USDA Forest Service, Intermountain Forest and Range Experiment Station General Technical Report INT-143. (Ogden, UT).

Sampaio OB, Soares RV, Viegas DX (1998). Calibration and test of four fire danger methods in Agudos Region, State of Sao Paulo, Brazil. In *`Proceedings III International Conference on Forest Fire Research. 14th Conference on Fire and Forest Meteorology', 16-20 November 1998.*Luso. Portugal. (Ed. DX Viegas), Vol. 1, pp 1249-1262. (University of Coimbra, Portugal).

Shunk C, Wastl C, Leucher M, Menzel A (2013). Forest fire danger in complex topography-results from a case study in the Bavarian Alps in autumn 2001. *Natural Hazards Earth Systems Science* 13, 2157-2167.

Stocks, BJ, Lawson, BD, Alexander ME, Van Wagner CE, McAlpine RS, Lynham TJ, Dubé DE (1989). The Canadian Forest Fire Danger Rating System: an overview. *Forestry Chronicle* 65(6), 450-457.

Taylor SW (2001). *Considerations for applying the Canadian Forest Fire Danger Rating System (CFFDRS) in Argentina. Natural Resources Canada. Canadian Forest Service,* Pacific Forestry Centre Contract Report Ref. No. 14020-90/02. (Victoria, BC.).

Taylor SW, Alexander ME (2006). Science, technology and human factors in fire danger rating: the Canadian experience. *International Journal of WildlandFire* 15, 21-135

Tortorelli LA (1947). *Los incendios de bosques en la Argentina.* (Dirección Forestal. Ministerio de Agricultura de la Nación Argentina).

In: Current International Perspectives
Editors: Brigitte Leblon and Martin E. Alexander
ISBN: 978-1-63463-682-7

Chapter 4

LACES: ADDING AN "*A*" FOR *ANCHOR POINT(S)* TO THE *LCES* WILDLAND FIREFIGHTER SAFETY SYSTEM

***Martin E. Alexander*[*1] *and William R. "Rob" Thorburn*[2]**
[1]University of Alberta, Edmonton, Alberta, Canada
[2]Alberta Environment and Sustainable Resource Development (Retired), Edmonton, Canada

ABSTRACT

As an acronym, *LACES* stands for *Lookout*(*s*) – *Anchor point*(*s*) – *Communication*(*s*) – *Escape routes* – *Safety zone*(*s*) and has gradually become a guideline for wildland firefighter safety in various regions of Canada over the past 15 years or so. *LACES* constitutes a slight modification, to the popular *LCES* wildland fire safety system as developed by the late Paul Gleason in 1991 and used by U.S. wildland firefighters to protect themselves from being killed or injured by free-burning wildfires and other fireline hazards (e.g., lightning, rolling rocks and logs, fire-weakened timber) – namely the addition of an "*A*" for *Anchor point*(*s*). An anchor point serves as a barrier to fire spread and is employed as a starting point to begin fire suppression work in order to minimize the chance of being outflanked by a fire while line is being systematically constructed, thereby decreasing the likelihood of being entrapped or burned-over by a spreading wildfire. The case is argued here that *LACES* is a significant and essential improvement over *LCES*.

Keywords: Alberta, Canada, fire behavior, fire entrapment avoidance, firefighter fatalities, fire suppression, risk management, safe work practices, situational awareness, wildfire case studies, wildland firefighting

[*] Corresponding Author address: Department of Renewable Resources and Alberta School of Forest Science and Management, University of Alberta, Edmonton, Alberta T6G 2H1. Email: mea2@telus.net.

At best, fire fighting is hazardous. But it is a job which much must be done. We do not want to develop the feeling that because fires are dangerous, we should back away from them without good reason. We must, however, respect the aggravated hazards in fire control and do our best to control them. At the same time, we can instill confidence in our men by letting them know that their welfare is our first consideration when laying out our fire-suppression plans.

Jackson (1950)

INTRODUCTION

The abbreviation *LCES* stands for *Lookout(s) – Communication(s) – Escape routes – Safety zone(s)* and represents a system developed by the late Paul Gleason, a veteran U.S. Forest Service interagency hotshot crew superintendent, for operational firefighter safety in the wildland fire environment (Gleason 1991a). *LCES* considers the four basic natural safety hazards that confront a wildland firefighter on the fireline (Jackson 1948; US Forest Service California Region 1954; Fuquay 1984; Beighley 1995; Mangan 1995; Valdez and Style 1996; Leuschen and Frederick 1999; Manning et al. 2000; Alexander and Stam 2003), namely:

- lightning;
- rolling rocks and logs;
- fire-weakened timber; and
- entrapment by running fires.

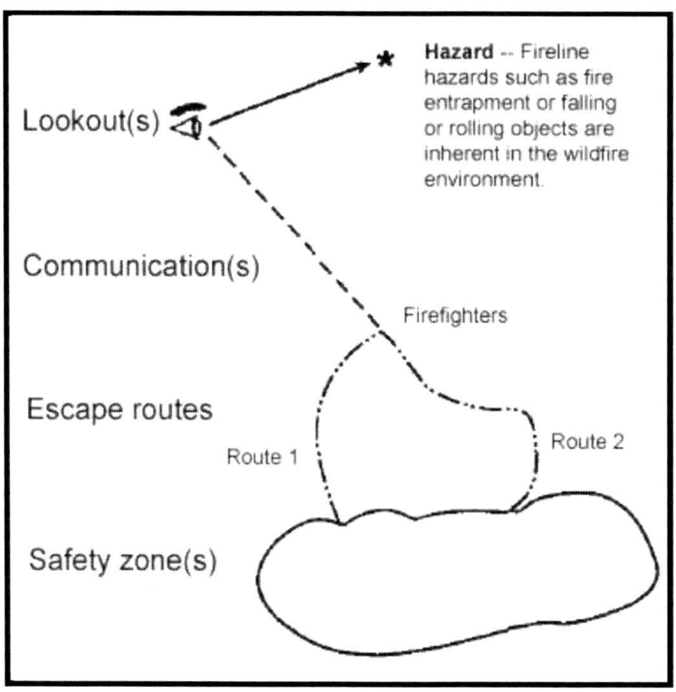

Figure 1. The components and concept of the ***LCES*** wildland firefighter safety system (from Gleason 1991a). Note that the "***E***" stands for ***Escape routes*** not ***Escape route****(s)*.

Together, the elements of *LCES* form a safety system used by wildland firefighters to protect themselves from these fireline hazards. *LCES* represents a procedure that is put in place prior to fighting a wildfire (Figures 1 and 2). It is built on two basic guidelines: (1) each firefighter must know how the *LCES* system will be implemented on a given incident and (2) the *LCES* system must be re-evaluated continuously as the fire environment changes.

LCES

LCES must be established and known to ALL firefighters BEFORE it is needed.

Lookout(s)

- Experienced, competent, trusted
- Enough lookouts at good vantage points
- Knowledge of crew locations
- Knowledge of escape and safety locations
- Knowledge of trigger points
- Map, Weather Kit, Watch, IAP

Communication(s)

- Radio frequencies confirmed
- Backup procedures and check-in times established
- Provide updates on any situation change
- Sound alarm early, not late

Escape Route(s)

- More than one escape route
- Avoid steep uphill escape routes
- Scouted for loose soils, rocks, vegetation
- Timed considering slowest person, fatigue, and temperature factors
- Marked for day or night
- Evaluate escape time vs. rate of spread
- Vehicles parked for escape

Safety Zone(s)

- Survivable without a fire shelter
- Back into clean burn
- Natural features (rock areas, water, meadows)
- Constructed sites (clearcuts, roads, helispots)
- Scouted for size and hazards
- Upslope?
- Downwind? → More heat impact → Larger safety zone
- Heavy Fuels?

Escape time and safety zone size requirements will change as fire behavior changes.

Figure 2. The ***LCES*** wildland firefighter safety system checklist as it appears in the U.S. National Wildfire Coordinating Group's Incident Pocket Response Guide (NWCG 2014: 7).

The need for an "*A*" to be added to *LCES*, denoting *Anchor point*(*s*) – in other words *LACES* – was identified and adopted by the Province of Alberta's fire management organization as a direct result of a wildfire entrapment and subsequent wildland firefighter fatality and serious injury that occurred in north-central Alberta on June 2, 1995. At the time, had there been a better understanding and application of the concept of an anchor point(s), the unfortunate incident may have never taken place. Wildland firefighter safety awareness training in Canada has steadily emphasized the *LACES* concept since that time (ETC and CIFFC 2000; Thorburn et al. 2000; Alexander and Thorburn 2001).

Although several Canadian wildland fire management agencies have considered and supported the concept of *LACES* as opposed to *LCES*, it has yet to be adopted as a national standard. The purpose of this chapter is to provide a rational basis for the addition of an "*A*" for *Anchor point(s)* to the *LCES* wildland firefighter safety system in the hopes that all wildland fire management agencies, at least in Canada, will consider adopting *LACES* as their fundamental wildland firefighter safety system. It is worth noting that *LACES* and other wildland firefigher safety guidelines can also be applied to the safety of recreationists and emergency medical personnel who may be in the vicinity of a wildland fire (Alexander et al. 2012).

BACKGROUND TO THE ORIGIN OF *LACES*

The *LCES* fireline safety system as originally developed by Gleason (1991a) was a result of the tragic loss of life on wildland fires and a concern for the cognitive overload of all the rules and procedures that had to be remembered by a wildland firefighter at any one time. More specifically, the *LCES* system as developed by Gleason (1991a) was a direct response, in part, to the firefighter fatalities that occurred on the Dude Fire in northern Arizona on June 26, 1990 (Gleason 1994a; Thomas 2003) and from his personal involvement in the rescue and recovery operations (Gleason 1991b; Chamberlin 2000).

Table 1. The ten standard firefighting orders as currently used by wildland firefighters in the U.S. (from http://www.fs.fed.us/fire/safety/10_18/10_18.html)

1. Keep informed on fire weather conditions and forecasts.
2. Know what your fire is doing at all times.
3. Base all actions on current and expected behavior of the fire.
4. Identify escape routes and safety zones and make them known.
5. Post lookouts when there is possible danger.
6. Be alert. Keep calm. Think clearly. Act decisively.
7. Maintain prompt communications with your forces, your supervisor, and adjoining forces.
8. Give clear instructions and insure they are understood.
9. Maintain control of your forces at all times.
10. Fight fire aggressively, having provided for safety first.

LCES checklists accordingly evolved (e.g., NWCG, 1993, 2014) as shown in Figure 2 and an *LCES* workshop (Chambelin 1998, 2000) in the late 1990s (Matthews 2000). This was followed by a formalized training course (S-134) developed by the U.S. National Wildfire Coordinating Group, first released in 2003 (NWCG 2013).

When someone is killed or seriously injured on a wildland fire, agencies are quick to respond, committing full attention to safety and adding yet another mix of rules and procedures into over 60 that exist today. These include for example, the *10 Standard Firefighting Orders* (McArdle 1957; Moore 1959) and the *18 Situations That Shout "Watch Out!"* (Wilson and Sorenson 1978; Morse and Monesmith 1987), which wildland firefighters have pursued in the name of safety for many years (NWCG 1980; Chandler et al. 1983; Braun et al. 2001; Ziegler 2007). These are commonly referred to as the 10 Orders (Table 1) and 18 Watchouts (Table 2). Weir (2009: 83) has adapted both to prescribed fire situations.

Each fire order or group of orders has a specific purpose (USDA Forest Service Division of Fire Control 1965; Gaylor 1974). The first three of the 10 Orders (i.e., 1, 2, and 3) are fire behavior related. The next three orders (i.e., 4, 5, and 6) deal with fireline safety issues. The following three orders (7, 8, and 9) are centered around organizational control. The last order (i.e., 10) is a simple reminder of the objective of the other nine orders (Figure 3). The 18 Watchouts on the other hand are more specific and cautionary to firefighter safety than the Orders. The Watchouts serve to expand upon the points of the orders. Humour has been used to illustrate both the 10 Orders and the 18 Watchouts (Figure 4) in the past (Murphy 2004).

Table 2. The eighteen situations that shout "watch out!" as currently used by wildland firefighters in the U.S. (from http://www.fs.fed.us/fire/safety/10_18/ 10_18.html). Watchouts 1, 3, 5, 8, and 10 were added to the original list of 13 in 1987

1. Fire not scouted and sized up.
2. In country not seen in daylight.
3. Safety zones and escape routes not identified.
4. Unfamiliar with weather and local factors influencing fire behavior.
5. Uninformed on strategy, tactics, and hazards.
6. Instructions and assignments not clear.
7. No communication link with crewmembers/supervisors.
8. Constructing line without safe anchor point.
9. Building fireline downhill with fire below.
10. Attempting frontal assault on fire.
11. Unburned fuel between you and the fire.
12. Cannot see main fire, not in contact with anyone who can.
13. .On a hillside where rolling material can ignite fuel below.
14. Weather is getting hotter and drier.
15. Wind increases and/or changes direction.
16. Getting frequent spot fires across line.
17. Terrain and fuels make escape to safety zones difficult.
18. Taking a nap near the fire line.

This chapter is not viewed as the appropriate medium to debate the relative merits of the "10 and 18" or any other wildland firefighter safety guidelines. Suffice it to say that there has been considerable criticism of the changes made to the 10 original orders over the years (e.g., Anon. 1988; Steele and Krebs 2000; Putnam 1995a, 2001; Brauneis 2002; Krebs 2003). In the mid 1980s, for example, the original orders were recast so as to make them easier to recall (Morse and Monesmith 1987; Monesmith 1988). Furthermore, the origin of the original 13 watchouts that comprise the current list of 18 remains unclear (Ziegler 2008); the addition of five other watchouts occurred in 1987 (Morse and Monesmith 1987; Monesmith 1988; Morse 1990). There is the tendency to continue adding to these watchouts through formal and informal processes (e.g., Valdez and Style 1996; Bennett 1998). Many firefighters, for example, have their own "19th watchout" based on personal experience (Mayhew 2010).

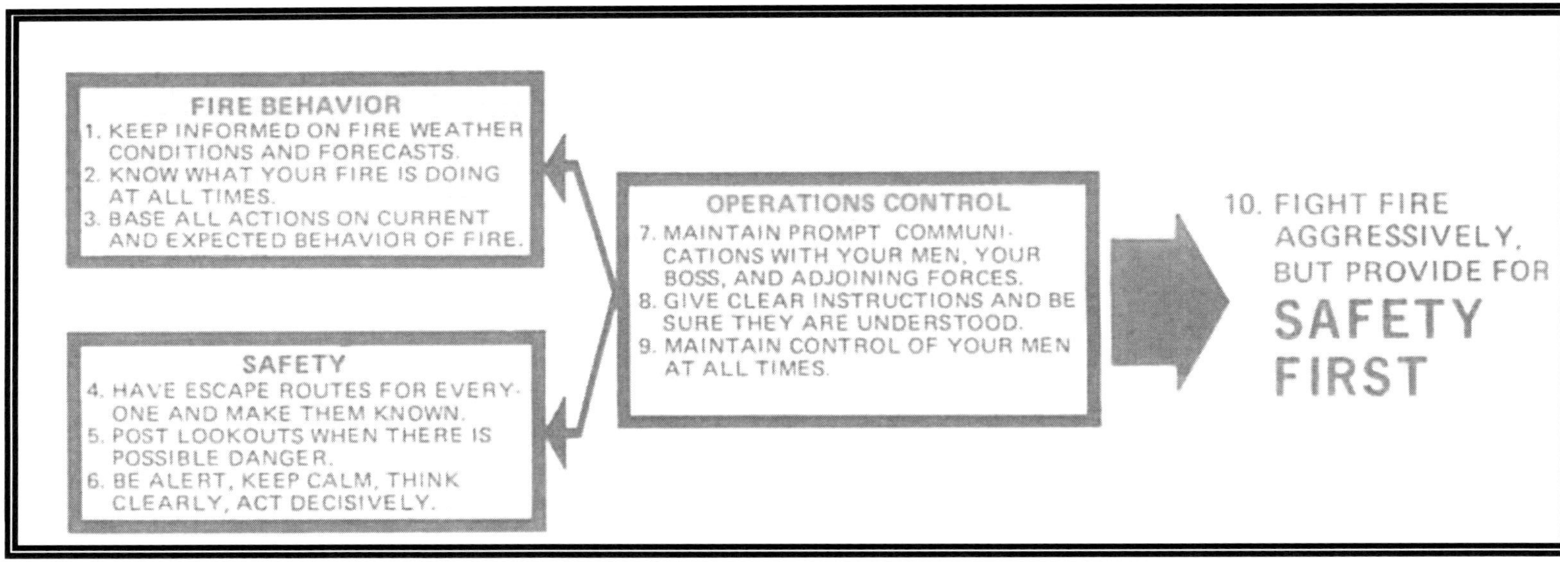

Figure 3. Flow chart of the original Ten Standard Fire Fighting Orders or rules emphasizing their specific purposes (from Gaylor 1974).

Figure 4. (Continued).

Figure 4. Maine Forest Ranger Rick Henion's (circa 1994) popular "18 Situations That Shout Watch Out" wall poster.

It could be argued that the use of the "10 and 18", no matter how well intentioned, may be somehow fatally flawed (Braun 2007). There has over the years, for example, been no systematic, independent peer-review of these guidelines. While the concepts of lookouts, communications, escape routes and safety zones have been formally recognized elements of wildland firefighter safety for over 55 years (McArdle 1957; Moore 1959), little attention has been directed towards specifics. Some effort, for example, was made with respect to lookouts and communications many years ago (Bell 1971) and there has been steady progress in radio communications (Gray 1982). But there has been surprisingly very little published quantitative work on firefighter travel rates using escape routes (Butler et al. 2000; Dakin 2001, 2002; Ruby et al. 2003; Baxter et al. 2004; Alexander et al. 2005, 2013) or on safety zone sizes (Butler and Cohen 1998a, 1998b, 2000; Alexander et al. 2009; Butler 2014) until about 15 years ago.

Based on his previous involvement with and study of several fatality fires, Paul Gleason, a veteran wildland firefighter, analyzed the many firefighter safety rules and procedures that had been developed over the years (Harris 1998). He determined that firefighters were simply overloaded with all that had to be remembered at any one time. His solution was the development and introduction of his *LCES* concept.

In supporting Paul Gleason's concern, Putnam (1995a) pointed out that the research conducted into cockpit resource management has indicated that no matter how many factors are important, the human mind normally can handle only a limited number. It has long been accepted in psychology that the amount of information the average person can maintain in working memory is approximately seven items, give or take two (Miller 1956). Cockpit resource management shares a lot of the characteristics in common with the rapid decision making sometimes required on the fireground. There is also strong evidence to suggest that this number is even lower when a person becomes anxious or stressed, which is most likely to occur when the situation becomes intense (McLennan et al. 2013).

As Gleason (1991a) points out, "*LCES* refocuses on the essential elements of the 10 Orders and 18 [Watch Out!] Situations." Its simplicity is a welcome solution to the problem of having to manage many elements at one time.

LACES Is Conceived: Alberta Adopts/Modifies the *LCES* Concept

Alberta's wildfire management organization adopted the elements of Gleason's (1991a) *LCES* as its basic wildland firefighter safety system in the early to mid 1990s. However, staff believed that an important element was missing from *LCES*, namely an "*A*" for *Anchor point(s)* – and thus the concept of *LACES* was born. The feeling was that an anchor point(s) must be an essential component of the *LCES* system, especially when burning conditions and potential fire behavior reach a critical stage. This philosophy slowly evolved over time within the provincial wildfire management organization. It became firmly entrenched as a result of a firefighter fatality that occurred in June 1995 when three rappel crewmembers became unexpectedly entrapped and had to be extracted using a helicopter mounted static line as described in the CD-ROM based training course *Wildland Fire – Safety on the Fireline* (ETC and CIFFC 2000). In the process, one crewmember was fatally injured while a second crewmember sustained serious injuries. The third crewmember, unable to be hooked on to the static line in time, escaped uninjured by running away from the fire's path.

On a historical note, the *LCES* workshop training initiative, as developed by Chamberlin (1998, 2000), was undertaken in Alberta. Paul Chamberlin facilitated an *LCES* workshop at the Hinton Training Centre in January 2001, which was very well received. While he did acknowledge that the addition of an "*A*" for *Anchor point(s)* to *LCES* was a worthy addition, he didn't formally sanction it.

RATIONALE FOR *LACES* VERSUS *LCES*

As Luke and McArthur (1978) forthrightly note, an anchor point is an essential component of wildland fire suppression safety:

> Whatever the course of action decided, suppression must at all-time start from an anchor point. This may be a natural firebreak such as a road or it may be the edge of a fire burning in light fuels which can be easily and safety suppressed. By proceeding in an orderly fashion, the safety provided at the anchor point is maintained and extended as work progresses along the fire perimeter. This provides an essential escape route if unexpected fire behaviour develops.

Admittedly, mandatory anchor points may in fact not be applicable in initial attack situations involving very small fires and low fire behavior potential.

Figure 5. Graphic representation of an anchor point at the road on the left flank of a wind-driven fire in cured grass, defining the logical position from which to begin safe wildland firefighting operations (from ETC and CIFFC 2000). A secondary anchor point could also be established at the road on the fire's right flank.

Merrill and Alexander (1987) defined an "anchor point" as illustrated in Figure 5 as "An advantageous location, usually a barrier to fire spread, from which *to start* or finish construction of a control line" [bold italics added for emphasis by authors]. The definition given in the U.S. "Glossary of Wildland Fire Terminology" (NWCG 2012) is similar in that it defines an anchor point as "An advantageous location, usually a barrier to fire spread, from which to start constructing a fireline. The anchor point is used to minimize the chance of being outflanked by the fire while the line is being constructed."

The descriptors "barrier to fire spread" and "minimize the chance of being outflanked by the fire", found in the above definitions, are noteworthy and key to the rationale for including an "*A*" for *Anchor point(s)* in *LACES*. Interestingly enough, neither *LCES* nor *LACES* are mentioned in the current Canadian glossary of forest fire management terms (CIFFC 2003). *LCES* is defined in the U.S. National Widlfire Coordinating Group's "Glossary of Wildland Fire Terminology" (NWCG 2012), although the abbreviation is not given.

If the intent of an anchor point is to serve as a barrier to fire spread and thus minimize the chance of being outflanked by the fire while the line is being constructed, would it not seem appropriate then to add an "*A*" for *Anchor point(s)* to *LCES*? In arriving at a conclusion, one should bear in mind the fact that anchor points are indeed established to prevent circumstances that result in wildland fire entrapments and burn-overs. An additional benefit of including the "*A*" would be to suggest that tactics and a review thereof must be considered carefully when engaging or re-engaging in suppression operations on a wildland fire. Furthermore, the selection of an anchor point is normally dictated by the assurance that a barrier to fire exists, thus avoiding a burn-over or entrapment that would likely threaten a crew while working on the line. The likelihood that firefighters would be forced into a compromising position where one would have to depend on the use of escape routes and a safety zone(s) would therefore be minimized by lessening the chances of a burn-over or entrapment situation.

Escape routes often utilize the constructed fireline leading back to safety zones such as the "black" or suitably burned-over area. In order to avoid compromising firefighter safety by seeking out escape routes and safety zones when time is of the essence, it makes good sense to establish anchor points that can by default also use the established fireline and the adjacent burned area. Both the constructed fireline and burnt-over area may be used as possible escape routes and safety zones. This is not to imply that escape routes and safety zones can be ignored in the presence of an anchor point. An anchor point selected as an escape route can be obstructed by falling trees that have been weakened by fire, making escape slow and difficult. A burned over area that has had an anchor point established against it, may not necessarily be a safety zone due, for example, to falling trees or rolling rocks and logs. Alternate, yet suitable escape routes and safety zones must be established as added insurance.

Arguments for the addition of an "*A*" for *Anchor point(s)* to *LCES* also come from a review of the 10 Orders and 18 Watchouts. No specific mention is made of anchor points in the 10 Orders. Note though Standard Firefighting Order 10: Fight fire aggressively, having provided for *safety first* [bold italics added for emphasis by authors]. We would regard the establishment of an anchor point as preparatory step *well before* in engaging in any firefighting operations. Of the 18 Watchouts, situation number 8 (which was only added to the original list of 13 in 1987) does specifically mention anchor points (Table 2): *Constructing line without safe anchor point* (Figure 6). Because anchor points are addressed indirectly in the 10 Orders and more specifically in the 18 Watchouts, and given as *LCES* attempts to manage the essential elements of the 10 Orders and 18 Watchouts, the argument is put forth that an "*A*" for *Anchor point(s)* should be incorporated within a condensed safety model or mnemonic device such as *LACES*.

Alberta's adoption of the *LACES* concept did not imply that the 10 Orders and 18 Watchouts should be ignored. They are still very important components of wildland firefighter safety awareness training in Canada (ETC and CIFFC 2000; Thorburn et al. 2000; Alexander and Thorburn 2001) and as operational fire safety guidelines.

Considering what was stated earlier about adding to the mix of rules and procedures, it can be strongly argued that the addition of a fifth component or element to the *LCES* abbreviation (thus changing it to *LACES*), still falls within the range of seven factors that one can manage successfully. Furthermore, it is believed that as an acronym, *LACES* fits more easily into the firefighter's lexicon, given that one periodically looks down to their boot "laces" (Figure 7) as they begin their day's work Teie (1994). This should serve as a reminder that line staff need to address and re-assess the situation as required. As a result, *LACES* may be far easier to remember or recall than *LCES*, which is in fact an abbreviation and not an acronym.

Figure 6. Cartoon depicting Watchout Situation Number 8 – Constructing Line Without Safe Anchor Point – from Henion's (ca. 1994) popular wall poster for the 18 Watchouts (http://www.novascotia.ca/natr/forestprotection/wildfire/bffsc/lessons/lesson6/lace.asp).

Figure 7. The *LACES* icon pin developed by the Alberta Environment and Sustainable Resource Development in the mid 2000s to promote the Lookout(s) – Anchor point(s) – Communication(s) – Escape routes – Safety zone(s) wildland firefighter safety system concept.

Cook (1995), who developed a five-step risk management model for wildland firefighting, also had concerns with respect to the numerous rules and procedures that exist in regards to firefighter safety guidelines. In the third step of his risk management process – "Hazard Control" – he associates the establishment of an anchor point and *LCES* in a single task. This relationship would suggest that perhaps anchor points and *LCES* should be combined.

Additional Background Information in Support of the Anchor Point and the *LACES Concept*

The Alberta initiative regarding *LACES* has not been the first attempt to add an "*A*" to *LCES*. Some contend that the "*A*" in *LACES* should stand for "Awareness" (e.g., Teie 1994; AFAC 2008a, 2008b) or in fact have dual meanings – i.e., both "Anchor point" and "Awareness" (Pearce et al. 2004; Stacey 2012). On the other hand, some have suggested that the letter "A" should denote "Attitude" and perhaps even other meanings (Kurtz 2003a, 2003b).

It is unfortunate that the lack of a national and/or international standard pertaining to the use of *LCES* versus *LACES* does not exist because it only confuses the issue and introduces the need to train imported personnel from other agencies before they can begin firefighting. One "close call" or "near miss" that occurred in Alberta during the 2001 fire season on the Chisholm Fire (Quintilio et al. 2001) was perhaps due, in part, to the interpretation by the different fire management agencies involved regarding the use of *LCES* versus *LACES* (Beck et al. 2001). Certainly the popularity of the *LACES* concept has steadily grown in Canada since that time (Figure 8). Nevertheless, it is time for the adoption of one national and possibly one international standard for wildland firefighters exported outside of their local jurisdiction.

Figure 8. The personalized ***LACES*** car license plate of Connie Bradburn, a Forest Protection Officer with the Saskatchewan Ministry of Environment's Wildfire Management Branch in Prince Albert. See Alexander (2013) for more information on this development.

As an adjunct to this issue, in Alberta anchor points are not always routinely implemented. For example, when rapid initial attack on small fires is conducted and a "hot-spotting" style of attack is used (Quintilio et al. 1990; Murphy et al. 1991), anchor points may not be feasible; hot-spotting is a suppression method or technique used to check the spread and intensity of a fire at those points that exhibit the most rapid spread or that otherwise pose some special threat to control of the situation in contrast to systematically working all parts of the fire at the same time, or progressively, in a step-by-step manner (Merrill and Alexander 1987). This is largely due to the small size of fires being actioned during initial attack, which also results in minimal opportunities to "anchor" – for example, a lightning strike in the middle of a large, continuous area of flammable fuels. The necessity to go in and without delay, knock down flames in areas of greatest threat to the immediate escape of the fire, often precludes the immediate establishment of anchor points. However, during hot-spotting tactics, the other four components of *LACES* are strictly observed without prejudice.

The following guidelines, documented in the province's wildfire management policy manual (Anon. 2001), govern when anchor points must be established on the basis of fireline intensity classes (Alexander and Cole 1995) as produced by the Canadian Forest Fire Behavior Prediction System (Forestry Canada Fire Danger Group 1992; Taylor et al. 1997):

- When fire behavior potential indicates that head fire intensity (HFI) is up to an HFI Class 3 (i.e., 500-2000 kW/m), wildfire resources should work from an anchor.
- At HFI Class 4 (i.e., 2001-4000 kW/m) or greater, wildfire resources must work from an anchor point.

Beck et al. (2002) have outlined how HFI forecasts are undertaken on an operational basis in the province of British Columbia.

In considering the above policy guidelines, one should keep in mind that certain cautions need to be observed in determining anchor points. One should also not assume that anchor points are necessarily safety zones, largely because their size amongst other factors may restrict their effectiveness (Steele 1997, 2000; Butler and Cohen 1998a, 1998b; Butler 2014). They are similar in that both must offer protection from a burn-over or entrapment. However, the "black", burned over area directly against an anchor point, although offering protection from a burn-over or entrapment, may not serve as a safety zone due to the presence of other basic safety hazards, especially falling trees or rolling rocks and logs. Furthermore, a recently burned-over area may not immediately serve as a safe refuge due to the burn-out time of woody fuels and duff (Sullivan et al. 2002) in contrast to grassland fuels in which a person can enter the recently burned area in a matter of a few seconds (Cheney and Sullivan 2008). And of course the effectiveness of a burned area will depend on the degree and completeness of the fuel consumption (Pearce et al. 2004).

Wildfire Case Studies in Support of *LACES*

A number of past wildfire incidents in which firefighters were either killed, injured or threatened, lend support to the adoption of *LACES* in lieu of *LCES* as a basic wildland firefighter safety system. Had an anchor point(s) been established or had the safety advantage afforded by an anchor point not been abandoned, these incidents may not have occurred at all. The 1949 Mann Gulch Fire in northwestern Montana (Maclean 1992; Rothermel 1993) represents a case in point. Had for example, foreman Wag Dodge and his smokejumper crew anchored to the fireline previously constructed by the local fire prevention guard Jim Harrison, there would have been a safety zone to go to when the fire blew up (i.e., the "black" or previously burned area). Then quite possibly, the tragedy that eventually unfolded would have never happened. Admittedly this opinion is based on the benefit of 20:20 hindsight.

CONCLUSION

As Barrows (1974) has stated, "... *there is one overriding challenge to fire management: that of maintaining full respect for the power of fire and the effects of this power on both wildland environments and the people who live and work in these environments.*" Wildland firefighter fatalities is a global wildland fire issue (Viegas 2009; Alexander and Buxton-Carr 2011; Cook 2013). Fatalities have occurred on both wildfires (Wilson 1977) and prescribed fires (Alexander and Thomas 2006).

It is not the intent of this chapter to imply that simply by adding an "*A*" for *Anchor point*(s) to *LCES* resulting in *LACES* would serve as a "sure-fire" solution to the prevention of all wildland firefighter injuries and fatalities, or that other fire safety guidelines should be ignored. "It cannot be made entirely safe" (Davis 1959). There are many other factors that enter into the equation when one examines past incidents involving injuries or fatalities. For example, the severity of the burn injuries sustained by California firefighter Kelly York during a burn-over, highlighted the fact that it was not just the lack of an anchor point that contributed to the nature of her accident but the lack of proper personal protective equipment and a lack of communications (Anon. 1996).

Chandler et al. (1983) notes that, "... *mnemonic devices ... have been used to instill the principles of safe firefighting practices. In the long run, however, only a firm understanding of fire behavior together with experienced leadership and good crew discipline can prevent fire tragedies*" (Budd et al. 1996). Firefighter fatalities can be related both directly and indirectly to fire behavior (Alexander 1998).

It is readily acknowledged that the addition of an "*A*" to *LCES* for Anchor point(s) may simply lead to a plethora of logical additions to *LCES* such as awareness (as in situational), accountability, attitude, avoidance (entrapment) abort, accountability, and access (Kruz 2003a, 2003b). For example, Putnam (1995a) has already suggested changing *LCES* to "I-LCES" where the "I" stood for "Inside, Inner and Interpersonal" in recognition of the importance of human factors in firefighter safety (Putnam 1995b). And Mayhew (2009) had advocated adding an "F", the Greek letter Delta (Δ) and two arrows to increase the effectiveness of *LCES* system (Figure 9).

Some members of the global wildland fire community are vehemently opposed to addition of any form of an "*A*" to *LCES* althogether. Why should Canada at least adopt *LACES*, where the "*A*" stands solely for *Anchor point(s)* as a national standard even if other parts of the world don't yet accept the acronym in its entirety? Well, why not? There is admittedly a myriad of well-intentioned wildland fire safety acronyms and guidelines (e.g., Chandler et al. 1983; Queen 1993), including the long forgotten "Accident Check List for Forest

Fire Fighters" (US Forest Service California Region 1954). In addition to the 10 Orders, 18 Watchouts, and the *LCES* checklist, there is also the following guides to wildland firefighter safety:

- Wildland-Urban Watch Outs (NWCG 1993, 2014);
- Tactical Watch Outs (NWCG 2014);
- Downhill Line Construction Guidelines (NWCG 1989);
- Structural 'Watch-Out' Situations (NWCG 1989; Goodson and Adams 1998);
- Eight Firefighting Commandments (NWCG 1980; Chandler et al. 1983);
- 13 Prescribed Fire Situations That Shout Watch Out! (Maupin 1981);
- Common Denominators of Fire Behavior on Tragedy and Near-miss Wildland Fires (Wilson 1977; Wilson and Sorenson 1978; NWCG 1996); and
- Look Up - Look Down - Look Around list of fire environment factors/indicators (NWCG 1993); "Look Out" for each other could also be added (Alexander and Stam 2003).

Figure 9. The F-LCES-Delta Tool icon (from Mayhew 2009). The "F" stands for fire behavior prompting one to ask the question: what's the worst this fire can do? The Greek letter Delta (**Δ**) in turn prompts you to look for change and in turn the two circular arrows indicates it is an ongoing process. ***LACES*** is afterall a dynamic and not a static checklist to be administered only once.

Other good suggestions have also appeared from time to time (e.g., Greenlee and Greenlee 2003), the most recent being the "Two & Seven Tool" (Mayhew 2008).

However, "anchor and flank" is viewed by many as a basic safe work practice in wildland firefighting operations (e.g., Luke and McArthur 1978; Cheney et al. 2001; Alexander and Fogarty 2002; Panko 2009). Adding the "*A*" for *Anchor point(s)* to *LCES*, resulting in *LACES* isn't likely to infringe on firefighter memory overload and on the contrary, may be easier for firefighters to remember.

Fire suppression actions that are based on observed or perceived fire behavior versus a potential fire behavior relative to an assessment of the fire environment has in the past, and will in the future, continue to lure wildland firefighters into a false sense of security (Gleason 1994b). Perhaps, as a few members of the wildland fire community have suggested (e.g., Morse 1990), we should look to adopt some measure of fuel dryness to gauge whether we put firefighters on the line or not and if so where and for what purpose. In Canada, at least, such criteria would logically be based on critical threshold values of the Fine Fuel Moisture Code (FFMC) and Buildup Index (BUI) components of the Canadian Forest Fire Weather Index System (Van Wagner 1987) along the lines of the generalized fire suppression guidelines developed by Alexander and Cole (1995), for example. This would enable fire suppression personnel to more objectively determine whether firefighters can continue to work safely on the fireline.

ACKNOWLDEGMENTS

The review and comments of Judi Beck, Hugh Boyd, Bruce MacGregor, Bruce Mayer, Brad Mayhew, Mary Omodei, Craig Quintilio, Dennis Quintilio, and Terry Van Nest on earlier drafts of this chapter are gratefully appreciated.

DEDICATION

This chapter is dedicated to the memory of the late Mr. Paul A. Gleason (1946-2003), the first "student of fire" (Cook and Tom 2003; Keller 2003). We very much appreciate Paul's widow, Karen Miranda-Gleason, for giving us her blessing to undertake this dedication. Thanks is also extended to Kelly Close for providing the following photo of Paul.

Paul Gleason on the Anaconda Fire, Glacier National Park, northwestern Montana, September 1, 1999. Photo by Kelly Close, Fire Behavior Analyst, Pourde Fire Authority, Fort Collins, CO.

REFERENCES

Alexander ME (1998) A wildland fire behavior researcher's perspective of firefighter safety in Canada. In *'Proceedings of Canada/US Wildland Fire Safety Summit'*, 29 September – 2 October 1997, Rossland, BC. (Ed JM Greenlee). pp. 37. (International Association of Wildland Fire: Fairfield, WA)

Alexander ME (2013) LACES license plates: a tool to promote wildland firefighter safety. *Forestry Chronicle* 89, 581, 584.

Alexander ME, Ackerman MY, Baxter GJ (2009) An analysis of Dodge's escape fire on the 1949 Mann Gulch Fire in terms of a survival zone for wildland firefighters. In *'Proceedings of 10th Wildland Fire Safety Summit',* 27-30 April 2009, Phoenix, AZ. (CD-ROM) (International Association of Wildland Fire: Birmingham, AL)

Alexander ME, Baxter GJ, Ackerman MY (2009) Is a wellsite opening a safety zone for a wildland firefighter or a survival zone or neither? In *'The '88 Fires: Yellowstone and Beyond'* 22-28 September 2008, Jackson, WY. (Eds RE Masters, KEM Galley and DG Despain) Tall Timbers Research Station, Tall Timbers Miscellaneous Publication No. 16, p. 110. (Tallahassee, FL)

Alexander ME, Baxter GJ, Dakin, GR (2005) Travel rates of Alberta wildland firefighters using escape routes. In *'Proceedings of Eighth International Wildland Fire Safety Summit: Human Factors - Ten Years Later',* 26-28 April 2005, Missoula, MT. (Eds BW Butler and ME Alexander) (CD-ROM) (International Association of Wildland Fire: Hot Springs, SD)

Alexander ME, Baxter GJ, Dakin GR (2013) How much time does it take for a wildland firefighter to reach a safety zone? *Wildfire* 22(4), 12-13.

Alexander ME, Buxton-Carr P 2011. Wildland fire suppression related fatalities in Canada, 1941-2010: a preliminary report. In '*Proceedings of the 11th International Wildland Fire Safety Summit*', 4-8 April 2011, Missoula, MT. (Ed RL Fox) (CD-ROM) (International Association of Wildland Fire: Missoula, MT)

Alexander ME, Cole FV (1995) Predicting and interpreting fire intensities in Alaskan black spruce forests using the Canadian system of fire danger rating. In 'Managing Forests to Meet People's Needs, Proceedings Society of American Foresters/Canadian Institute of Forestry Convention' 18-22 September 1994, Anchorage, AK. Society of American Foresters, SAF Publication 95-02, pp. 185-192. (Bethesda, MD)

Alexander ME, Fogarty LG (2002) A pocket card for predicting fire behavior in grasslands under severe burning conditions. Natural Resources Canada, Canadian Forest Service, Forest Research, and National Rural Fire Authority, Fire Technology Transfer Note No. 25. (Ottawa, ON, Rotorua, New Zealand and Wellington, New Zealand)

Alexander ME, Mutch RW, Davis KM, Bucks CM (2012) Wildland fires: dangers and survival. In *'Wilderness Medicine'*. (Ed PS Auerbach) pp. 240-280. 6th edn. (Elsevier, Philadelphia, PA)

Alexander ME, Stam JC (2003) Safety alert for wildland firefighters: fuel conditions in spruce-beetle-killed forests of Alaska. *Fire Management Today* 63(2), 25.

Alexander ME, Thomas DA (2006) Prescribed fire case studies, decision aids, and planning guides. *Fire Management Today* 66(1), 5-20.

Alexander ME, Thorburn RW (2001) Fireline safety training course available on CD-ROM. *Fire Management Today* 61(2), 45.

Anonymous (1988) Health and safety. National Fire Protection Association, Quincy, MA. *Wildland Fire Management Briefings* 1(2), 22.

Anonymous (1996) 'A Firefighter's Return From a Burnover: The Kelly York Story'. (VHS videotape, 21 minutes) (LCTV Advertising: Lakeport, CA) [Also available at: http://www.youtube.com/watch?v=0AiyrH2uHWQ]

Anonymous (2001) Forest protection policy manual, FPD 6.13 – P1. Alberta Sustainable Resource Development, Forest Protection Division. (Edmonton, AB)

AFAC (2008a) Lookouts, awareness, communications, escape routes, safety zones (LACES) – discussion paper. Australasian Fire Authorities Council. (East Melbourne, Victoria)

AFAC (2008b) AFAC position on the use of lookouts, awareness, communications, escape routes, safety zones (LACES) system for wildfire firefighters safety on the fireground. Australasian Fire Authorities Council. (East Melbourne, Victoria)

Barrows JS (1974) The challenges of forest fire management. *Western Wildlands* 1(3), 3-5.

Baxter G, Alexander M, Dakin G (2004) Travel rates of Alberta wildland firefighters using escape routes on a moderately steep slope. *Forest Engineering Research Institute of Canada Advantage* 5(25), 1-15.

Beck JA, Alexander ME, Harvey SD, Beaver AK (2002) Forecasting diurnal variations in fire intensity to enhance wildland firefighter safety. *International Journal of Wildland Fire* 11, 173-182.

Beck J, Van Nest T, Hutchinson B, Bachop S (2001) Chisholm wildfire entrapment investigation. Forest Engineering Research Institute of Canada, Alberta Sustainable Resource Development and British Columbia Ministry of Forests,Victoria, BC. (Hinton, AB, Edmonton, AB and Victoria, BC)

Beckley B (2001) Flagging for firefighting escape routes and safety zones. USDA Forest Service, Missoula Technology and Development Center, Tech Tip 00151-2399-MTDC. (Missoula, MT)

Beighley M (1995) Beyond the safety zone: creating a margin of safety. *Fire Management Notes* 55(4), 21-24.

Bell AG (1971) Air horn helpful in fire emergencies. *Fire Control Notes* 32(1), 9, 15.

Bennett B (1998) Safety alert: Watch out for aircraft turbulence! *Fire Management Notes* 58(4), 20-21.

Braun CC (2007) Standard failures. *Wildfire* 15(5), 20-25.

Braun CC, Gage J, Booth C, Rowe AL (2001) Creating and evaluating alternatives to the 10 Standard Orders and 18 Watch-out Situations. *International Journal of Cognitive Ergonomics* 5, 23-35.

Brauneis K (2002) Fire orders: do you know their original intent? *Fire Management Today* 62(2), 27-29.

Budd G, Brotherhood J, Hendrie L, Cheney P, Dawson M (1996) *'Safe and Productive Bushfire Fighting With Hand Tools'*. (Australian Government Publishing Service: Canberra, ACT)

Butler BW (2014) Wildland firefighter safety zones: a review of past science and summary of future needs. doi.org/10.1071/WF13021.

Butler BW, Cohen JD (1998a) Firefighter safety zones: how big is big enough? *Fire Management Notes* 58(1), 13-16.

Butler BW, Cohen JD (1998b) Firefighter safety zones: a theoretical model based on radiative heating. *International Journal of Wildland Fire* 8, 73-77.

Butler BW, Cohen JD (2000) Field verification of a firefighter safety zone model. In 'Proceedings of the 2000 International Wildfire Safety Summit' 10-12 October 2000, Edmonton, AB. (Eds BW Butler, K Shannon) pp. 54-61. (International Association of Wildland Fire: Montana City, MT)

Butler BW, Cohen JD, Putnam T, Bartlette RA, Bradshaw LS (2000) A method for evaluating the effectiveness of firefighter escape routes. In 'Proceedings of the 2000 International Wildfire Safety Summit' 10-12 October 2000, Edmonton, AB. (Eds BW Butler, K Shannon) pp. 42-53. (International Association of Wildland Fire: Montana City, MT).

Chamberlin P (1998) LCES workshop. *Wildfire* 7(5/6), 17-18.

Chamberlin P (2000) LCES workshop, do basic things very very well. In 'Proceedings of the 2000 International Wildfire Safety Summit' 10-12 October 2000, Edmonton, AB. (Eds BW Butler, K Shannon) pp. 62-88. (International Association of Wildland Fire: Montana City, MT)

Chandler C, Cheney P, Thomas P, Trabaud L, Williams D (1983) *'Fire in forestry. Volume II: forest fire management and organization.'* (Wiley: New York, NY)

Cheney P, Gould J, McCaw L (2001) The dead-man zone – a neglected area of firefighter safety. *Australian Forestry* 64, 45-50.

Cheney P, Sullivan A (2008) *'Grassfires: Fuel, Weather and Fire Behaviour'*. 2nd edn. (CSIRO Publishing: Collingwood, Victoria)

CIFFC (2003) The glossary of forest fire management terms. 2003 edn. Canadian Interagency Forest Fire Centre, Winnipeg, MB.

Cook J (1995) Fire environment size-up: human limitations vs. superhuman expectations. *Wildfire* 4(4), 49-53.

Cook JR (2013) Trends in wildland fire entrapment fatalities ... revisted. Available at: http://www.nifc.gov/wfstar/downloads/Wildland_Fire_Entrapment_Fatalities_Revisited_JC.pdf

Cook J, Tom A (2003) Interview with Paul Gleason. *Fire Management Today* 63(3), 91-94.

Dakin G (2001) The determination of ground travel rates by fire crews using escape routes in variable fuel types and terrain factors. In 'Proceedings of 2001 International Wildland Fire Safety Summit'. 6-8 November 2001, Missoula MT. (Eds BW Butler, D Mangan) (CD-ROM) (International Association of Wildland Fire: Fairfax, WA)

Dakin G (2002) Ground rates of travel by fire crews using escape routes: an interim report. *Forest Engineering Research Institute of Canada Advantage* 3(15), 1-17.

Davis KP (Ed) (1959) *'Forest Fire: Control and Use'*. (McGraw-Hill: New York, NY)

ETC, CIFFC (2000) 'Wildland Fire – Safety on the Fireline'. (CD-ROM) (Environmental Training Centre and Canadian Interagency Forest Fire Centre, National Training Working Group: Hinton, AB and Winnipeg, MB)

Forestry Canada Fire Danger Group (1992) Development and structure of the Canadian Forest Fire Behavior Prediction System. Forestry Canada, *Information Report* ST-X-3. (Ottawa, ON)

Fuquay DM (1984) Living with lightning. *Fire Management Notes* 44(1), 18-23.

Gaylor HP (1974) *'Wildfires: Prevention and Control'*. (RJ Brady Company: Bowie, MD)

Gleason P (1991a) LCES – a key to safety in the wildland fire environment. *Fire Management Notes* 52(4), 9.

Gleason P (1991b) L.C.E.S. – the key to safe procedures. *Wildfire News & Notes* 5(2), 1, 4.

Gleason P (1994a) Appendix G: lookouts, communications, escape routes, safety zones – "LCES". Original document, June 1991. In 'An Analysis – Learning From Our Mistakes; Lessons Gleaned From the 1994 Shelter Deployments in the Southwest Area and the South Canyon Fire-Colorado'. (Ed WL Russell) pp. 77-82. USDA Forest Service, Southwest Region, Albuquerque, NM.

Gleason P (1994b) Unprepared for the worst case scenario. *Wildfire* 3(3), 23-26.

Goodson C, Adams B (1998) *'Fundamentals of Wildland Fire Fighting'*. 3rd edn. (Oklahoma State University, Fire Protection Publications: Stillwater, OK)

Gray GC (1982) Radio for the fireline: a history of electronic communication in the Forest Service, 1905–1975. USDA Forest Service. (Washington, DC)

Greenlee J, Greenlee D (2003) Trigger points and the rules of disengagement. *Fire Management Today* 63(1), 10-13.

Harris JP (1998) LCES in the interface: learn it, remember it, use it every time. *Wildfire* 7(9), 10-16.

Henion R (circa 1994) *'18 Situations That Shout Watch Out.'* (Poster) (Maine Forest Service, Fire Control Division and International Association of Wildland Fire: Gouldboro, ME and Fairfield, WA)

Jackson S (1948) Fire line safety. *Fire Control Notes* 9(2&3), 1-4.

Jackson S (1950) Death on the fire line. *Fire Control Notes* 11(3), 26-27.

Keller P (2003) "Gleason Complex" puts up huge "plume": a tribute to Paul Gleason. *Fire Management Today* 63(3), 84-90.

Krebs J (2003) Original ten standard orders. http://www.nifc.gov/safety_study/annual-refresh/hot_topics/10_fireorders.html.

Kurtz E (2003a) LCES vs. LACES: is there a need to change? *Wildland Firefighter* 6(11), 16, 18.

Kurtz E (2003b) Give your crews an 'A' on the fireground. *Wildfire* 12(3), 14-15.

Leuschen T, Frederick K (1999) The consumption strategy: increasing safety during mopup. *Fire Management Notes* 59(4), 29-33.

Luke RH, McArthur AG (1978) *'Bushfires in Australia'*. (Australian Government Publishing Service: Canberra, ACT)

Maclean N (1992) *'Young men and fire'*. (University of Chicago Press: Chicago, IL)

Mangan R (1995) Investigating wildland fire entrapments. USDA Forest Service, Missoula Technology Development Center, Technical Report 9551-2845-MTDC. (Missoula, MT)

Manning T, Taudin-Chabot, P. Bunleavey M, Rowe D, Densmore N (2000) Dangerous tree assessment – British Columbia's wildland fire safety module. In 'Proceedings of the 2000 International Wildfire Safety Summit' 10-12 October 2000, Edmonton, AB. (Eds BW Butler, K Shannon) pp. 137-142. (International Association of Wildland Fire: Montana City, MT)

Matthews M (2000) LCES. *Wildland Firefighter* 6(8), 44-48.

Maupin J (1981) Thirteen prescribed fire situations that shout watch out! *Fire Management Notes* 42(4), 10.

Mayhew B (2008) The two & seven tool. *FireRescue Magazine* 26(8), 82, 84-85, 87-88.

Mayhew B (2009) The intent of LCES. *FireRescue Magazine* 27(4), 96, 100, 102, 104, 106.

Mayhew B (2010) The 19th Watchout – treating a WUI fire like a structure fire. *FireRescue Magazine* 28(4): 103-104, 106.

McArdle RJ (1957) Standard fire fighting orders. *Fire Control Notes 18*(4), 151.

McLennan J, Strickland R, Omodei M, Suss J (2013) Wildfire safety-related decisions and actions: lessons from stress and performance research. In 'Proceeding of 12th International Wildland Fire Safety Summit', 25-26 October 2012, Sydney, NSW. (Ed RL Fox) pp. 33-45. (International Association of Wildland Fire: Missoula, MT)

Merrill DF, Alexander ME (Eds) (1987) Glossary of forest fire management terms. 4th edn. National Research Council of Canada, Canadian Committee on Forest Fire Management, Publication NRCC No. 26516. (Ottawa, ON)

Miller GA (1956) The magical number seven, plus or minus two: some limits on our capacity for processing information. *Psychological Review* 63, 81-97.

Monesmith J (1988) Standards for survival. *Fire Management Notes* 49(3), 30-31.

Moore WR (1959) Training in the ten standard fire fighting orders. *Fire Control Notes* 20(2), 58-60.

Morse GA (1990) A trend analysis of fireline "watch out" situations in seven fire-suppression fatality accidents. *Fire Management Notes* 51(2), 8-12.

Morse GA, Monesmith JL (1987) Firefighter safety: a new national emphasis. *Fire Management Notes* 48(4), 3-5.

Murphy K (2004) Visualizing the Ten and Eighteen – with humor. *Fire Management Today* 64(2), 4-5.

Murphy PJ, Woodard PM, Quintilio D, Titus SJ (1991) Exploratory analysis of the variables affecting initial attack hot-spotting containment rate. *Canadian Journal of Forest Research* 21, 540-544.

NWCG (1980) Preliminary Report of Task Force on Study of Fatal/Near-fatal Wildland Fire Accidents. (National Wildfire Coordinating Group: Washington, DC)

NWCG (1989) Fireline handbook, NWCG handbook 3. National Wildfire Coordinating Group, National Fire Equipment System Publication NFES 0065. (Washington, DC)

NWCG (1993) Fireline safety reference. National Wildfire Coordinating Group, National Fire Equipment System Publication NFES 2243. (Boise, ID)

NWCG (1996) Common denominators of fire behavior on tragedy and near-miss wildland fires. National Wildfire Coordinating Group, National Fire Equipment System Publication NFES 2225. (Boise, ID)

NWCG (2014) Incident response pocket guide. National Wildfire Coordinating Group, National Fire Equipment System Publication NFES 1077. (Boise, ID)

NWCG (2012) Glossary of wildland fire terminology. National Wildfire Coordinating Group, National Fire Equipment System, Publication PMS 205. (Boise, ID)

NWCG (2013) NWCG National Fire Equipment System catalog part 2: publications. National Wildfire Coordinating Group, National Fire Equipment System Publication PMS 449-2. (Boise, ID)

Panko B (2009) Want to be a great media spokesperson? Remember LCES and other fundamentals. *Fire Management Today* 69(1), 24-25.

Pearce HG, Hamilton RW, Millman RI (2004) Fire behaviour and firefighter safety implications associated with the Bucklands Crossing Fire burnover of 24 March 1998. Forest Research in association with National Rural Fire Authority, FRI Bulletin No. 197,

Forest and Rural Fire Scientific and Technical Series Report No. 4. (Rotorua and Wellington, NZ)

Putnam T (1995a) The collapse of decisionmaking and organizational structure on Storm King Mountain. *Wildfire* 4(2), 40-47.

Putnam T (1995b) Findings from the wildland firefighters human factors workshop. Updated July 1996. USDA Forest Service, Missoula Technology Development Center, Technical Report 9551-2855-MTDC. (Missoula, MT)

Putnam T (2001) The ten standard firefighting orders: can anyone follow them? In *'Proceedings of 2001 International Wildland Fire Safety Summit'.* 6-8 November 2001, Missoula MT. (Eds BW Butler, D Mangan) (CD-ROM) (International Association of Wildland Fire: Fairfax, WA)

Queen PL (1993) *'Fighting fire in the wildland/urban interface'.* (Fire Publications, Inc: Bellflower, CA)

Quintilio D, Lawson BD, Walkinshaw S, Van Nest T (2001) Final documentation report Chisholm Fire (LWF-063). Alberta Sustainable Resource Development, Forest Protection Division, Publication No. I/036. (Edmonton, AB)

Quintilio D, Van Nest TA, Murphy PJ, Woodard PM (1990) Determining production rates of initial attack crews. In *'The Art and Science of Fire Management, Proceedings of the First Interior West Fire Council Annual Meeting and Workshop',* 24-27 October 24-27, 1988, Kananaskis Village, AB. (Tech Coords ME Alexander, GF Bisgrove) Forestry Canada, Northern Forestry Centre, Information Report NOR-X-309, pp. 105-113. (Edmonton, AB)

Rothermel RC (1993) Mann Gulch Fire: a race that couldn't be won. USDA Forest Service, Intermountain Research Station, General Technical Report INT-299. (Ogden, UT)

Ruby BC, Leadbetter GW III, Armstrong DW, Gaskill SE (2003) Wildland firefighter load carriage: effects on transit time and physiological responses during simulated escape to safety zone. *International Journal of Wildland Fire* 12, 111-116.

Stacey R (2012) *European glossary for wildfires and forest fires.* European Forest Fire Networks (EUFOFINET) Project. Available at: http://www.northumberland.gov.uk/default.aspx?page=4604.

Steele J (1997) Safety zones: concepts vs. standards. *Wildfire* 6(3), 32-34.

Steele J (2000) Effective firefighter safety zone size: a perception of firefighter safety. In 'Proceedings of the 2000 International Wildfire Safety Summit' 10-12 October 2000, Edmonton, AB. (Eds BW Butler, K Shannon) pp. 173-178. (International Association of Wildland Fire: Montana City, MT)

Steele J, Krebs J (2000) Revisiting the ten standard orders. *Wildfire* 9(5), 21-23.

Sullivan AL, Knight IK, Cheney NP (2002) Predicting the radiant heat flux from burning logs in a forest following a fire. *Australian Forestry* 65, 59-67.

Taylor SW, Pike RG, Alexander ME (1997) Field guide to the Canadian Forest Fire Behavior Prediction System. Natural Resources Canada, Canadian Forest Service, Northern Forestry Centre, Special Report 11. (Edmonton, AB)

Teie WC (1994) *'Firefighter's Handbook on Wildland Firefighting'.* (Deer Valley Press: Rescue, CA)

Thomas D (Issue Coord) (2003) Dude Fire staff ride. *Fire Management Today* 62(4), 1-59.

Thorburn RW, Alexander ME (2001) LACES versus LCES: Adopting an "A" for "anchor points" to improve wildland firefighter safety. In '*Proceedings of 2001 International*

Wildland Fire Safety Summit'. 6-8 November 2001, Missoula MT. (Eds BW Butler, D Mangan) (CD-ROM) (International Association of Wildland Fire: Fairfax, WA). [Reprinted as Canadian Interagency Forest Fire Centre Wildfire Safety Tip 005]

Thorburn RW, MacMillan A, Alexander ME (2000) The application of interactive multimedia CD-ROM technology to wildland fire safety training. *Forestry Chronicle* 76, 953-959.

US Forest Service California Region (1954) Accident check list for forest fire fighters. *Fire Control Notes* 15(3), 14-15.

USDA Forest Service Division of Fire Control (1965) 10 standard fire fighting orders. a programmed text for training fire fighters to know and apply the ten standard orders to fire situations. (U.S. Government Printing Office: Washington DC)

Valdez M, Style JR (1996) Shout "watch out – snag!". *Fire Management Notes* 56(3), 26-27.

Van Wagner CE (1987) *Development and structure of the Canadian Forest Fire Weather Index System.* Government of Canada, Canadian Forestry Service, Forestry Technical Report 35. (Ottawa, ON)

Viegas DX (Ed) (2009) Recent forest fire related accidents in Europe. European Commission, Institute for Environment and Sustainability, Joint Research Centre, JRC Scientific and Technical Report EUR 24121 EN. (Ispra, Italy)

Weir JR (2009) *'Conducting prescribed fires: a comprehensive manual.'* (Texas A&M Univeristy Press: College Station, TX)

Wilson CC (1977) Fatal and near-fatal forest fires: the common denominators. *The International Fire Chief* 43(9), 9-10, 12-15.

Wilson CC, Sorenson JC (1978) *Some common denominators of fire behavior on tragedy and near-miss forest fires.* USDA Forest Service, Northeastern Area State and Private Forestry, Publication NA-GR-8. (Broomall, PA)

Ziegler JA (2007) The story behind an organizational list: a genealogy of wildland firefighters' 10 Standard Fire Orders. *Communication Monographs* 74, 415-442.

Ziegler JA (2008) How the "13 Situations that Shout 'Watch Out'" became the "18 Watch Out Situations". Available at: http://works.bepress.com/jennifer_ziegler/7.

In: Current International Perspectives
Editors: Brigitte Leblon and Martin E. Alexander
ISBN: 978-1-63463-682-7

Chapter 5

WILDFIRE MAPPING AND CLOUD-BASED GIS: SILVER FIRE 2013

***Cassandra Hansen*[*,1], *Paul J. Doherty*[2], *Samuel Lanier*[3] and *Spencer R. Adkisson*[3]**

[1]FireWhat, Inc., Dunsmuir, California and University of Nevada, Reno, Nevada, US
[2]Sierra Nevada Research Institute, University of California, Merced, California, US
[3]FireWhat, Inc., Dunsmuir, California, US

ABSTRACT

Wildfires in the Western United States are a natural and inevitable occurrence, yet evolving conditions have increased the risk to human populations in the wildand-urban interface (WUI). The information systems used to respond to the increasing threats must also evolve to meet the various challenges. Since the 1990's, the operational response to wildfire has relied heavily on the use of traditional geographic information systems (GIS) for planning and operations. With the advent of increasingly sophisticated information systems, GIS can now be used as a platform for sharing geographic information in new and innovative ways. Here we provide a case study of the Silver Fire of 2013, which will highlight the emerging innovation. The case study will serve as an example of the value added by cloud-based GIS to strategic descision makers in the field, raise awareness about future use implications to emergency response, and discuss best practices moving forward. The Silver Fire started on August 7, 2013, outside the towns of Banning and Cabezon, California. As with any emergency response, split-second decisions based on the best available data can prove pivotal to the degree of success to which the situation (in this case, wildland fire) is handled. For this reason the delivery of near real-time fire information, particularly thermal infrared images, to the incident commander (IC), and the incident management team (IMT) prior to arrival on the scene is critical. In this instance, the use of data analysis through GIS and compression of time to turn the data around into a useful product for strategic decision makers was a historic milestone. The ability to integrate infrared thermal hot spot data, with other relevant spatial data into a useful map, was effectively shortened from 12 hours, to 40 minutes via dynamic cloud-based GIS maps. This fact represents a significant evolutionary step for emergency

[*] Corresponding Author address. Email: cassie@firewhat.com.

responders, and provides a major advantage in the ability to make strategic decisions regarding the allocation of finite resources.

We will describe the impact and show examples of cloud-based GIS, on the timeliness of information sharing for planning, operations, and public information. Incident commanders and emergency responders in future cases will need accurate spatial data as close to real-time as possible, in order to make the best decision for allocating resources to save lives, and property. We aim to provide an informative framework for future research and development.

Keywords: Geographic information systems, operations research, cloud-based GIS, wildfire, wildland urban interface

INTRODUCTION

Geographic Information Systems (GIS) have been used for nearly 30 years in wildfire operations. Printed maps are a vital component of any incident action plan, briefing, and public information documention. These paper maps work well in the field, where digital media may have limitations. However, due to the dynamic nature of wildfires, paper maps are often expired as soon as they are printed. The paper map gives a snapshot of how the fire was behaving in the past, but does not reflect how the fire is currently behaving. Moreover, wildfire conditions and zones of increasing threat of wildfire are changing dramatically in response to our changing climate, and the encroachment of human settlement into wildfire prone landscapes (Mell et al. 2010). This ubran sprawl into the wildland areas has recently become more prevelant. Fire and Insurance services refer to this space as the Wildland Urban Interface (WUI), an area where ihuman development conflicts with both naturally occuring and human-caused wildfire. For these reasons new technology is needed in order to keep pace with the landscape changes. We need to change our perspective on geographic information and incorporate dynamic maps into planning and operations. It is now possible to integrate "remote sensors" that detect and describe wildfire geography with dynamic maps that can be distributed via the Internet (cloud-based GIS). In this chapter we will discuss the recent advancements of wildfire information sharing, and how it will likely change the way geographic information is disseminated and utilized in the management of wildfire operations.

Wildland fire sensors vary from traditional (e.g., a fire spotter in a tower, a thermal detection satellite in space), to non-traditional (e.g., a citizen observer simply taking a photo and sharing it across a social media platform) - but both of these types of sensors can be integrated and transformed into informative geographic information if done in a way that supports the direct needs of the community. For instance, the use of WebGIS as a platform in wildfire management can decrease the time lag between sensor receiving data to the intended audience receiving the message. This paradigm shift leads to actionable information being delivered in a shorter time frame for incident managers, where timely and accurate decisions can be made. This will likely require a period of transition where WebGIS can be adapted to operational workflows. First, let us describe WebGIS at a high level.

WebGIS is an information system that uses web technology to communicate between a server and a client (Fu and Sun 2010). Benefits of WebGIS consist of:

- Having a global audience
- Multiple users at any given time, simultaneously allowing for collaborative mapping efforts
- Works across commonly used web browsers, Internet Explorer, Mozilla FireFox, Apple Safari, Google Chrome etc, allowing for a large number of users
- Use of mobile cloud-based GIS applications that are both, device agnostic, and browser agnostic
- Real-time information data feeds, ensuring that the maps are displaying current information
- The diverse application of cloud-based GIS maps allows for a wide range of users and workflows

In this chapter we will highlight a case study that demonstrates the use of cloud-based GIS maps during the Silver Fire, a catastropic wildland fire, in Southern California on August 7, 2013. In order to highlight the significant advancement of GIS by the web platform, we must first reflect on a brief history of wildland fire GIS.

Wildland Fire GIS Background

Maps have had a long and significant role as an invaluable tool for firefighters. The ability to take unique information on-scene and transcribe it onto a flat map has been crucial for the distribution of fire information. As technology has advanced and increased the mobility of computers, GPS units, and remotely sensed data the poteniitial application of these tools have increased exponentially. Regardless, the fire fighting industry has experienced challenges adopting these changes, favoring the familiar analog methods for data collection and geographic analysis. We have discussed this process with several industry experts and summarized our findings here. This section will provide a historic outline of the gradual adoption of GIS technology and highlight pivotal moments and contributions that GIS has made to the firefighting industry.

Since 1964 the National Infrared Operations (NIROPS) has been flying over the western United States recording thousands of hours of night time thermal infred images. These images have proven invaluable for wildland fire thermal detection in remote regions with complex terrain (US Forest Service, 2013). However the information product of choice was printed images up until recent years.

The use of GIS in the wildland fire indutrstry gained momentum in the late 90’s, due to the persistent need to have consistency between wildland fire GIS specialists (GISS) from different governmental agencies. The fire season of 2000 was the worst in 90 years and was characterized by large agresive fires across the western U.S. Nearly 7 million acres burned in 2000 (twice the national average), 15 firefighters perished, and total land and economic losses approached $10 billion. As a result of the castrophic fire season of 2000, changes in the amount of allocated funds for GIS fire support were enacted (U.S. Fire Admin, 2001). In an effort to promote the access to data, access to public online fire perimeter information from the Geospatial Multi-Agency Coordination Group (GeoMAC) was eased. Around this time Selective Availabity was turned off on May 1, 2000, improving the accuracy of GPS globally (Christophe et al. 2002). While this greatly increased the use of GIS in wildland fire operations, there were still no standard protocols designed to reconcile the variability associated with the proper methodology for applying GIS, lack of metadata information, and the absence of standardized maps. The 2000 fire season clearly illustrated the need for such a framework and served as a call to action for all agencies associated with wildland fire response but did not lead to systematic adoption of GIS for operations (Triplett, 2013).

Figure 1. National Infrared Operations (NIROPS) thermal interperation used for identification of fire perimeter hand drawn on the mylar layers overlaying the topomap (NIROPS, 2001).

Continuous implementation of GIS in the fire industy prompted the beginning of Fire GIS classes using Ventura Tools along with Environmental Systems Research Institute's (Esri) ArcView 3.0. Ventura Tools represented the standard application used by emergency responders for GIS/Mapping situations specifically designed for California. These extentions were used for data creation, manipulation and display of emergency response information. This mapping software allowed for symbology, provided consistency, and began to allow for the automation of mapping (Triplett, 2013). In conjunction with mapping software, advancements in computer processors and plotters increased the speed and capability of GIS technicians, which reinforced their strategic value to a fire response team. It also created the need to train and develop dedicated GIS technicians, rather than using firefighters for GIS duties in the field. In 2004 a comprehensive GIS training program was established by the National Wildfire Coordinating Group (NWCG). Adoption of this GIS training and standardization program has had a positive and lasting effect on the industry as it continues to evolve in both breadth and depth. Even though GIS and related mapping systems have become more prevalent as an information systems tool, the majority of situational awareness and incident information exchange are slow to keep pace, and are still expressed using the traditional Incident Command System (ICS) forms (NWCG 2013). The ICS was developed in the early 1970's after a series of catastropic wildland fires in Southern California. The system was an adaptation of a military command and control structure to coordinate resources with common goals, terminology, and communication. The ICS is primarly paper based and is employed to plan, organize, staff, control, evaluate, and direct an incident for successful execution. It is against this backdrop that the events of the Silver Fire can be fully appreciated to indicate the next evolutionary step forward.

CASE STUDY

Incident Information

The Silver Fire started on August 7, 2013 and was, in comparison to the conditions and severity of other western wildfires, a relatively commonplace wildfire incident. However, the use of data analysis through GIS, and compression of time to turn the data around and into a useful product for strategic decision makers made this a historic event. The Silver Fire was an extremely agressive fire with a rapid rate of spread (ROS) near the towns of Banning and Cabazon, California. The fire destroyed 48 structures and burned approximately 82 square kilometers (20,292 acres). At the peak of the fire, there were over 2,106 fire personel on scene. Residents were given approximately 15 minutes to evacuate their homes (Figure 2).

a) b)

Figure 2. a) A burnt car is the only remains after the fast moving Silver Fire and b) evacuated resident displays Silver Fire photos from his smart phone (Photos taken by Nathan Johnston, 2014).

As with any emergency response, split-second decisions based on the best available data can prove pivotal to the degree of success to which the situation (in this case, wildland fire) is handled. For this reason the delivery of near real-time fire information (particularly thermal infrared images), to the hands of the incident commander (IC), and their IMT prior to arrival at the incident marks a major technological advancement for the GIS in the wildland firefighting industry. The ability to integrate infrared thermal hot spot data, with spatial geographic data, into a useful map, shortened the standard 12-hour turnaround time down to 40 minutes via dynamic cloud-based GIS maps. The process is a major leap forward for emergency responders and provides the IC and their IMT with unprecedented power in strategic fire response (Figure 3).

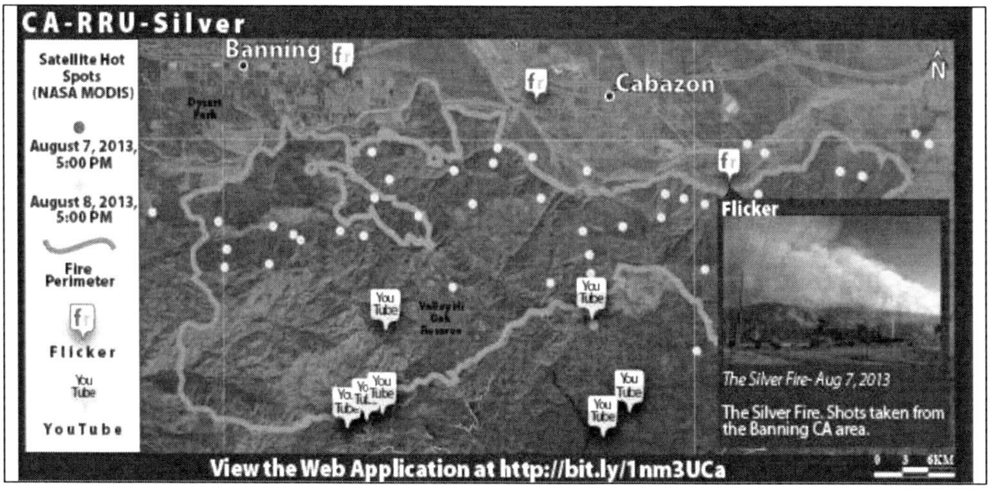

Figure 3. Silver Fire WebGIS situation map dispaying the fire perimeter, MODIS satellite thermal detection, along with geotagged social media (Wildlandfire 2013). RRU signifies the Riverside Unit of CALFIRE, in Riverside County, California. Visit http://bit.ly/1nm3UCa to interact with the Silver Fire situation map shown above.

Data

Public data about fire perimeter is near real-time information updated daily from fire incident intelligence sources. GPS data is obtained from field personnel uploading track files to the GIS System. The track files come from firefighters, bulldozer operators, helicopter pilots, and field observers walking, driving, or flying the fire perimeter. Infrared (IR) imagery is obtained from both fixed-wing and satellite platforms providing fire perimeter data from remote locations (Walters et al. 2011). All fire maps incorporate the GeoMAC fire perimeters and are available via an ftp site (USGS 2013). The perimeters are loaded into GeoMac after the first operational period. An operational period typically runs from the time the fire ignited, to 0700 the following morning. The Silver Fire incident was unique in that the fire perimeter data was obtained, analyzed by a GIS technician, and redistributed to the IC prior to being uploaded to the GeoMAC resulting in a dramatically reduced wait time for useable information.

Moderate Resolution Imaging Spectroradiometer (MODIS) thermal data are available from the Terra (EOS AM) and Aqua (EOS PM) satellites (NASA 2013). Both satellites are on different orbital paths collecting global data every 1-2 days. The Terra satellite orbital sequence passes from north to south and passes over the equator in the morning. The Aqua satellite passes over the equator in the afternoon. MODIS satellites that orbit 705 kilometers above the earth can recognize the thermal signature of a fire which can be used to identify "hotspots".

NOAA surface wind data provides real-time wind velocity (direction and speed) (NOAA 2013). Both MODIS and surface wind data sets use rich site summary (RSS) data feeds from NOAA. RSS data layer feeds are continually updating the map providing quantifiable data that reflects the rate of change over time.

In partnership with the Esri Disaster Response Program (www.esri.com/disaster), all of these datasets are made freely available as map services (Esri 2013) on secure cloud-hosted infrastructure. These map services use the publically available datasets from government sites but re-host the services to reduce the impact during high volume web traffic and to help cloud-based GIS users integrate the geographic content for display and analysis. The user does not have to download these data sources, they are simply transfered to a map interface via http. These map services can be added to an agency's web mapping application or consumed by the end user in a ready to use web mapping application.

Methodology

Web-based cloud GIS maps can facilitate a multi-use platform to create specialized applications from single web-map. In this case a single map application was created for a public information purposes while another map application was designed specifically for the internal use of the fire service. This section outlines the detailed sequence of how specific data became available, and how it was specifically used for the distribution of fire information.

The Silver Fire was reported on August 7, 2013 at 1430 hrs. local time in the San Jacinto Mountains of Riverside County, California south of the towns of Banning and Cabazon (ICS 209 Form, 2013) (See Figure 3). Surface observations regarding this fire were posted on a social media forum at 1430 describing the column of smoke near Banning, CA (Wildlandfire.com 2013). At the initial report estimated that the size of the fire was 300+ acres (ICS 209 Form, 2013). However, due to the rugged geographic location of this fire, high winds, low relative humidity and high surface temperatures, the fire moved faster than anticipated (Wildlandfire.com, 2013).

At 1545 an IMT Type 1 team was ordered (Wildandfire.com, 2013) and it was estimated that the fire had reached 22 sq. km (5000 acres) with zero percent containment and that the fire would be going "extended" (ICS 209 Form, 2013). Extended refers to a fire that has no known or projected termination time. A Fire is upgraded to a Type 1 designation when the logistical needs exceed the capabilities of local jurisdiction (NWCG, 2013). An IMT Type 1 team consists of several high level staff with specific skill sets required to manage complex incidents. It takes, on average, 12-24 hours to organize a team and up to 48 hours for an IMT to mobilize to an incident, depending on the geographic location of the incident vs travel time to the incident. A fire burning into the next operational period elevates the fire into an extended attack, requiring additional resources, and shifts of suppression personnel necessary to staff the incident operations.

Later in the evening of Aug. 7th, National Infrared Information (NIROPS) aerial imagery was flown, collected and utilized to obtain the thermal fire perimeter (US Forest Service 2013). Once the fire perimeter was flown and the aerial information was post-processed, the GIS perimeter was sent to the data interpreter at 0300, the morning of Aug.8. The post-processed data was available for cloud-based GIS analysis at 0425 and was uploaded onto cloud-based GIS map. The time sensitive information was made available to Silver Fire Type 1 IC at 0438, several hours prior to the IC arriving on scene (Figure 4).

The fire perimeter data was accompanied with real-time information on surface wind speed and direction, along with MODIS satellite thermal "hotspot" locations updated twice daily. At 0800, the fire had burned over 26 sq.km. (6,500 acres) and was still 0% contained. At this time the fire split into different directions threatening the communities of Poppet Flats, Silent Valley Club, Twin Pines and Cabazon (ICS 209 Form). Mandatory evacuations were sent out to residents using the Early Warning Notification System (County of Riverside 2013). As the fire moved closer to residential areas, census data was used to generate enriched polygons of the cities of Banning and Cabazon for evacuation purposes (Figure 5).

As this fire was in a classic WUI area, the WUI census data consisted of the following information about: 1) population in the area (children under the age of 15 and median income), 2) housing units (number of structures owned and rented), 3) business types (hospitals, grocery stores, and gas stations), and 4) acres (precentage of water, barren, and developed land).

At 1000 the time sensitive fire perimeter data was made public on GeoMac. At this time the cloud-based GIS map's URL was made public and distributed on social media fire forums such as Wildlandfire.com. Firefighters, media networks and people who were evacuated had access to this dynamic near real-time map. Social media public tools such as twitter and flicker were also used to capture geographic surface conditions (Figure 3).

Date/Time	Silver Fire Incident Information	Cloud Base GIS
Aug 07, 2013 1700	Initial Fire Report	--
Aug 07, 2013 2000	Mandatory evacuation are in affect for approximately 1500 residents of Silent Valley and Poppet Flats	
Aug 08, 2013 0300		NIROPS postprocessed perimter data
Aug 08, 2013 0425		Fire perimeters uploaded to WebGIS Maps
Aug 08, 2013 0730	22 sq. km. (5500 acres), 0% contained. Fire has critical rate of spread and has grown rapidly.	
Aug 08, 2013 1340	40 sq.km. (10000 acres), 10% contained	
Aug 08, 2013 2000	56 sq. km. (14,000 acres), 20% contained	
Aug 09, 2013 0800	72 sq. km. (18,000 acres) 40% contained	
Aug 10, 2013 0745	76 sq. km (19,000 acres) 47% contained Governor Edmond G. Brown Jr. issued an emerergancy proclamation for River Side County	
Aug 10, 2013 2000	76 sq. km (19,000 acres) 70% contained	
Aug 11, 2013 1800	78 sq. km (19,436 acres)	
Aug 12, 2013 0600	82 sq. km (20,292 acres) 90% contained	
Aug 12, 2013 1800	82 sq. km (20,292 acres) 100% contained	

Figure 4. Silver Fire Timeline based on the Silver Fire Incident Report (ICS 209 Form, 2013).

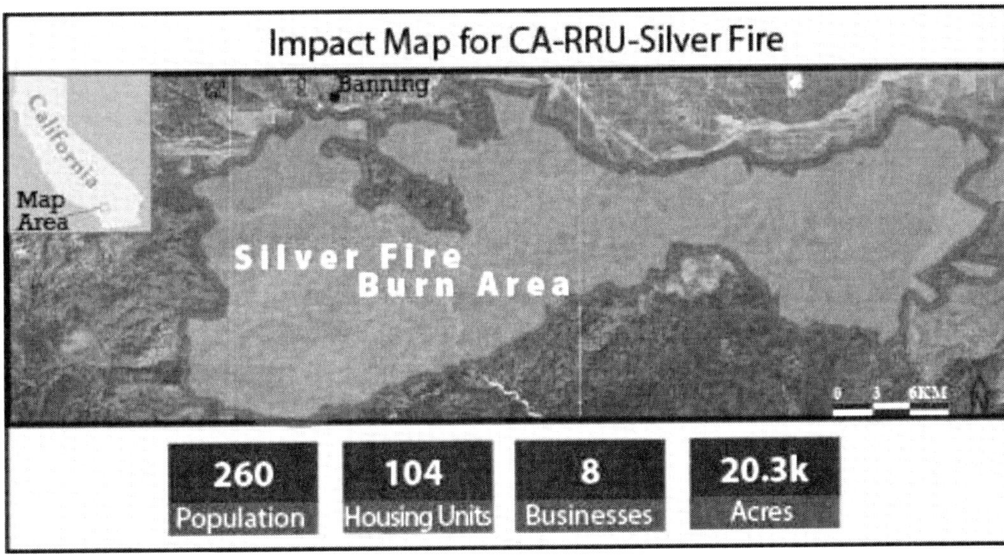

Figure 5. Silver Fire WebGIS impact map dispaying the fire perimeter along with an analysis of the potential impact in terms of population, households, and businesses. (Wildlandfire, 2013). Please visit http://bit.ly/1CngzGB to view the Impact Map Live.

RESULTS

The timing of the Standard Operating Protocol (SOP) varies from each fire and geographically available resources. However, it is certain that accurate near real-time maps are pertinent for split second decision making by fire ICs. The type 1 IC arrived on the scene at the Silver firing knowing that the fire acreage had increased over night to approximately 11,000 acres. If not presented with this in hand cloud-based GIS map, it was assumed that the fire still remained at approximately 6,000 acres. With this knowledge more air support and ground support was ordered and deployed.

At the time the WebGIS map was made public it received over 500 user hits from around the world within the first hour. This cloud-based GIS map was also disseminated to multiple media outlets from the Social Media Forum Wildlandfire.com. Over the next six days approximately 82 square kilometers (20,292 acres) burned before becoming 100% contained. There were 13 injuries reported (12 from fighters and 1 civilian). At the peak of the fire there were 2,106 personel on the ground with, 201 engines, 63 fire crews, 9 helicopters, 16 dozers, and 18 water tenders costing an estimate of $9,999,000 US dollars (Inciweb, 2013).

CONCLUSION

During the Silver Fire, the construction and disseminating of this dynamic WebGIS map from start to finish was unprecedented. While GIS has been historically used for producing paper assignment and briefing maps – here we highlighted how cloud-based GIS could be used as communication tool to give all responders a common map to discuss. Discussions of the potential for use of real-time cloud-based GIS have been ongoing as far back as 2002, according to Tom Patterson (retired Fire Management Officer and Technology Specialist, Esri). Patterson states " for the first time, that I know of, an Incident Management Team responding to a major fire were armed with the most up-to-date firemap available as they were travelling and had a good perspective of the situation upon arrival".

The internet has proven to be a fast and useful source for obtaining critical information. While the IMT was responding to the Silver Fire it could obtain this information while in a moving vehicle. ArcGIS Online offers a secure site for agencies to collaborate from different locations (Incident Command Post, Agency Headquarters, Fire Stations, etc.). Technology can enhance situational awareness and provide for firefighter safety but this was the first incident in which it was actually done during the developing phase of an ongoing emergency operation. When an IMT assumes command of a major wildfire, the first operational period is spent determining what has already happened, what is the potential to happen, where the fire is currently located, direction of spread, values at risk, etc. During the Silver Fire this team was provided all those answers before they arrived and could more quickly begin their operational roles.

Using this case study we have highlighted the paradigm shift of Web GIS for wildland fire in a similar way to the use of Desktop GIS for emergency management in the previous decade (Cutter 2003). In the early part of the 2000 - 2010 decade, GIS was emerging as a critical information system that would eventually embed itself into the emergency response cycle during major disasters, such as the World Trade Center disaster on 11 September 2001. By the middle part of the decade, GIS was used on large natural disasters such as Hurricane Katrina and as a result great advances were made to the field of Public Safety GIS. The key components that led to this shift were understandable user interfaces data quantity, quality, and integration as-well-as real-time data and information. With cloud-based GIS – these key components can now become fully realized by wildland fire responders as demonstrated in the Silver Fire.

Disseminating Public Information Maps through social media networks was clearly the fastest mode of distribution for the Silver Fire. As cloud-based GIS maps become more frequently used, it is of the utmost importance to ensure that valid information is being used and distributed. This will require that Public Information Officers (PIO) and media liasons have access to easy-to-use web map authoring tools. Timely and relevant data layers from creditable sources is one of the strongest characteristics of cloud-based GIS maps that sets these digital maps apart from any other Web Maps. Likewise, the dissemination of cloud-based GIS maps for decision-makers must be authored using dynamic information from real-time sensors – therefore platforms that can aggregate these live feeds will continue to grow in importance. Moreover, in order for a cloud-based GIS platform to be successful – the end user web mapping applications need to be well designed. From this incident we learned that web mapping applications should be:

- cross-platform
- responsive in design (works well on all devices)
- customizable for the audience
- easy to understand

It is also essential that all operational data in the cloud-based GIS platform can be printed on paper maps, loaded onto mobile devices in the field, and work in disconnected environments when needed.

It is important to anticipate how this will change our perspective on wildfire prevalance, safety conerns, and impacts to the environment. As populations grow people are moving farther and farther out into what was previously « wild » land. As a result, the WUI -- the intersection between a destructive wildfire event and people, their property and infrastructure, also increases. Technology offers the only means for compresing distance and time with regard to disseminating critical information to residents and stakeholders in the event of a wildland fire. One resident of Poppett Flats noted that his property had narrowly avoided three major fires within a 12 year period. He also stated that he had no information on the Silver Fire as it burned closer and closer to his home. He was warned and told to leave by his daughter through social media, not neccessarily a formal government notification system.

WebGIS offers the ability for residents of rural areas to have access to important information regarding the location of a fire, rate and direction of spread, possible escape routes, evacuation centers, in a simple hand-held device, your mobile phone. This represents a major step forward in fire safety, and allows people to make more informed decisions in a time of crisis. No longer will fire maps be made with yesterday's expired data. Once technology is used in such an efficient and effective manner as witnessed in the Silver Fire, there is no going back. IMT's and the public will demand increasingly accurate dynamic maps will full unrestricted public access to the information. Mobile GIS technology has advanced fire maps and technology over a short period of time, but is still in its infancy.

With regards to «*Wildfires Prevalence, Safety Concerns and Impacts on the Environment*», we feel WebGIS will play a defining role in the advancement of public fire safety in the 21st century. Moreover, WebGIS can also benefit all professionals and trying to understand wildfire prevalence and the impact of this changing fire regime on the environment. We hope that our example here can be expanded upon and utilized by the greater community concerned with science, policy, and law of wildfire.

"Knowing what your fire is doing is necessary in deciding what to do. By simply glancing at a map and determining hazards to personnel, areas that need evacuation and routes to shelters can all be pre-planned, or in this case, strategies can begin to be considered while enroute to the scene. Instead of having everyone respond to the same location, using this technology will enable fire officials to determine staging areas closer to the threat for faster deployment of incoming resources" (Patterson, 2014).

ACKNOWLEDGMENTS

The authors would like to thank (1) Sean Triplett for providing information on the historical aspects of GIS technology in the fire industry, (2) Tom Patterson for his professional experience of wildland fire mapping and (3) to all the residents affected by the Silver Fire.

References

Cutter SL (2003) GI science, disasters, and emergency management. *Transactions in GIS* 7, 439– 445.

Christophe A, Girard I, Gendner JP, Janeau G (2002) Global positiong system (gps) location accuracy imporvement due to selective availabity removal. *Comptes Rendus- Biologies* 2, 165-170.

ESRI (2013) Esri Live Feed Map Services [online]. Available from: http://tmservices1. esri.com/arcgis/rest/services/LiveFeeds.

Fu P, Sun J (2010) Web GIS: Principles and Applications. ESRI Press. 24-48. (Redlands, CA)

Inciweb (2014) Incident Information Web [online]. Available from: http://inciweb.nwcg.gov/

Johnston N (2014) Silver Fire Photos. Taken on March 11, 2014.

Mell WE, Manzello SL, Butry D., and Rehm, RG (2010) The wildland–urban interface fire problem – current approaches and research needs. *International Journal of Wildland Fire* 2, 238-240.

National Interagency Coordination Center (2013) Silver Fire ICS 209 Report. [online]. Available from: http://fam.nwcg.gov/fam-web/hist_209/hist_r_print_209_ head_2013?v_number=CA-RRU-079781&v_report_date=08/07/2013&v_hour=1950&v_gaid=SO

NIROPS (2001) Incident Informatoin Template: How are these map made? [online]. Available from: http://nirops.fs.fed.us/docs/about-more/08-How do we Make those maps.pdf

NASA (2013) NASA LANCE-FIRMS MODIS Hotspot/Active Fire Detections. [online]. Available from: https://earthdata.nasa.gov/data/near-real-time-data/firms

NOAA (2013) National Oceanic and Atmospheric Administeration (NOAA) Meteorological Aviation Report (METAR) [online]. Available from: http://www. aviationweather. gov/adds/metars/

NWCG (2013) Incident Command System (ICS) forms [online]. Available from: http:// www.nwcg.gov/pms/forms/ics.htm.

Patterson T (2014) Personal Interview. 10 Feburary, 2014.

Riverside County (2013) Riverside Early Warning Notification System [online]. Available from: http://www.countyofriverside.us/emergencies/ews.html.

Triplett S (2014) Personal Interview. 9 January, 2014.

US Forest Service (2013) National Infrared Operations [online]. Available from: http://nirops.fs.fed.us/pages/about.

US Geological Survey (2013) GeoMAC Outgoing [online]. Available from: http://rmgsc.cr. usgs.gov/outgoing/Geomac/.

Walters SP, Schneider NJ, Guthrie, JD (2011) Geospatial multi-agency coordination (GeoMAC) wildland fire perimeters. U.S. Geological Survey. U.S. Geological Survey Data Series, 612, 6-8.

Wildlandfire (2013) CA-RRU-Silver Forum [online]. Available from: http://wlfhotlist. com/threads/36203-CA-RRU-Silver-(contained).

In: Current International Perspectives
Editors: Brigitte Leblon and Martin E. Alexander
ISBN: 978-1-63463-682-7

Chapter 6

WILDFIRES AND SAFETY ISSUES IN GREECE

Gavriil Xanthopoulos*
Hellenic Agricultural Organization "Demeter",
Institute of Mediterranean Forest Ecosystems
and Forest Products Technology, Athens, Greece

ABSTRACT

Greece is a European country with Mediterranean climate that faces serious challenges with wildfires, as the combination of high forest vegetation flammability and hot and windy summer weather, often resulting in very intense fire behavior. As a result, large forest and agricultural areas are burned every year. Villages and other wildland-urban interface areas are commonly found in the path of the fires, resulting in serious damages and sometimes loss of human life. The aim of this chapter is to examine the safety aspect of wildfires in Greece. Loss of life of firefighters and other citizens due to forest fires in Greece is historically tracked. It is shown that, with one exception, no deaths were recorded before 1977. Worsening fire behavior after the end of the 1970s, as reflected in the increasing yearly burned area trend, gradually made fire-related fatalities a regular phenomenon. A few of the most interesting accidents are described in brief and the factors contributing to this trend are discussed in order to learn and be better prepared for the future.

Keywords: Wildfire, safety, fatality, Greece

INTRODUCTION

Greece is a country in southern Europe nearly surrounded by the Mediterranean sea. A typical Mediterranean climate, with cool wet winters and hot and dry summers, prevails over most of the country. Similarly to all other countries on the north coast of the Mediterranean

* Corresponding Author address: Hellenic Agricultural Organization "Demeter", Institute of Mediterranean Forest Ecosystems and Forest Products Technology, P. O. Box: 14180, Terma Alkmanos, 11528, Athens, Greece
Email: gxnrtc@fria.gr.

sea, it faces serious challenges with wildfires, as the combination of forest vegetation flammability with the hot and windy summer weather often results in very intense fire behavior, causing the loss of large areas of forest and agricultural lands. Villages and other wildland-urban interface areas are commonly in the path of these fires, resulting in serious damages and sometimes in loss of human life.

The aim of this chapter is to examine the safety aspect of wildfires in Greece. Loss of life of firefighters and other citizens due to forest fires in Greece is historically tracked and discussed in an effort to identify trends and to provide insights through analysis of specific events.

Evolution of the Forest Fire Problem in Greece

Understanding wildfire safety issues requires some knowledge of the evolution and the characteristics of the forest fire problem in Greece. About 60% of the country's area is considered as forest land. The forest vegetation is very diverse, with the elevation differences and proximity of the sea playing a major role. Low to mid-elevation forest ecosystems, up to 800 m, are typically Mediterranean, while at higher elevations, especially in the northern part of the country, the vegetation is similar to mid-latitude temperate European forests. Wildfires mainly occur in the lower elevation Mediterranean vegetation, which is also adapted to higher fire frequency.

Mediterranean vegetation in Greece mainly consists of low, often thorny, shrubs called phrygana, of taller typical Mediterranean evergreen shrubfields (maquis), and of flammable low-elevation pine forests (mainly *Pinus halepensis* Miller and *Pinus brutia* Ten., usually with evergreen shrub understory). Deciduous and semi-deciduous oak (*Quercus spp.)* forests are also found below 800 m elevation.

Fire in phrygana is characterized by low to medium fire intensity, but can reach very high rates of spread under high wind conditions. Fire behavior in evergreen shrubfields varies significantly because, depending on species composition, site characteristics, and intensity of livestock grazing, the fuel height ranges from 0.5 m to more than 2.5 m and fuel load varies accordingly. Mediterranean pine forests are notorious for extreme fire behavior, especially in regard to high intensity crown fires that occur under high fire danger conditions. The shrub layer in the understory facilitates crowning and reduces the number and size of vegetation gaps that could reduce the rate of spread. Oak forests generally burn with less intensity than the other types of Mediterranean vegetation.

Another characteristic of the forest lands in Greece is that they are interspersed with villages where most of the population of the country resided until recently. Accordingly, and especially around villages, there were many cultivated fields, olive groves, vineyards, etc. interrupting forest continuity. Forest fuels were controlled through forest biomass utilization for timber (construction and furniture), for cooking and heating, and for animal feed. Agricultural cultivations were kept free of fine fuels in the summer in order to reduce risk of fire damage.

Until the 1960s forest fires were considered a relatively minor problem. Forest protection was in the responsibility of the Forest Service, which did not even have fire engines until 1970. Forest fires were mainly fought by the people in the villages, who were often seasoned forest workers, under the guidance of the local Forest Service officers. A few fire engines of the (urban) Fire Service were occasionally available. Firefighting was quite successful, as the people used local knowledge about the topography and the road network, including vegetation breaks and previous burns (Xanthopoulos 2007a). They also knew and understood local wind and moisture patterns. Use of fire to fight fire was one of the ways in which they increased their chances of controlling serious fires in spite of the limited, mainly agricultural tools they had (Kontos 1933; Karras 1954).

Starting in the 1950s, when Greece was a devastated country after the Second World War and the Civil War that followed, and picking up in the late 1960s, a wave of immigration abroad and migration to the big cities in the country led to a sharp decrease in the population of the villages, especially those in mountain areas. This and the gradual reduction of wood utilization for heating and cooking in the villages in favor of using alternative energy sources such as electricity, resulted in significant forest fuel built-up. Forest fuels increased in quantity while abandonment of agricultural activities around mountain villages resulted in higher spatial fuel continuity.

Population migration resulted in quick and poorly planned expansion of cities. Subsequently, and in parallel with the economic growth that came in the last three decades, many people tried to find better living conditions than those in the crowded city centers. This led to the development of extensive wildland-urban interface (WUI) areas around the largest cities (Athens, Thessaloniki). Also, the effort of many people to escape the summer heat in the cities, combined with a substantial growth in tourism, led to the development of WUI areas along the coast where villages and tourist installations are often found in contact with forest vegetation. Starting with a large fire in the high-class northern suburbs of Athens in 1981, fires in the WUI gradually became a major issue nearly every summer.

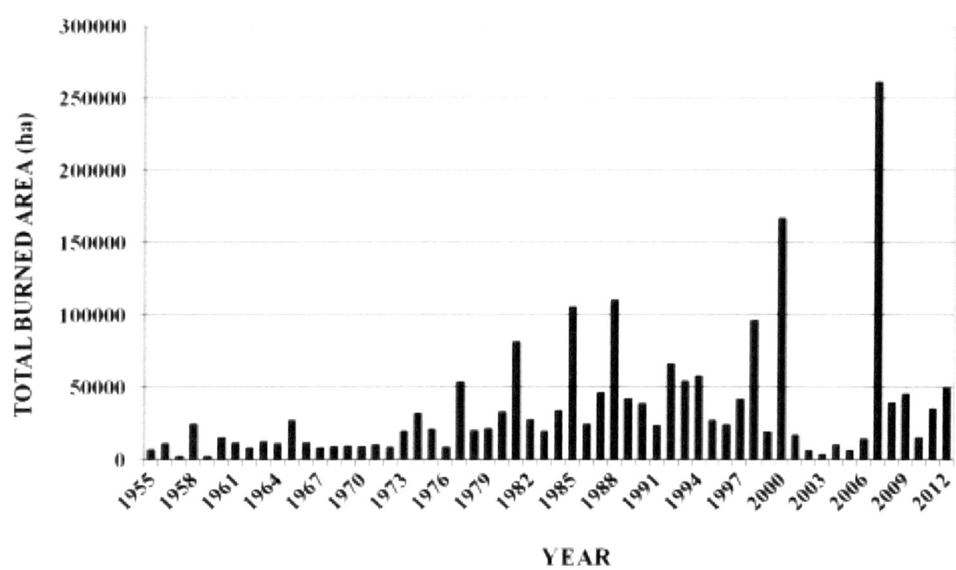

Figure 1. Evolution of the total yearly burned area by wildfires in Greece in the 1955-2012 period.

All these changes had an immediate effect on the appearance, characteristics and effects of forest fires, gradually making this phenomenon one of the most important natural hazards that Greece has to face every year. The number of fires increased and the yearly burned area broke previous records multiple times in the last forty years (Figure 1). Efforts to fight the fires more effectively, spending more money on firefighters and resources, and even transferring the responsibility of forest firefighting from the Forest Service to the Fire Service in 1998, did not solve the problem. Furthermore, the damages caused by these fires became worse. Nowadays, fuel continuity allows fires to reach and enter WUI areas and even villages. In addition to damaging homes and infrastructure, wildfires have now become an important safety issue both for the firefighters fighting them and for the citizens who get trapped in their path. A look into the evolution of wildfire-related deaths in Greece provides an understanding of the safety problems and the causative factors that the country currently faces (Figure 2).

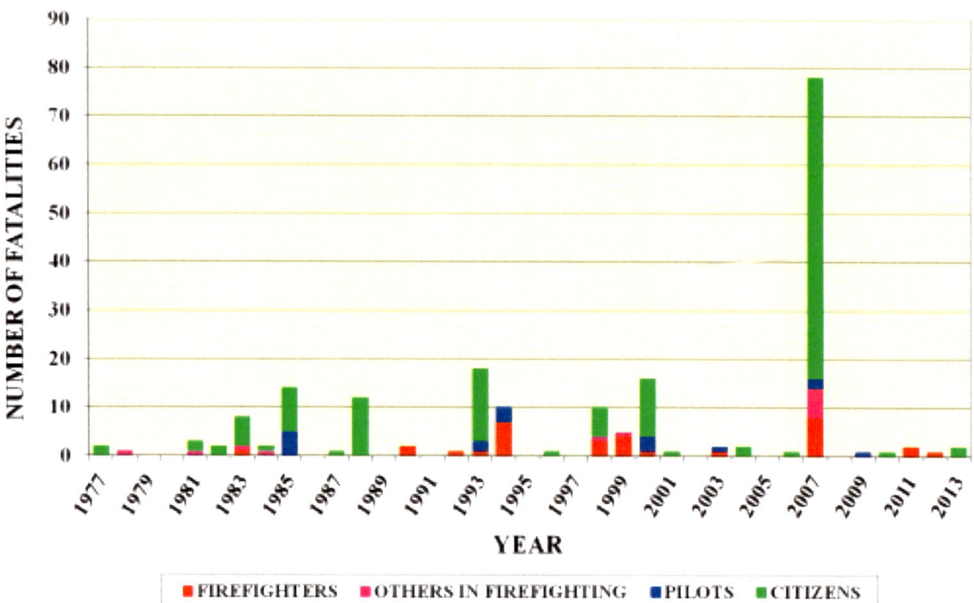

Figure 2. Evolution of wildfire deaths in Greece in the 1977-2013 period.

Evolution of Wildfire-Related Deaths in Greece

The risk to life posed by wildfires was not a cause for serious concern until the last two decades of the 20th century. Actually, the only record of wildfire-related deaths before 1977 was in 1916, as described later in this chapter. While, due to the lack of an official effort to record and count such events, it cannot be ruled out completely that there may have been a few such deaths in the first part of the 20th century, no information could be located in articles on forests and forestry history, in newspapers, or in other journals. In 1977, the first firefighting-related death in recent forest fire history occurred in the same general area, where the 1916 fire and deaths took place. Following that, deaths of firefighters, pilots, and civilians started becoming a relatively regular phenomenon, with occasional peaks and a general upward trend (Figure 2). Examination of these deaths with emphasis on some of the most noteworthy accidents, taking into account the change in conditions described earlier, provides some understanding about the causes of this trend.

DEATHS RELATED TO GROUND FIREFIGHTING

According to Kailidis (1990), there were no forest fire-caused fatalities in Greece in the 1950-1976 period. From 1977 on, forest fires in Greece started having victims on a more or less regular basis. However, few of these fatalities were related to firefighting. In 1978, during a fire in the forest of Tatoi near Athens, a soldier whose unit was participating in firefighting was killed when hit by a tree broken by the water drop of a Canadair CL-215 water bomber. According to an account of this accident, narrated to this author by a witness, the army unit was ordered to enter into the firefighting zone, where the drops were being made, by a minister who was present in the area at the time, in spite of the objections of the Forest Service officer in charge.

In 1981, a young volunteer was killed on the island of Rhodes while firefighting (Kailidis 1990).

In 1983, a Fire Service professional firefighter who participated in firefighting in an area of low shrubs with fine fuels (phrygana) received extensive burns and died. According to an account of this accident, narrated to the author twenty years later by Fire Service officers, he was surprised by the change in fire behavior and the quick spread of the fire in these flashy fuels.

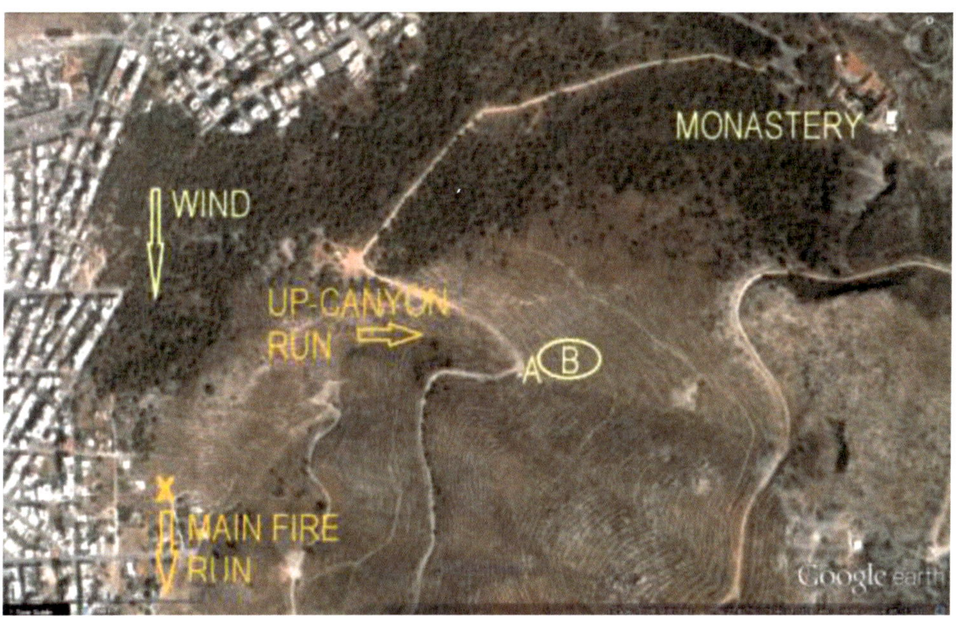

Figure 3. The site of the fatality at Kareas on a satellite image captured in 2001 (source: Google Earth). The approximate ignition area is noted with "X". Point (A), at the apex of the turn on the forest road, is where the firefighters stopped their engine and got ready to fight the up-canyon fire run. Area (B) is where their bodies were found.

In 1984, a soldier was killed during firefighting on Rodopi Mountain in Northern Greece, and another was seriously injured on the island of Rhodes.

In 1985 a forest guard was trapped during a large fire on Thassos island that lasted 7 days and burned 12000 ha. He chose to cross the fire front and enter into the black. He received serious burns, but survived. Six other citizens died in that fire (Kailidis (1990).

In 1988, one of the worst fire seasons in which 12 people died, the chief officer of the Forest Service office in Sperhiada, Central Greece, received serious burns when, according to descriptions narrated to the author, the vehicle he was using for patrolling an intense fire was trapped on a forest road, in a V-shaped draw, with the fire coming from below.

On 26 July 1990, a fire in the area of Paleovarvasena, Pyrgos, Peloponnese, killed two young seasonal firefighters. The fuel was mainly tall dry grass with few shrubs and sparse Aleppo pine (*Pinus halepensis* Miller) trees. According to the account of a forester who worked at the Forest Service office of Pyrgos at the time, the two firefighters reached the location of the fire in their fire engine. They left the engine on the forest road and went down a steep slope (about 50%), from their engine towards the flank of the fast moving fire below them. They started hosing the flank when a change of wind direction, probably an aerodynamic effect of the movement of wind around the hill, spread the fire below and then upslope around them. They were overpowered by the fire, abandoned the hose, and run upslope towards the road but they could not make the last few steep meters to the road surface. The nozzle at the end of the hose was found about ten meters downslope from them, indicating that they did not have much time to run from the moment they realized they could not extinguish the front reaching them and tried to make a run upslope to reach safety on the forest road. When other firefighters arrived a few minutes later, the fire was burning their fire engine.

Two months after the Fire Service undertook the responsibility for forest firefighting, on July 22, 1998, at 3 p.m., a fire erupted in an Aleppo pine forest at the base of Hymettus Mountain, next to the suburb of Kareas, at the east edge of the city of Athens. It was fanned by a strong north wind called "meltemi" which occurs regularly in the summer in the eastern part of Greece. The fire started moving along the slope of the mountain that runs in a north-south direction. A fire engine and a command vehicle started moving on a forest road in the middle of the slope. Their movement was along the east flank of the fire that was burning southwards (Figure 3). However, near the starting point of the fire, there is a small canyon that has a west-east direction. As the fire reached the bottom of the canyon, to the surprise of the firefighters, it started spreading as a crown fire up the canyon, perpendicular to the general direction of the main fire and the wind, in a clear demonstration of the so-called "chimney effect" (Figure 4).

The firefighters, fearing they might be cut-off by this finger, turned back, trying to get out of the area, towards the monastery of Saint John of Kareas. The command vehicle escaped, but three firefighters and two volunteers in the fire engine stopped at exactly the U-point of the road in the center of the canyon. There, they turned the front of their engine towards the advancing fire in a position that they could move in either direction on the forest road. According to one of the two volunteers who stayed with the engine, they started unreeling the hose to fight the fire. The driver of the engine disembarked, joining the other two firefighters and the second volunteer. Exposed to the heat, smoke and gases of the advancing fire in the canyon they soon had to abandon their effort. Unfortunately, instead of trying to escape on the road on either side of the U-point, or remaining in the cabin of their truck, they started climbing upslope, probably trying to reach another road that exists higher on the slope, at a distance of approximately 250 m (Figure 3). With the fire behind them, they were soon overwhelmed, as they were breathing the smoke and hot gasses channeled in the canyon. Their bodies were found on the slope less than 150 m from the truck. The first volunteer, who had remained on the road near the truck, left it, and, running on the road, passed through the smoke and reached the unburned forest at a distance of less than 200 m from the truck (Figure 3). The fire truck suffered relatively little damage. One of the back tires caught on fire when a nearby tree started burning. The tank had more than 1 ton of water. The cabin was nearly intact. After changing the tire, it was possible to drive it to the Fire Station. It can be assumed with some certainty that the four victims could have survived if they were trained to anticipate the fire behavior and had avoided stopping in the canyon (Figure 5) or had taken similar actions as the surviving volunteer, instead of climbing up the slope with the fire below them. Even staying in the truck would have been a better survival strategy (Xanthopoulos 1999).

On July 15, 1999, two young firefighters were killed when their fire truck got off the road and fell in a large natural cave on Hymettus Mountain near Athens. A third firefighter managed to open the door and jump out of the truck before it fell in the cave. She escaped unharmed.

Figure 4. A view of the Kareas fire on Hymettus mountain showing on the right the main fire front moving to the south and on the left the head of the up-canyon fire run at the moment it reached the top of the canyon after moving nearly perpendicular to the direction of the main fire and the wind.

In the evening of July 28, 1999, a fire ignited in the area of Cape Melanios on the island of Chios. It could not be controlled before dark. The forces fighting the fire in the night included a ground crew with hand tools. The fire was spreading in phryganic vegetation, consisting of small low shrubs, mainly *Sarcopoterium spinosum* L., which is known as a flashy fuel. The crew was sent downslope to extinguish small parts of the perimeter of the fire that were still burning. The fire appeared calm at the time. A sudden wind gust re-kindled the fire, which, due to the slope and the flashy nature of the fuels, started an upslope run that surprised the firefighters. Most of the crew managed to retreat upslope to the road, making a narrow escape. However, a female 27 year-old firefighter, who had been educated as assistant-forester, had received firefighting training, and had four years of experience in firefighting with helitack crews, suddenly collapsed as they were retreating. The reason for this remained unclear. Another young firefighter and the Fire Service officer who was in charge paused trying to help her. All three were overtaken by the flames. The young firefighter suffered extensive burns (>60% of his body) and, although he walked out of the burned area to the ambulance on his own, died a week later in the hospital. The Fire Service officer, who escaped with intense burns over 40% of his body, survived after long treatment in the hospital and multiple operations (Xanthopoulos 2000a).

In September 1999, one civilian driver of a water-truck belonging to the town of Poros in Peloponnese, was killed when his truck fell off a forest road. He had just refilled Fire Service fire trucks that were fighting a small wildfire and was returning to his base (Xanthopoulos 2000a).

Figure 5. The site of the fatality at Kareas, photographed in 2002 from the point where one of the bodies was found. The box-canyon topography is evident. On the left is the monument built in memory of the victims at the spot where the firefighters had parked their firetruck. The tree stand to the right shows where the flank of the up-canyon fire was stopped, a distance of roughly 200 m from the firetruck. The surviving trees on the left side, less than 80 m from the firetruck location in the direction from which the firefighter truck came, are an indication that fire intensity there was relatively low.

On June 14, 2000, a Fire Service officer, commander of the Fire Service in Sparta, Peloponnese, was injured by a large rock that came down a burning slope while coordinating the suppression of a large fire on Taygetos Mountain west of Sparta. He was taken to a hospital where he died on July 23, 2000.

The fire season of 2007 was the worst on record (Xanthopoulos 2007b). On July 11, 2007, an eight-person ground crew of seasonal firefighters was operating in low, sparse shrub fuels in a rocky area in a gorge near Doxaro village, in the prefecture of Rethymno on the island of Crete (Figure 6). Fire behavior was not intense and four of the firefighters advanced on one of the slopes trying to extinguish a flank. A sudden and turbulent wind gust, probably caused, according to some witnesses, by the pass of a Kamov-32 firefighting helicopter over the gorge, fanned the fire, which suddenly started to spread actively in the gorge. The four firefighters, realizing that the fire was below them, tried to escape running in the gorge with the flames behind them. Three of them were overrun by the flames and perished. The fourth found some shelter behind a rock outcrop. He suffered serious burns, but survived after a long treatment in the hospital.

Figure 6. The gorge near Doxaro in which the three seasonal firefighters lost their lives.

On August 24, 2007, fire danger conditions were extreme. Many lives were lost due to the numerous wildfires that erupted in the South part of Greece. In the morning of that day, a wildfire near the town of Areopolis, in the areα of Mani, in south Peloponnese, claimed the lives of six people. One of them was a 43 year old seasonal Fire Service firefighter in the village of Xifaria, who died of heart attack while fighting the fire.

In the afternoon of that same day, another wildfire claimed 24 lives near the village of Artemida in the prefecture of Ilia, in Peloponnese. Three of the fatalities were seasonal Fire Service firefighters. The fire started accidentally at 14:40 at the village of Palaiohori. The temperature was 40°C, relative humidity varied between 10% and 17%, and the wind was blowing from a northeast direction at 12 km/h, with gusts reaching 29 km/h. The fire immediately started spreading very fast in the diverse vegetation of the area that consisted mainly of olive groves with grassy fuels on the ground intermixed with small stands of Aleppo pine (*Pinus halepensis* Miller) (Figure 7). With the help of heavy spotting, its rate of spread exceeded 5 km/h. The firefighters coming from the direction of the town of Zacharo were moving with their fire truck towards the village of Artemida when they met a convoy of cars with people fleeing the fire. Seeing that they could not do much at the village, which was about to be hit by the fire, they turned to leave, followed by the convoy, towards the town of Zacharo. A little further, at a curve of the asphalt road, they found a car coming from Zacharo that had left the road and fallen in a swallow ditch. They stopped there, partially blocking the road, and got off the truck in an attempt to fight the fire. They either did not have the time or were overwhelmed by panic, because they were not able to get water to the nozzle.

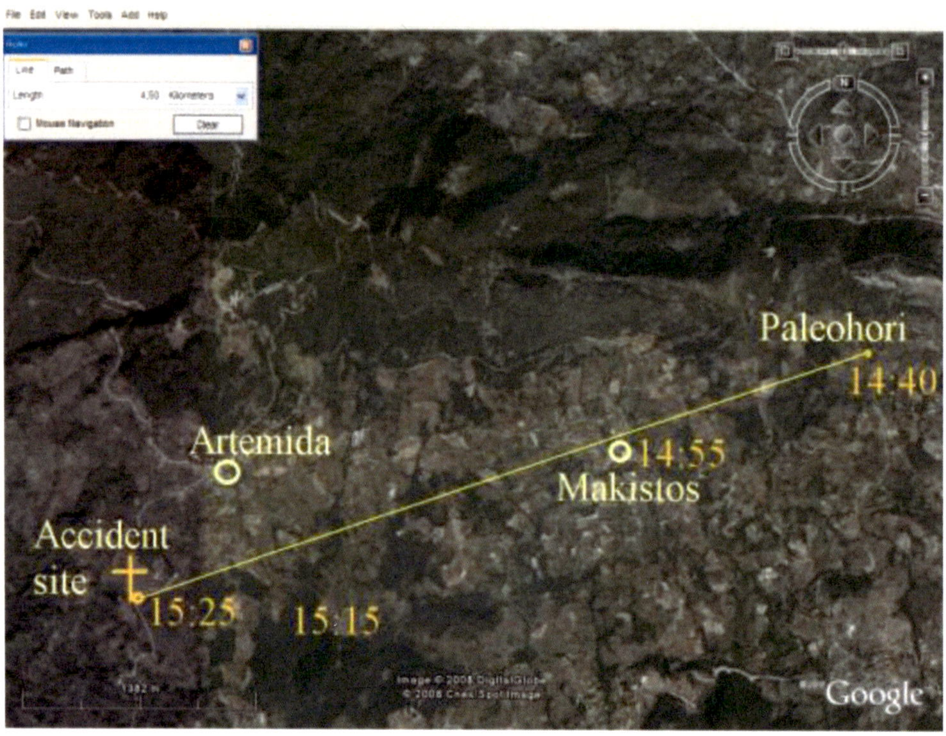

Figure 7. The general area of the fire run that killed 3 seasonal firefighters and 21 other civilians near Artemida, in Ilia, Peloponnese. The times shown refer to the approximate location of the fire front. The fire spread about 4.5 km in 45 min reaching a rate of spread of 6 km/h. Image Source: Google Earth.

The fire front reached them in a few minutes (about 15:25). Their bodies were found in the olive grove next to the road curve. Two of them were next to the hose nozzle. Some of the civilians were found close to the firefighters in the olive grove, obviously hoping for protection, while others, who made a run uphill, died next to a pine stand, where they were trapped by a spot fire that had started on the other side of the hill. A young man, who was an athlete, tried to outrun the fire running uphill, but was finally trapped and died two kilometres from the site. Two others, one of them the deputy mayor of Zacharo, managed to get into an open field (noted by B in Figure 8) and covered themselves with dirt. After the passage of the fire front, they walked back on the road having burns on the unclothed parts of their bodies, mainly caused by sparks. The deputy mayor died in the hospital a few days later due to infection of his burns. A third person, who initially had stayed behind in the village of Artemida, received a call through mobile phone from his sister who was caught with the others and was asking for help. He left the village on foot towards the accident site, but was caught by flames near a pine stand next to the road (noted by C in Figure 8) (Xanthopoulos et al. 2009).

Figure 8. The fatality site near Artemida, marked with A. The letter B shows approximately the field where two people lied on the ground in the open covering themselves with dirt. The area marked with C shows the location where a person lost his life coming from Artemida to help his sister who was caught with the others in A. Vegetation shown in this 2003 Google Earth image is the one that existed at the time of the fire.

On August 26, 2007, a fire near the village of Mistros on Evia Island claimed the lives of five people, aged 19 to 33, who tried to control its front with chainsaws and handtools. The fire started around 15:30 and spread as a crown fire in rough topography, burning a mature *Pinus halepensis* Miller forest. Very soon, many people left the village aiming to fight the fire. The high intensity of the fire made most of them stop at a distance. However, a group of four young men from the village proceeded to fight the fire. One of them, a shepherd and resin collector, also worked in the summer as seasonal firefighter for the Fire Service. He had just finished his shift, but rushed to the fire without second thought. The second man was also a shepherd and the third a factory worker. These three were fit and had experience in firefighting. The fourth person worked as a driver, was in poor physical shape, and did not wear proper clothes and shoes. The four men started building a fireline with chainsaws and handtools in an effort to stop the fire front. Their past experience and their zeal made them underestimate the speed and intensity of the fire, which was channeled towards them in a ravine. Even when they realized that their efforts were in vain and they were being overcome by the flames, they did not immediately try to escape. Instead, they tried to help their unfit friend, losing precious time. The seasonal firefighter realized that they were running out of time. Being familiar with the topography and the trails, he run a few hundred meters to an opening in the forest where there was a fold for livestock and a pool of water. He saved himself by taking shelter in the water. The unfit man was overtaken by the flames first. One of the other two men was soon incapacitated. The remaining individual tried carrying this man on his shoulder, but it soon became evident that the effort was futile. Soon, at the request of his incapacitated friend to leave him and go get help, he laid his companion on the ground. Realizing that he could not outrun the fire uphill, he chose to turn towards the fire front and run through the wall of flames to get into the black. The larger group of people from the village who had been following the rear of the flame front trying to find their friends, suddenly saw him appearing through the flames. He had been seriously burned over 30% of his body. He was hospitalized and survived, losing, however, much of his ability to work. When the rescue patrol reached the accident site. it found the man who was left on the ground with extensive burns (over 70% of his body). He died in the hospital a week later. Two other young men, also seasonal firefighters for the Fire Service, and a 19-year old boy, perished as well. According to the reports broadcast on TV, they were trying to find and help the four person group when they were overtaken by the fire in a narrow ravine (Rizospastis 2008).

On August 22, 2011, a fire in mountainous topography near the village of Vathy, in the area of Mani in Peloponnese, claimed the life of the local Fire Service head officer. The fire was burning in olive groves with grass in the understory and in light shrub fuels (phrygana). The officer and his driver reached the area by car. Scouting the fire, the officer realized the need for more resources. He asked for reinforcements on the radio and asked his driver to leave with the car to meet the arriving firetrucks and lead them back to his location. He stayed on the spot to keep an eye on the fire below and prepare an attack plan. When the driver returned the officer was dead, overtaken by the fire. It was deduced that he was overtaken by the fire, which had accelerated due to flashy fuels and the steep topography.

On November 19, 2011, a firefighter lost his life fighting a grass fire in steep alpine terrain at an elevation of 2000 m, in the area of Kallithea in the prefecture of Florina, in northern Greece. As a driver, he was backing up a fire truck on a dirt road when the edge of the road surface failed and the vehicle overturned and rolled down the steep slope, killing him.

On June 20, 2012, a young firefighter, driver of the head officer of the region, exited the Jeep he was driving and tried to help other firefighters who were battling a fast moving fire in the area of Aspropyrgos a short distance from Athens. Fuels were mainly light grasses and phrygana shrubs, but there were also plastics and tires emitting thick black smoke. Firefighting was intense as the fire threatened the Greek Air Force base at Elefsis. Exposed to toxic smoke, the young firefighter lost consciousness and his heart soon failed. Although he was taken immediately to a nearby hospital, the doctors could not revive him. Four other firefighters suffered injuries or faced respiratory problems in that fire.

DEATHS RELATED TO AERIAL FIREFIGHTING

The fleet of national firefighting aircraft includes Canadair CL-215s, Canadair CL-415s, and PZL M-18 Dromaders, operated by the Greek Air Force. Between 1984 and 2005, three C-130 aircrafts were used for retardant delivery through a Modular Airborne FireFighting System (MAFFS). From 1994 to 1998, Army helicopters (Chinook and UH-1H "Huey") were used to carry helitack crews to new fires and to drop water. In 1993 and from 1999 until present, a fleet of contracted helicopters of various types (Erickson S-64 "Air-Crane", MIL MI-26, Kamov KA-32, etc.) has been used for firefighting. In the last few years, the Fire Service also operates its own fleet of three Eurocopter BK117C and two Super Puma AS332 L1 helicopters.

As aerial firefighting, which is inherently a risky operation, kept growing in Greece, it was inevitable that accidents would happen. Five CL-215 pilots were killed between 1973 and 1990. Four CL-215's were destroyed during the 1980s. Two more pilots were killed when their CL-215 crashed on a slope due to failure of one of the engines while executing a water drop on steep topography near the village of Ano Rodini in 1993. In the difficult fire season of 2000, two more pilots were lost on July 15, 2000, on Pelion mountain in central Greece when, while doing a low water drop, one of the wings of their CL-215 hit the top of a tree (Xanthopoulos 2000b). In the extreme fire season of 2007, on July 23, two more pilots lost their lives when, after making a very low drop with their CL-415, they did not manage to clear the slope. The plane crashed and broke up on the hill.

Three pilots lost their lives while firefighting with a PZL single-engine single-pilot plane. The first was lost on August 7, 2000, during firefighting on the island of Corfu. The second pilot crashed on August 25, 2003, on the island of Lesvos. He was on a fire prevention patrol when he spotted a starting fire. He crashed as he was trying to perform initial attack on the fire. On August 27, 2009, another pilot lost his life when his PZL crashed in an inhabited area while firefighting on the island of Kefallonia. According to witnesses the PZL hit high voltage power lines.

Three Army pilots and a helitack crew of seven firefighters were killed in 1994 in the prefecture of Kavala in northern Greece when their UH-1H "Huey" helicopter hit power lines on its way back from a fire. The accident happened at dusk, because the pilot had waited to pick-up the last persons of the returning crew and took-off for the base at a time that he should have already been there.

DEATHS OF CITIZENS NOT INVOLVED IN FIRE SUPPRESSION

The first wildfire causing civilian deaths was at Crown's Estate at Tatoi, northwest of Athens, on July 12, 1916. A fire that started at 10:30 in the morning of that day under drought conditions, fanned by a strong wind, burned most of the royal forest, which consisted mainly of Aleppo pine, the summer palace, and other buildings, and numerous deer and wild boars that were fenced in the royal forest. About one thousand soldiers were dispatched by train to the site, but could not accomplish much. King Constantine the 1st and his guards tried to escape the flames in a convoy of four cars, but they were trapped. The cars burned and the king was saved at the last moment by two soldiers who carried him on their backs. Three high ranking army officers, the king's driver, the soldiers who tried to save the king, and a number of civilians who were working in the management of the royal facilities, perished. The total death toll was 20 people. Another 300 people, mainly soldiers, were injured (Empros 1916).

There are no records of further loss of human life due to forest fires until 1976. In 1977, however, the death of a nun trapped by a wildfire in a monastery on Mount Parnassos signalled the beginning of a new era, in which the loss of human life due to wildfires became common. There is no official record of all civilian fatalities that have occurred, but a search of various sources (books, newspapers, internet sources, etc.) showed that at least thirty five civilian deaths were recorded in the 1980s and at least 22 in the 1990s. From 2000 to 2013, the number exceeded 81 people. Some notable cases are briefly described below.

In 1993, Greece was shocked by the death of thirteen people on the island of Ikaria in the NE part of the Aegean sea. The fire started at 14:30 on July 30th. There were two fire starts north of the town of Agios Kirikos, attributed to arson. According to witnesses, it took firefighting forces more than an hour to arrive on the scene. The wind was blowing at about 8 Beaufort scale force (the usual NE wind called "meltemi" that blows at this time of the year in the Aegean), and soon reached the Glaredes and Kountouma settlements, where houses are intermixed with forest and agricultural vegetation. People evacuated in panic. On one occasion, five people perished when they tried to flee in their car. After an apparent collision, these people were caught by smoke and flames as they were trying to leave their damaged car. In another case, three young men tried to help three older persons evacuate their home. All were caught in the open by flames and perished. The three young men were heralded as heroes in Ikaria, and a monument has been built to commemorate their sacrifice. However, the outcome might have been different had all remained in a house rather than trying to flee. The final toll was 13 people dead, many more injured, 35 destroyed houses, and 400 ha of pine forest and agricultural land burned.

On July 4, 1998, on a high fire danger day during which more than 15000 ha burned in southern Greece, a woman was killed by a fast moving fire on Salamina Island near Piraeus. She had fled her home in panic and the fire caught her in the open.

On August 5, 1998, the 7500 ha Penteli mountain fire that had started on August 2, killed a 67-year old man near the National Weather Observatory in the village of Penteli. He was was trying to fill his car with some of his belongings and evacuate (Xanthopoulos 2002). He was found on the asphalt next to the car, his death being the result of inhaling smoke and hot gases. His wife who had found refuge in a neighbour's home survived.

Another disastrous event occurred on August 24, 2000, near the Greek-Albanian border. A fire had been burning for many days in Albania. On that day, the wind, coming from the north, exceeded 50 km/h. The fire accelerated, crossed the border, and entered Greece in a mountainous area that is generally moist and where the forest is normally less flammable than in the rest of the country. However, in that extremely dry year, forest stands were as dry as they had ever been. Entering a long wide valley near the village of Haraugi, the fire spread over a distance of nearly 17 km within four hours, burning mainly dense broadleaved vegetation. Many small villages in the shrubby oak forests were soon surrounded by fire. Most people fled or were evacuated by the authorities, but some did not make it or may have stayed to protect their homes and perished. By the next morning, seven people were dead and dozens of homes were destroyed. The fire continued for three more days, finally burning about 8000 ha. The final death toll reached eleven people, all of them older than 67 years old.

In July 7, 2004, a wildland-urban interface fire in the Thrakomakedones suburb of Athens, in the foothills of mount Parnis, burned Aleppo pine stands mixed with homes, damaging many homes and claiming the life of a 74-year old man who had left his home trying to escape the fire. He was caught by the fire and died due to smoke inhalation, 200 m from his home. His house did not suffer any significant damage.

On August 9, 2006, in the area of Astakos, in the prefecture of Aitoloakarnania, a shepherd and his wife lost their lives when their car drove off a cliff as they were trying to escape the flames that threatened their sheepfold.

On August 21, 2006 a large, lightning-caused fire, in the peninsula of Kassandra in northern Greece burned through a mosaic of forest, agricultural land, and small towns that forms an extensive wildland-urban interface with significant tourist development. Thousands of people, including many tourists, were evacuated towards the sea. Evacuation in some cases was performed using boats since the fire blocked the main road connecting the towns of Kassandra peninsula with each other. In one incident, obviously in a state of panic, a German tourist trying to get on a boat to leave the area, fell in the sea and drowned in front of his family. In another case on that same fire a civilian suffered a stroke and died.

In 2007, during the worst fire season on record, 64 civilians perished. The first two perished on June 27, near the town of Agia in the prefecture of Larissa. They were driving as a three-person group in a car when they realized that their route was blocked by a fast moving wildfire that was burning shrub vegetation. Initially, the three men left the car in panic trying to escape on foot. One of the three soon felt overwhelmed by the conditions and returned to the car shouting to the others to join him. He survived in the car while the others, who tried to outrun the fire uphill, did not survive.

One of the largest fires of 2007 occurred in the area of Aigialia, in the prefecture of Achaia in Peloponnese. It started on July 24 and ran until July 27, burning about 30,000 ha and killing three people. One of them was a 75 years old shepherd who refused to abandon his sheepfold and his flock near the village of Mamousia. His wife followed evacuation orders and survived. The other two fatalities were two elderly women who lived in the village of Ano Diakopto. One of them was unable to move and the other attended her. They died when their home caught on fire because they were unable to escape.

Most of the deaths in 2007 occurred between August 24 and August 27 when the country had very high fire danger conditions, resulting in many simultaneous fires with extreme behavior. The first reports of deaths came in the morning of August 24 from the fire at Areopolis that was mentioned earlier. Five people perished in addition to the seasonal firefighter that suffered a heart attack. The two owners of a guest house, a girl who worked for them, and two guests died of smoke and hot gases. They had left the house and were trying to escape on foot. A few hours later, news of the large number of deaths of civilians and firefighters in Artemida became the main news item for that day. Many other deaths took place on that and the following four days, but only minimal information was provided by the mass media. Among the fatalities were many shepherds who were caught by the fires in the open. Most of them were trying to save their flocks.

DISCUSSION

The information presented above can be viewed in two ways. First, it can reveal long term trends and can lead to conclusions about the evolution of the wildfire problem in Greece. This may help to anticipate the future of wildfire safety in the country and to take appropriate mitigation measures. Second, this information can serve to identify common errors in firefighting and in citizen response that have led to disasters. These errors can be compared to the findings of similar studies in other parts of the world, ultimately leading to conclusions and lessons learned for dissemination to firefighters and the public.

Long Term Trends

The absence of recorded firefighter and other citizen fatalities before 1977, with the exception of 1916, is very interesting and cannot be considered as coincidental. One can speculate that the 1916 fire in the Crown's Estate at Tatoi burned the only forest in the country where, at the time, fuelwood collection by the people was not permitted (Xanthopoulos 1988). As long as mountain villages maintained a thriving population which lived near the forest and exploited the forest for timber, fuelwood, and other products, fire intensities remained relatively low. Although roads were few, the people knew the existing trails very well. They were also very familiar with the topography and the vegetation mosaic. They knew where there were agricultural cultivations such as vineyards, olive groves, vegetable gardens, etc. within the forests. These were kept free of fine fuels and often served as fuel breaks but could also be used as safety zones in case of risk of entrapment. Another contributing factor to the lack of fatalities may have been the method of attack. The local people were the main firefighting force. They would immediately mobilize for initial attack while the fires were small and easy to control. However, when the fires escaped initial attack, having no fire engines available, they would resort to indirect attack, would work on the flanks, or would take advantage of the night and early morning hours. Backfiring and burning out were among their firefighting tools.

This changed as fuel loads increased following the abandonment of villages in the 1960s. It took roughly 10 years until the fuels built up. As much of this built-up occurred around villages, fires could now reach them and sometimes spread through them. At the same time, firefighting tactics started changing with the acquisition by the Forest Service of the first fire engines in 1971 and the first Canadair CL-215 amphibian water bombers in 1974. Direct attack gradually became the nearly exclusive method of firefighting at a time that fire behavior was worsening.

The year of 1977 was the first exceptional year in regard to burned area and the first year marked by a forest fire fatality. This was the start of a trend with more bad fire seasons in the 1980s and an increasing number of fatalities. At about that time, extensive WUI interface areas appeared, which contributed to the fire safety issue, further compounding the wildfire management problem. Lack of knowledge of the people living in such areas on what to do in case of a wildfire is a key factor leading to more disasters.

For the future, the trend of fatalities is likely to continue as long as fire intensity due to fuel built-up gets worse. Fire weather extremes caused by climate change may further contribute. If firefighting tactics are not adapted to these realities, by for example re-introducing indirect attack, the number of firefighter fatalities will probably increase. The safety of civilians is also likely to be affected by the occurrence of periods with extreme fire behavior. The only way to avoid future catastrophes is to put an emphasis on wildfire prevention and to introduce careful changes in firefighting strategies. Prevention should focus in three areas:

1. Fire oriented forest management in order to maintain healthy forests, to reduce fuel loads, and to develop the necessary vegetation mosaics that will help fire suppression efforts. The objective should be to control the potential for megafires.
2. Public education/awareness campaigns that will make people aware of what they should do before and during wildfires.
3. Improvement in the layout and safety of structures in WUI areas.

Firefighting changes should include re-introduction of the use of indirect fire suppression when appropriate, intensive high quality firefighter training with emphasis on safety, and procurement, and increased use of equipment that can improve safety (personal protection equipment, GPS, GIS, telecommunications, etc.).

Lessons Learned

The conditions that have led to the accidents discussed here are generally similar to those identified in other parts of the world, such as the USA and Australia (Mangan 2007; Haynes 2010). In regard to firefighting, practically all cases listed in the international literature (Mangan 2007, Rothermel 1993, Simeoni and Dold 2009, Viegas and Caballero 2009) such as burnovers, vehicle accidents, aircraft accidents, heart attacks, falling trees and rocks, etc. are represented within the relatively small number of fatal firefighter accidents in Greece. Flashy fuels, steep canyon topography (Viegas and Pita 2004), wind direction changes, and even unexpected influence of aerial resources on fire behavior have all been identified in the list of causes. The number of ground firefighter fatalities is not excessive by international standards. On the other hand, the death toll in accidents related to aerial firefighting is much higher, especially considering the small number of people involved in it. Probable causes of these accidents are: bending the rules, as in the case of the UH-1H helicopter crash in Kavala, very intensive work load on pilots for long periods of time in difficult fire seasons that, although within the limits of international regulations, builds routine and probably excessive pilot self-confidence, and potentially, in some cases, poor aircraft maintenance.

In regard to civilian deaths, their causes are very similar to those noted in other countries around the world. Flashy fuels are the cause of many deaths, especially related to agricultural stubble burning. Leaving home and trying to escape in panic becomes another major cause of death when people find themselves in the smoke of the advancing fire. It should be noted that houses in Greece are generally built with non-flammable materials so, as a rule, it would make sense to stay in them during the passage of the fire front instead of attempting a risky escape. It is quite clear that an effort is needed for educating people on how to prepare for the case of a fire close to their WUI home, and what to do to survive and save their home when a fire actually occurs. It must be stressed that the information on past fire incidents involving deaths, injuries, or near-hits in Greece is sketchy. The results of the investigations that follow such events were never publicized. Furthermore, even if one tries to obtain the documents through legal procedures, the confidence on the findings can be quite weak. The reason is that the people performing the investigation usually do not possess specialized knowledge on fire behavior etc. Also, there is often an effort to cover-up the true responsibilities and mistakes protecting specific individuals who were probably at fault because of the general inadequacy of their organization (e.g., lacking training, clear guidelines, appropriate dependable tools, etc.). There is little effort to this day devoted to investigation of the accidents in order to draw conclusions and guide future training. On the contrary, declaring the victims as heroes, building monuments for them, and providing some compensation to their families appears to be the preferred approach.

CONCLUSION

The incidents described here provide the necessary information that leads to the basic conclusion of this work: the causes of each fatal accident are not unique. They have been identified before in some other places and time, and they fall into broad categories for which warnings have been established in international literature. The question that has to be answered then is whether the victims possessed, at the time of the accident, the necessary knowledge to avoid putting their life in danger. A negative answer to this question necessarily highlights the ever-existing need to educate firefighters and the public, especially when the latter lives in wildland-urban interface areas, on what they should do, what they should not do, and which are the situations that should "watch-out".

ACKNOWLEDGMENTS

This work was carried out in the context of the "*Prevention of Forest Fires in 2012-2013 with the INCA methodology*" project, and was funded by the Greek Special Secretariat for Forests through the Green Fund of the Ministry for the Environment, Energy and Climate Change. This support is gratefully acknowledged.

REFERENCES

Empros newspaper (1916) The funeral of the victims of the tragedy at Tatoi. Athens, Greece (newspaper aricle published on July 2, 1916 (old calendar), in Greek)

Haynes K, Handmer J, McAneney J, Tibbits A, Coates L (2010) Australian bushfire fatalities 1900–2008: exploring trends in relation to the 'Prepare, stay and defend or leave early'policy. *Environmental Science and Policy* 13, 185-194.

Kailidis D (1990) 'Forest Fires.' (3rd edition), 510 p. (Giahoudi-Giapouli Publishers: Thessaloniki, Greece) (in Greek)

Mangan R (2007) Wildland firefighter fatalities in the United States: 1990–2006. National Wildfire Coordinating Group, Safety and Health Working Team, National Interagency Fire Center. NWCG PMS 841. 28 p. (Boise, ID)

Karras T (1954) 'Forest fires in Greece.', 24 p. (Ministry of Agriculture, General Direction of Forests, Forest Propaganda Office: Athens, Greece) (in Greek)

Kontos P (1933) 'Against forest fires.' 40 p. (M. Triantafyllou and Co. Publishers: Thessaloniki, Greece) (in Greek)

Rizospastis newspaper (2008) Once upon a time there was a beautiful village. (Newspaper article published on 2008-8-3, last accessedted online on 2014-5-11). Available from: URL: http://www2.rizospastis.gr/page.do? publDate= 3/8/2008&pageNo=20&direction= 1 (in Greek)

Rothermel RC (1993) Mann Gulch fire: a race that couldn't be won. 10 p. (USDA Forest Service, Intermountain Research Station, Gen. Tech. Rep. INT-299, Ogden, UT, USA)

Simeoni A, Dold J (2009) The Accident of Palasca (France). In 'Recent Forest Fire Related Accidents in Europe' (Ed. D. X. Viegas) EUR 24121 EN. pp. 54-64. (European Commission, Joint Research Centre, Institute for Environment and Sustainability: Ispra, Italy)

Viegas DX, Pita LP (2004) Fire spread in canyons. *International Journal of Wildland Fire* 13, 253–274.

Viegas DX, Caballero D (2009) The Accident of Guadalajara (Spain). In 'Recent Forest Fire Related Accidents in Europe' (Ed. D. X. Viegas) EUR 24121 EN. pp. 9-15. (European Commission, Joint Research Centre, Institute for Environment and Sustainability: Ispra, Italy)

Xanthopoulos G (1988) Greek forest fires and property damage: A brief history. In Proceedings of the conference 'Protecting People and Homes from Wildfire in the Interior West', 6-8 October 1987, Missoula, MT, USA. (Eds W.C. Fischer, S.F. Arno) USDA Forest Service, Intermountain Research Station, Proceedings INT-251, pp. 199-200. (Missoula, MT)

Xanthopoulos G (1999) The 1998 forest fire season in Greece: A forest fire expert's account. *International Forest Fire News (ECE/FAO)* 20, 57-60.

Xanthopoulos G (2000a) Greece: the 1999 forest fire season. *International Forest Fire News (ECE/FAO)* 22, 25-28.

Xanthopoulos G (2000b) Fire situation in Greece. *International Forest Fire News (ECE/FAO)* 23, 76-84.

Xanthopoulos G (2002) The forest fires of 1995 and 1998 on Penteli mountain. In Proceedings 'Improving Dispatching for Forest Fire Control', 6-8 December 2001, Chania, Greece. (Ed. G. Xanthopoulos) Mediterranean Agronomic Institute of Chania, Chania, Greece., pp. 85-94. (Chania, Greece)

Xanthopoulos G (2007a) Forest fire related deaths in Greece: confirming what we already know. In Proceedings of the IV International Wildland Fire Conference, 13-17 May 2007, Seville, Spain. p. 339 (Book of Abstracts, Full paper on accompanying CD).

Xanthopoulos G (2007b) Olympic Flames. *Wildfire* 16,10-18.

Xanthopoulos G, Viegas DX, Caballero D (2009) The fatal fire entrapment of Artemida (Greece) 2007. In 'Recent Forest Fire Related Accidents in Europe' (Ed. D. X. Viegas) EUR 24121 EN. pp. 65-75. (European Commission, Joint Research Centre, Institute for Environment and Sustainability: Ispra, Italy)

In: Current International Perspectives
Editors: Brigitte Leblon and Martin E. Alexander
ISBN: 978-1-63463-682-7

Chapter 7

FIRE HISTORY, FIRE ECOLOGY AND MANAGEMENT IN ARGENTINE PATAGONIA: FROM ANCIENT TIMES TO NOWADAYS

Guillermo E. Defossé[1*], María Marcela Godoy[1], Lucas O. Bianchi[1], Natalia S. Lederer[1] and Carlos Kunst[2]

[1]Patagonian Andes Forest Research and Extension Center (CIEFAP-CONICET) and Forest Department, University of Patagonia San Juan Bosco, Esquel, Chubut, Argentina

[2]National Institute of Agricultural Technologies (INTA), Santiago del Estero Agricultural Experimental Station, Santiago del Estero, Argentina

ABSTRACT

Patagonia is a vast region in South America that covers the southern part of Argentina and Chile, from the 36° to the 56° S. In Argentina, the Patagonian region comprises about 750 thousand km^2 and goes east to west, from the Atlantic Ocean to the Andean Cordillera and north to south, from the Colorado River up to the Beagle Channel in Tierra del Fuego. Three main phytogeographic provinces occur in Patagonia: the Monte Province, the Patagonian Steppe, and the Andean Forests. Between the Patagonian Steppe and the Andean Forest, a narrow strip comprising about 2 millon ha is starting to be used for afforestation purposes, mainly with non-native conifers such as ponderosa pine, Douglas fir, and lodgepole pine. The whole region is dominated by a Mediterranean climate, cold and humid during late fall and winter, followed by a dry and temperate period during spring and summer. This Mediterranean type of climate has favored the occurrence of wildfires from October to early April (mid-spring to late summer in the southern Hemisphere). In the Andean Forests, there is evidence of fire 10,000 years B.P. in the early Holocene, where great magnitude fires took place in pure *Notahofagus* forests. Further in the mid-Holocene, between 6000 and 3000 years B.P., precipitations increased to the east and vegetation changed toward mixed *Nothofagus – Austrocedrus chilensis* (D.Don) Florin et Boutelje forests, shifting also the fire regime to more frequent

[*] Corresponding Author address: Patagonian Andes Forest Research and Extension Center (CIEFAP-CONICET) and Forest Department, University of Patagonia San Juan Bosco, Esquel, Chubut, Argentina, Email: gdefosse@ciefap.org.ar.

and less severe fires. Fire scars were also found in *Fitzroya cupressoides* (Molina) I.M. Johnst. (Mol.) and *A. chilensis* trees from 990 and 1437 AD, respectively. From 1800´s to early 1900´s, human set fires increased in the forest piedmont, inducing vegetational changes to open shrublands and grasslands. In 1930´s, a fire suppression policy began, and *Austrocedrus* forests started to recover toward the east. This trend, however, was not linear, and other disturbances such as logging for either timber or firewood, grazing, and increasing human population in the forest-steppe Wildland Urban Interface (WUI) have been shaping the landscape in a dynamic way. In the other two provinces (Monte and Patagonian Steppe), there are also evidences that fire played a major role in vegetation dynamics, and that the introduction of grazing may have disrupted the pristine role of fire. Studies about fire and fire effects in Patagonia done at present, have focused more on landscape or stand level than on individual species. Nevertheless, it is evident that most species have developed strategies to survive fires of different frequency and severity. These strategies varied from resprouting from their stem bases, producing seeds resistant to fire, or migrating seeds from adjacent unburned areas that would colonize open areas left by the fire. In relation to conifer afforestations, the main objectives are focused at finding the ways (through the development of appropriate silvicultural techniques) of reducing wildfire hazard, allowing then for greatest growth and help mitigate global CO_2 emissions.

Keywords: Andean Forests, Fire ecology and management, Monte region, Patagonia, Patagonian steppe

INTRODUCTION

Patagonia is a vast region located in the southern tip of South America, covering the austral territories of Argentina and Chile. In Argentina, Patagonia comprises about 750 thousand km^2, and is bordered to the north by the Colorado river, to the south by the 56° South Parallel, to the east by the Atlantic Ocean, and to the west by the high peaks of the Andean Cordillera, which delimits the frontier with Chile. The physiography of Patagonia includes vast deserts and semideserts that go from the Atlantic Ocean to the foothills of the Andes, in which an ecotonal area marks the transition toward the Andean forests. Although with huge differences in total annual precipitation among different Patagonian subregions, the whole territory is characterized by a Mediterranean climate (cold and rainy during fall, winter, and early spring, and temperate and markedly dry during late spring and summer). The dry period coincides with the fire season, which is also influenced by the dessicant effects of high westerly winds that blow continuously throughout the year. From the vegetational point of view, Patagonia has been divided into three main phytogeographical provinces, named the Monte Province (Cabrera 1976), the Patagonian Steppe (Soriano 1956), and the Subantarctic Andean Forest (Dimitri 1972) (Figure 1). Physiognomically, the Monte and Patagonian Steppe regions resemble in some ways the vast western rangelands of North America, including the sagebrush ecosystem, the mixed and short grass prairies of the Great Plains in the USA as well as the prairies of Manitoba, Saskatchewan, and Alberta in Canada.

Patagonia is crossed west to east by several rivers that go from the Andean Cordillera to the Atlantic Ocean. Along the shorelines of these rivers, fertile valleys have been developed based on intensive irrigated agriculture, producing fruits, vegetables, and forest products provided by *Salix* and *Populus* spp. plantations. In the narrow ecotone that connects the

Patagonian Steppe with the Andean Forests, dryland afforestations are currently being developed, mainly with North American conifer species such as ponderosa pine, Douglas fir, lodegepole, and radiata pine), and other valuable broadleaf timber species, (Godoy et al., 2007).

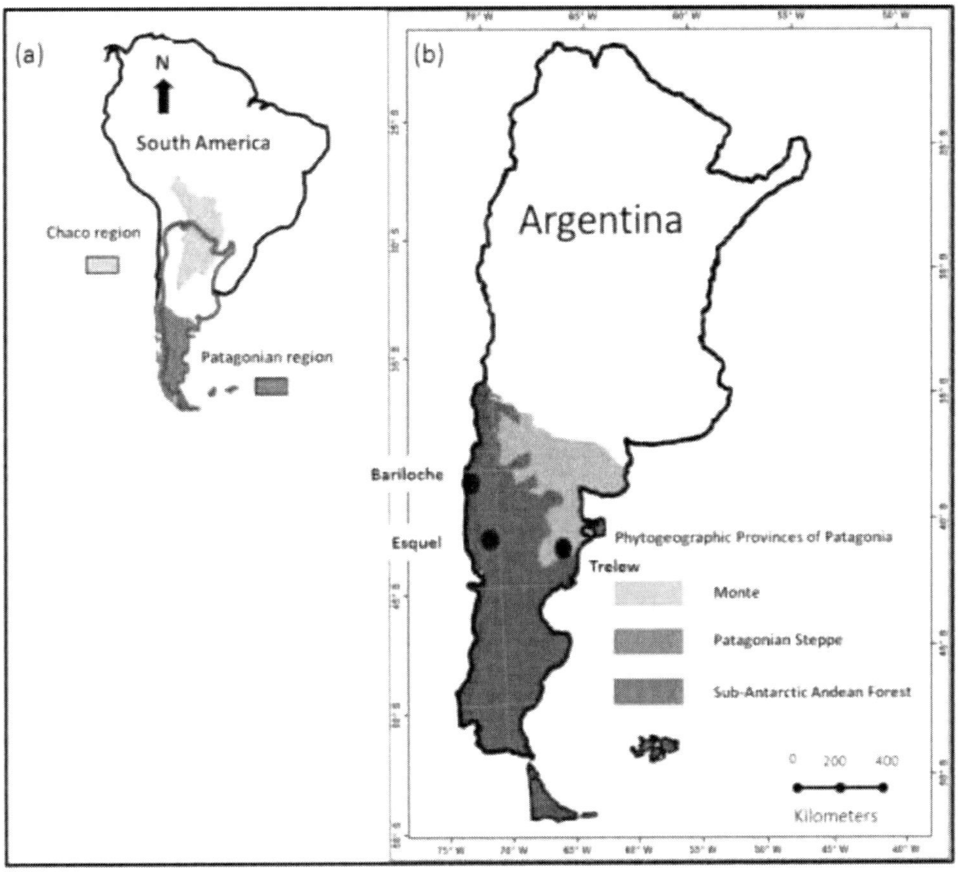

Figure 1. (a) Location of the temperate Patagonia (south) and the sub-tropical Chaco, which are among the most important biogeographical regions of South America (Cabrera and Willink 1973); (b) Phytogeographical Provinces of Patagonia. The cities of Trelew, Esquel, and Bariloche are located in the Monte, Patagonian Steppe and Andean Forest, respectively (modified from Cabrera 1976 and SAGPyA 1999).

In this chapter, we review the fire history, ecology, and management in the three main phyogeographic provinces found in Patagonia, including the ecotonal area in which afforestations are nowadays being fomented. We intendently excluded fire occurrence and management in irrigated valleys. The present and predicted impacts of increasing human populations around Wildland Urban Interfaces (WUI) of Andean cities, pine afforestations, and fire effects in Patagonia in a future scenario of climate change are also discussed.

FIRE HISTORY OF PATAGONIA (FROM ANCIENT TIMES TO NOWADAYS)

Before the Triassic, Patagonia was a humid and mild territory with exuberant vegetation, as it is attested by fossil trunks of the extinct *Araucaria mirabilis* (Spegazzini) Windhausen found today in several plateaus of arid Patagonia (Channing et al., 2007). Because of fire evidences are difficult to find for that period, the fire history of Patagonia can be traced back to the Cretaceous, when the Andes rose and began to take its present form. Since then, the Patagonian Andean Cordillera started to become a barrier for the humid westerly winds that once blew freely from the Pacific to the Atlantic Oceans. This new physiographical condition implied that winds, after discharging their moisture in the western side of the Andes in Chile, passed through the eastern Andean side, releasing there the rest of their remnant moisture. These winds crossed the Patagonian territory as cold and dry air masses until reaching the Atlantic coast. This remarkable change in ambient conditions produced the gradual disappearance of former ecosystems and facilitated the establishment of vegetation types adapted to the new environment (Barreda et al., 2007; Iglesias et al., 2011). During the subsequent Pliocene and Pleistocene (from 5×10^6 to 10,000 years ago), climatic change, vulcanism, and naturally occurring wildfires caused by lightning shaped and differentiated the vegetation of Andean from the extra-Andean Patagonia (Iglesias et al., 2011). These processes ended in the modern cool-temperate Andean forests, dominated by *Nothofagus* and other associated species to the west, and the arid and semiarid grass and grass-shrub Patagonian steppe and the Monte region to the east (Barreda et al., 2007; Iglesias et al. 2011). Dry and hot (Mediterranean) climate conditions started to prevail during the summer in either Andean or extra- Andean Patagonia. Fires of different magnitudes and frequencies helped shape the structural and functional characteristics of the different vegetation assemblages found from the mid- Holocene to presettlement times.

The evidences that fire has always been present in all three main phytogeographical provinces of Patagonia have been widely acknowledged (Claraz 1988; Goldammer et al., 1996; Veblen et al., 1999; Veblen et al., 2003; Defossé et al., 2003; Ghermandi 2006). These evidences, however, are more visible in the Andean forests than in the Monte or in the Patagonian Steppe. In the Andean forests, fire scars present in long lived trees give precise evidences, from which fire occurrence and frequency could be dated. By contrast, as in any grassland or shrubland, aboveground physical evidences do not last enough time in the Monte and the Patagonian steppe to allow estimating ancient fire events and make inferences about them with certain degree of accuracy (D´Alia 1988; Frost 1988; Defossé et al., 2003). Despite this fact, human presence in the Patagonian steppe since the Holocene (ca 10,000 years ago) probably increased fire occurrence, as confirmed by several charcoal records for this period found in sediments of lakes and lagoons of these steppes (Withlock et al., 2006; 2007),

Toward the west, great magnitude fires took place in pure *Nothofagus* forests during the same period (Withlock et al., 2006). Further in the mid-Holocene (from 6000 to 3000 years B.P.), precipitation increased toward the east, and then vegetation changed toward mixed *Nothofagus – Austrocedrus chilensis* (D.Don) Florin et Boutelje forests, altering also the fire regime to more frequent and less severe fires. Evidences of fires occurred during the last 1000 yr were also determined by charcoal records in river banks of the Andean forest-steppe ecotone dominated by *A.chilensis* (Goldammer et al., 1996). Fire scars dated from 990 and

1437 AD were found in modern times in alive *Fitzroya cupressoides* (Molina) I.M. Johnst and *A.chilensis* trees, respectively (Veblen et al., 1999). From 1700´s to early 1900´s, human-caused fires increased in the forest piedmont, inducing vegetational changes toward open shrublands and grasslands (Veblen et al., 2008). The existence of extensively burned forests in northern Patagonia was attested by the first explorers who arrived to Patagonia in the late eigthteenth and nineteenth centuries (Cox 1863; Musters 1871; Fonk 1896; Moreno 1897). The priest Francisco Menéndez mentioned having seen extensive burns in northwestern Patagonia, such as the explorers Fonck and Hess in 1856 (Fonck 1896). Various writers later reported that large forest areas had been burned in northwestern Rio Negro and southern Neuquen provinces prior to the initiaton of white settlement, which occurred around 1850 (Rothkugel 1916; Tortorelli 1947; Veblen and Lorenz 1988; Veblen et al., 1999). When mapping the region for forest purposes, Rothkugel (1916) described an evenaged 150 year old post-fire *N. dombeyi* (Mirb.) Oerst. stand at the western end of Lake Nahuel Huapi. He attributed the origin of this stand to burning by the Jesuits or Indians in the seventeenth century to create pasture for cattle.

These places become afterwards large evenaged cohorts of *Nothofagus – Austrocedrus* forests. From 1850, white settlement began in southern Chile and Indians displaced towards the east were forced to cross the Andes to Argentine Patagonia. In the eastern slopes of the Andes and in the adjacent steppe, the Indians domesticated horses and with the aid of fire, started to hunt rhea (*Pterocnemia pennata* D'Orbigny) and guanaco (*Lama guanicoe* Müller) in open areas.

In 1890's and early 1900, fire frequency reached its maximum in western Andean Patagonia with the arrival of European settlers coming from the Atlantic. These early European settlers used to burn extensive areas, primarily to create open areas for cattle grazing. Willis (1914)'s maps reported large extensions of recently burned forests of *Nothofagus antarctica* (Forster) Oerst. and Rothkugel (1916) mapped as burned the 26% and 59% of the forests of the provinces of Neuquen and Río Negro, respectively. Although fires during this settlement period were almost exclusively set by humans, National Park authorities started to record lightning-ignited fires in the Andean region between ca. 39° to 41° 30' S, in a period that went from 1938 to 1982. Whether the ignition source was human or natural, it is clear that there were sizable areas of burned forests in northern Patagonia prior to European settlement (Veblen and Lorenz 1988).

In the other two phytogeographical provinces, i.e., Monte and Patagonian Steppe (Figure 1), there were also evidences that fire had played a major role in vegetation dynamics, and that the introduction of grazing by settlers may have disrupted the pristine role of fire. For the Monte region, Claraz (1988), describing his journey across that region in the book "*Exploration trip to Chubut 1865-1866*", stated that different areas were usually burned by Indians, who used fires for hunting and communication purposes. Recent studies showed that summer lightning may have also played a role in igniting fires in this region (Dentoni and Cerne 1999; Defossé et al., 1999; Dentoni et al., 2001; Defossé et al., 2003; del Valle et al., 2004). However and contrary to what happened in the Andean region, early settlers arrived in the 1900´s did not use fire for management purposes neither in the Monte nor in the Patagonian steppe.

PHYSIOGNOMY, CLIMATE, AND FIRE ECOLOGY IN PATAGONIAN PHYTOGEOGRAPHIC PROVINCES

The Monte Province

Climate

The Monte Province is a vast temperate semidesert, in which several shrub species of the *Larrea, Prosopis, Chuquiraga, Schinus, Junelia,* and *Lycium* genera, arranged into two or more strata, are intermixed with grasses of the *Poa, Pappostipa, Elymus, Bromus,* and *Hordeum* genera (Cabrera 1976; León et al., 1998) (Figure 2). The Monte region extends beyond Patagonian limits and shows differences in climate as we go north of the Colorado River.

Figure 2. Landscape view of the Monte vegetation. The upper stratum (> 2.5 m tall) corresponds to the shrub *Larrea divaricate* Cav., while other shrubs such as *Prosopidastrum globosum* (Gillies ex Hook. & Arn.) Burkart, *Atriplex lampa* Gill.ex Moq., and *Chuquiraga avellanedae* Lorentz compose the medium (~ 0.8 m tall) stratum. The scattered grass cover in the lower stratum indicates that this paddock has been heavily grazed for many years (photo by A. Bisigato).

In its Patagonian part, however, the climate of the Monte region is typically Mediterranean, with rains mainly concentrated from late fall to early spring, followed by a drought period during late winter, spring, summer, and early fall. Annual precipitation averages about 200 mm and presents high interannual variability. Temperatures during the summer can reach up to 40 °C, and relative humidity can drop from 40 % to 15% or 10 % during hot days. This type of climate conditions the fire season, which extends from late August to early April (late winter to early fall in Southern Hemisphere) (Figure 3) (Defossé et al., 2003). Wildfires in the Monte are spatially heterogeneous and generally occur in patches of different sizes (Hardtke et al., 2011).

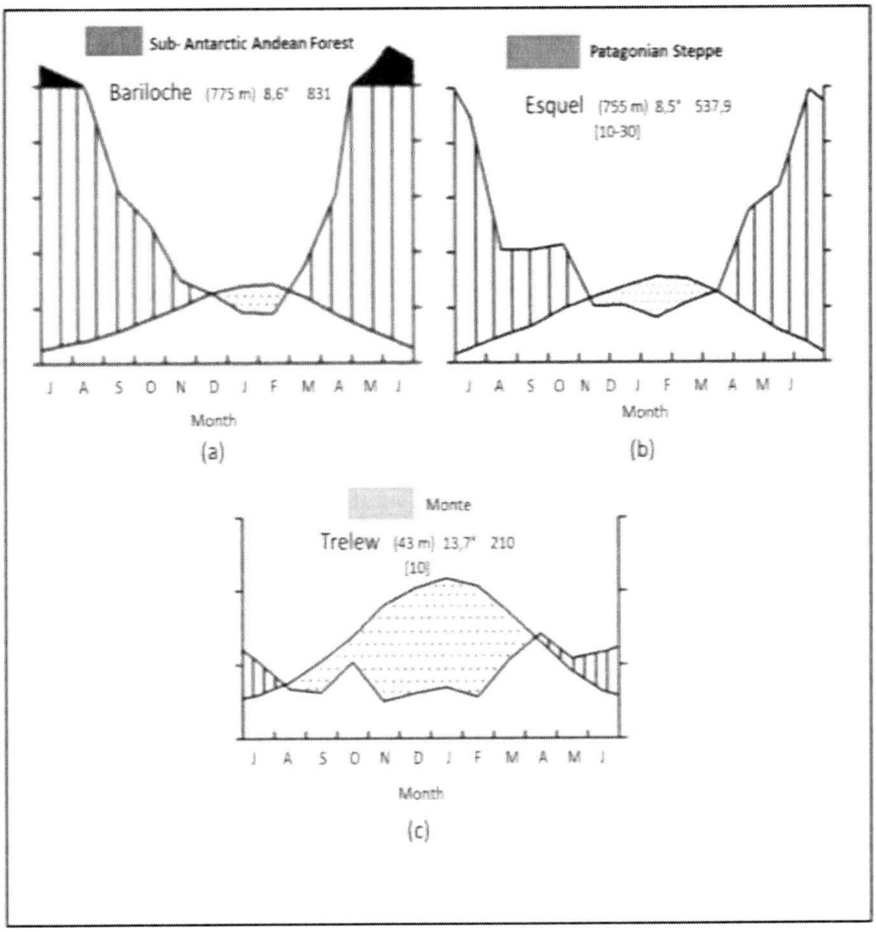

Figure 3. Climate diagrams for the cities of Bariloche (a), Esquel (b) and Trelew (c), located in the Sub-Antartic Andean Forest, the Patagonian steppe, and Monte regions, respectively. Each ordinate subdivission represents either 10 ºC or 20 mm precipitation. Dotted area is a drought period, matching also the length of the fire season; vertical bars denote humid season. Black area represents a per-humid season (when monthly precipitations exceeds 100 mm, the scale is reduced to 1/10).

Fire Ecology and Management

The most conspicuous perennial grasses in the Monte are species of the *Pappostipa* genus (*P. speciosa* (Trin. & Rupr.) Romasch, *P. tenuis* Phil.), *Nassella clarazii* (Ball) Barkworth, *Pappostipa sorianoi* (Parodi) Romasch), and also *Poa ligularis* Nees ex Steud., *Pappophorum caespitosum* Fries R.E., *Elymus patagonicus* Speg., and some *Hordeum* and *Bromus* spp. Several studies showed that these grasses developed strategies to survive fire disturbance, increasing their biomass related to unburned plants in only two years after the fire event (Rostagno et al., 2006; Bran et al., 2007; Kröpfl et al., 2007; Villagra et al., 2009). If the fire is not severe enough to burn the crown completely, most grass species will regrow from new meristems located at the base of the crown (Figure 4 a). However, and if the fire has been severe enough as to burn the crowns completely, these species would recolonize burned patches from seeds coming from adjacent unburned areas. In general and after a fire disturbance and in absence of grazing, the biomass of these grasses increases as compared to the unburned areas, and remain so until at least 10 year after the fire event (Rostagno et al., 2006; Figure 4 b).

(a) (b)

Figure 4. a) *Elymus patagonicus* Speg. two years after a wildfire. Note the burned branches of a shrub behind it. b) *Elymus patagonicus* and other grasses grown after a wildfire in the Monte region. Without grazing pressure, the grass component is the first to recover, while for shrubs recovery takes longer periods (photos by G. E. Defossé).

Shrubs in this ecosystem, instead, recover after fire at slower rates than do grasses (Figure 5 a and b). Studies showed that 10 years after a wildfire and in absence of grazing, shrub density for the *Chuquiraga avellanedae* Lorentz, *Nardophyllum chiilotrichioides* (J. Remy) A. Gray, *Larrea divaricata* Cav., *Prosopidastrum globosum* (Gillies ex Hook. & Arn.) Burkart, *Chuquiraga histrix* D. Don, *Junelia alatocarpa* (Troncoso) Moldenke, and *Lycium chilense* Miers species still remained significantly lower in burned as compared to unburned areas, while *Schinus polygamus* Cabrera Hardee is the only shrub that significantly increased in burned as compared to unburned areas (Defossé et al., 2003; Rostagno et al., 2006; Villagra et al., 2009).

From an ecological point of view, all grass or shrub species of the Monte region are well adapted to survive fires of different intensity and severity. During the first seral stages after a fire, the proportion of regrowing grasses is higher than that of shrubs, which gradually increases as succession proceeds (Kröpfl et al., 2007). The interaction between grazing and wildfire in the southern Monte was analyzed by Rostagno et al., (2006). These authors showed that shrub-dominated steppes appear as the stable, climax community in the Monte Region. Should a wildfire disturbance occurs, succession will set back the community to a pioneer stage in which annuals and tussok grasses dominate. The persistence of this stage will depend on the recovery rate of dominant shrub species. Without grazing pressure, the climax stage is reached in about 7 to 12 years. Continuous grazing pressure however, would reduce the grass component, and consequently, modifies the fire regime (Figure 2) (Rostagno et al., 2006; Bran et al., 2007).

(a) (b)

Figure 5. a) *Lycium chilense* Miers, *Junelia* sp. and *Chuquiraga avellanedae* Lorentz shrubs slowly recover after a wildfire in Monte region. b) *Prosopidastrum globosum* (Gillies ex Hook. & Arn.) Burkart resprouting from the stem base two years after a fire (photos by G. E. Defossé).

From a grazing management perspective, frequent fires would then favor the maintenance of a seral stage in which the grass component prevails (Willard 1973), meaning more forage available for grazers (Rostagno et al., 2006; Bran et al., 2007; Villagra et al., 2009). Although this fact is acknowledged and known by many technicians and land management agencies, prescribed burns to preserve the grass stage is still not used in the Monte region. Continuous grazing is then the main stress, driving vegetation into a disclimax stage in which the shrub component encroaches in detriment of grasses (Villagra et al., 2009; Figure 2). Under this scenario and since prescribed burns are not yet considered as a management tool, eventual wildfires followed by a pasture rest appear as the only way for returning the ecosystem to its natural successional path (Defossé et al., 2003).

There are several reasons, however, for which ranch owners do not implement prescribed burns to return the range to earlier successional stages and improve pasture condition. Among these reasons, we will point out two. The first one is due to the fact that most ranches in the region have only one, or at least two, allotments in which yearlong grazing is practiced. If a burning plan is applied, grazers would not be allowed to graze the burned pastures for at least two years after the burn, until the grass component has fully recovered. This implies the construction of new fences to increase the number of allotments to locate the animals until the burned pasture is ready to be grazed. Apart from being expensive, the construction of new allotments would change the traditional yearlong grazing management scheme sustained by ranchers for decades, and almost all are reluctant to make that change. The other reason has more to do with an emotional question that crossed the whole society that inhabits in cities located in the Monte region of Patagonia. In January 21, 1994, 25 volunteer firefighters lost their lives when trying to combat a wildfire in a ranch located in the outskirts of the city of Puerto Madryn, The whole society was so badly shocked, that still today the mere mention of "wildfire", or the possibility of conducting "prescribed burns", causes resistance and disapproval. Another factor that impedes the application of fire management practices in this phytogeographical region is that the fire authority relays on civil defense and structural volunteer firefighters. Their vision of fire as "the enemy", and their reluctance to accept fire as a natural force in rangelands and associated vegetation makes still difficult to promote fire management in the Monte region.

Related to the factors beyond human control, it is important to recognize that most models, which predict global climate change, forecast an increase in temperatures and decrease in precipitation during the summer for the Monte Region (i.e., Labraga 1998). According to this predicted scenario, climatic variability would increase the probability of electric storms during the summer, with the consequently increase in wildfire frequency (Villagra et al., 2009; Fischer et al., 2012).

The Patagonian Steppe

Climate

The Patagonian Steppe covers the central part of Patagonia (Figure 1) and has been divided into 5 different floristic districts (Soriano 1956). Although the vegetation of this steppe has similar structural characteristics all across, subtle differences in climatic and soil conditions have justified its division into floristic districts (Soriano 1956). In this chapter, we will focus in the Central and Occidental districts, which cover about 80% of the whole Patagonian Steppe. Within them, the Central District occupies about 66% of this steppe, and its vegetation is mainly composed of different grass species of the *Pappostipa, Festuca, Poa, Hordeum*, and other genera, interspersed with several shrubs, such as *Nassauria glomerulosa* (Lag. Ex Lindl.) D. Don, *Chuquiraga avellanedae* Lorentz, *Prosopis denudans* Bentham, *Berberis cuneata* K.Koch and cushion-like half-shrubs, like *Brachiclados caespitoides* (Phil) Speg., *Mulinum microphyllum* (Cav.) Pers., *Acaena platyacantha* Speg., and *A. splendens* Hook. & Arn. (Soriano 1956; Ares et al., 1990; Figure 6).

Figure 6. Landscape view of the Central District of Patagonia, in which the dominant *Chuquiraga avellanedae* Lorentz is surrounded by half-shrubs and tussock grasses (photo by G. E. Defossé).

The climate is similar to that of the Monte region, although with lower temperatures and higher winds. Rainfall in this district is less than 150 mm per year and evapotranspiration is very high. Below-freezing temperatures occur during the fall, winter and early spring, while mild to hot temperatures (average 15 °C and extreme of 43°C) may occur during the summer. Although the vegetation in this area is very sparse, wildfires are relatively frequent. This frequency is sustained by the combined effects of low relative humitity and strong westerly winds that dries up the fine fuels, and carries the fire across this sparse vegetation. Even though no specific studies have been conducted related to fire effects on vegetation of this vast District, evidences suggest that native plants are well adapted to summer wildfires.

Toward the west and near the piedmont of the Andes, the Occidental District of the Patagonian Steppe appears, incresing in species diversity and biomass as compared to the Central District. This district shows more grasses besides *Pappostipa* and *Festuca* spp., as *Poa huecu* Parodi , *P. ligularis* Nees ap. Steud., *Bromus macranthus* Hack., and other bushes, such as *Fabiana patagonica* Speg., *Fabiana imbricata* Ruiz & Pav., *Adesmia campestris* (Rendle) Rowlee, *Mulinum spinosum* (Cav.) Pers., and *Senecio filaginoides* DC (Figure 7).

Figure 7. The Occidental District of the Patagonian phytogeographic province, in which shrubs intermix with grasses. At the back, the eastern slopes of the Patagonian Andes (photo by G. E. Defossé).

Rains are higher and less variable than in the Central District, and temperatures are somewhat similar during the summer (Figure 3). As in the rest of Patagonia, wildfires are common from late October to early April. Toward the west, an ecotonal area marks the end of this district and the beginning of the Subantarctic Andean Forest. In this ecotone, diversity increases as compared to other districts, and if we include the Andean piedmont, it is the region in which the highest proportion of wildland and human-set fires occurs in the whole Patagonian region. As we will see later, this is also the area most suitable for dryland conifer afforestations, since its soil has a layer of volcanic ashes (Allophanes, located more than a meter deep), which stores moisture during the winter months. Since roots of native plants do not reach this deep soil profile, the accumulated water into the layer of volcanic ashes is then available for conifer growth during the summer drought (Colmet Daage 1993; Buduba 2006).

Fire Ecology and Management

In a similar way as in the Monte region, most shrub and grass species of the Patagonian steppe are well adapted to periodic wildfires. Two studies showed that vegetation entirely recover in about 4 to 6 years after a wildfire, and species richness and species representativeness are higher in burned than in the adjacent unburned areas (Gobbi 1994; Ghermandi 2006). Working in post-fire dinamics, Ghermandi (2006) found that the grass *Festuca pallescens* (St.-Yves) Parodi. (preferred by cattle) is more sensitive to fire injury, showing higher mortality rates (37%) than *P. speciosa* (24%), not preferred by cattle. This higher sensitivity could be due to the shape of the crown of *F. pallescens*, which is denser than that of the *P. speciosa*, and could have received more heat during the fire, damaging or killing the growing meristems.

Mulinum spinosum is an example of a shrub that quickly resprouts after a wildfire. This hemispheric, non palatable shrub species is very well adapted to fire and very persistent (Figure 8). It recovers its initial size in a few years after a fire event, showing almost null post fire mortality. Resprouting is the main strategy for post fire recovering, as observed by the rapid growth of fire damaged individuals and the almost null percentage of seedlings present after the fire (Ghermandi et al., 2006; Damascos and Ghermandi 2007).

Figure 8. Typical *Mulinum spinosum* (Cav.) Pers. shrub surrounded of Patagonian buchgrasses (photo by M. M. Godoy).

Another interesting shrub species that shows and adaptation to fire events in this steppe is *Fabiana imbricata*. This species forms contemporary bushes in mid-slopes, and produces allelopatic substances that impede the establishment of its own seedlings in its surroundings. The recruitment of new individuals will then depend on different factors. In the case of a severe fire, adult *F. imbricata* plants do not resprout and die. If post-fire conditions are benign (as a rainy spring), numerous new individuals will grow in the unburned edges of the dead bushes, where seeds are dispersed by water. This may explain why distribution patterns of *F. imbricata* at landscape scale are in part related to wildfires (Ghermandi et al., 2006). Another example of particular adaptations to fire is given by the cushion-like half shrub *Acaena splendens* Hook & Arn. that does not resprout after a fire. Its seeds, instead, can support high temperatures during a wildfire, remaining alive and able to germinate when environmental conditions are favorable. New *Acaena* plants can reach the same cover their ancestors had in only 4 years after the fire (Ghermandi et al., 2006).

Senecio bracteolatus Hook & Arn., another non-palatable shrub, not only resprouts as *M. spinosum* does, but also produces abundant seeds after a fire event (González 2011). Many *Senecio* spp., at seedling stage, need steppe grasses (*Festuca* spp. or *Pappostipa* spp.) as nurse plants for survival in summer, which in turn grow more after a fire event. Ghermandi et al., (2006) concluded that native shrubs in this steppe not only recover after wildfires but also can create new shrub centers, taking advantage of favorable conditions generated by the availaility of resources right after the fire. In general, all grass and shrub species grown in this steppe are capable to persist after a wildfire by developing two main classic strategies. In the case of a mild severity fires, the persistence strategy implies resprouting from their stem or crown base. If the fire is severe enough as to damage the crown or stem base, the persistence strategy implies the production of seeds resistant to fire, which will germinate when environmental conditions are aproppriate.

Figure 9. *Senecio filaginoides* DC surrounded by buchgrasses in the patagonian steppe. To the upper left, the carpet-like *Azorella patagonica* Speg., locally named "leña de piedra" (stone-firewood). Photo by M. M. Godoy.

As mentioned for the Monte, prescribed fires are still not used as a range management tool in the Patagonian steppe. The sporadic occurrence of extreme wildfires in ranches of this region may explain the reluctance of ranchers to use fire as a management tool. In 1999, a lightning-ignited fire, facilitated by previous hot and dry environmental conditions, presented extreme behavior and, in about a week, affected 18,700 ha, including grasslands, shrublands, and conifer afforestations, destroying also houses and barracks of the famous San Ramón Ranch ranch, located near the city of Bariloche (Dentoni et al., 1999). While this kind of extreme wildfires may occur from time to time and affect thousand of hectares, the steppe vegetation seems to be unharmed and quickly recovers, as demonstrated by several studies (Gittins 2001; Ghermandi and González 2009; Ghermandi et al., 2004; 2006; 2010).

In 1996, Argentina established the National Fire Management Plan (PNMF, National Fire Management Service today) in response to several extreme wildfires that affected the ecotonal areas between the Patagonian steppe and the Andean Forests. In 2000 and with the cooperation of Canadian scientists, PNMF started to develop and implement a FDRS for

Argentina, based on the Canadian Forest Fire Danger Rating System (CFFDRS) (Dentoni et al., 2014). The first phase consisted in the adjustement and implementation of the FWI component in diferent regions of the country, starting by the northern Andean region of Patagonia, because of the availabitity of fire information and the human capabilities in the region. Recently, some work has been done involving the development of fuel models to fit that system (Bianchi et al., 2012). As we will expand later for the Andean Forests, institutions related to fire prevention, fire management, and fire research are very well developed in this region, and constitute one of the most advanced in the field in Argentina.

The Sub-Antarctic Andean Forest

Climate

The eastern border of the Sub-Antarctic Andean Forest begins by intermingling some of its main species with those of the Patagonian steppe, in a narrow strip about 80 km wide and 2000 km long located in the foothils of the Andean Cordillera (Figure 1) (Dimitri 1972). Its vegetation traits rapidly change as one goes up to higher elevations, in one of the steepests precipitation gradients of the world (Veblen and Lorenz 1988). To illustrate the features of this gradient, it should be considered that at lower eleventions in the ecotone with the steppe, annual rains average from 400 to 600 mm, at mid elevations from 600 to 1,500 mm, and reach more than 3,000 to 4,000 mm at the timberline. It is interesting to note that in spite of the huge differences in total precipitation among these areas, all share a similar (Mediterranean) pattern of annual distribution of rains, which are mostly concentrated during the winter months and followed by a drought period during the summer. Rains at the piedmont of the Andean region are significantly higher and drought period is shorter than in the Patagonian Steppe (Figure 3). At lower elevations in the ecotone with the steppe (from the 36° to the 44° S), and under relatively xeric conditions, spinny shrubs (*Mulinun* spp., *Acaena* spp.), bunchgrasses (*Pappostipa* spp. and *Festuca* spp.), scattered trees of the *Maitenus, Lomatia, Schinus,* and *Diostea* genera combined with some *A. chilensis* prevail (Figure 10 a).

a) b)

Figure 10. a) Ecotone forest-steppe of Patagonia. At the back, a stand of *Austrocedrus chilensis* (D.Don) Florin et Boutelje. At the front, *Lomatia hirsuta* Diels ex J.F.Macbr. and *Bulnesia retama* (Gillies ex Hoock. & Arn.) Griseb. trees. Precipitation reaches 700 mm/year (Photo by L.O.Bianchi). b) Mixed forest of *Nothofagus dombeyii* (Mirb.) Oerst. and *A. chilensis*. Precipitation reaches 1000 mm/year. Photo by M. M. Godoy.

From that area and up to mid elevations, these species are gradually replaced by open stands dominated by *A. chilensis* trees. This species forms pure stands in some places, and in others it intermixes with *Nothofagus dombeyi* (Mirb.) Oerst., an evergreen *Nothofagus* which grows near water sources (Figure 10 b). At higher elevations, these species are replaced by ñire forests (*Nothofagus antarctica* (Forster) Oerst. Figure 11 a), and in areas in which yearly precipitation reaches from 1500 to 2200 mm, *Nothofagus pumilio* (Poepp. & Endl.) Krasser appears as the dominant species (Figure 11 b).

a) b)

Figure 11. Autumn views of Patagonian Nothofagus forests. a) *Nothofagus antarctica* (Forster) Oerst. forest (precipitation ~ 1200 mm/year (photo by A. H. Franke), and b) *Nothofagus pumilio* (Poepp. & Endl.) Krasser forest; precipitation: ~2000 mm/year (photo by J. O. Bava).

In the northern part of the Andean forests, from the 36° to the 40° S, small patches of the evergreen coniferous tree *Araucaria araucana* (Molina) K.Koch dominates the landscape. *Araucaria araucana* trees 800 years old have been dated, allowing long-term analysis of fire occurrence (Figure 12 a). From that place to higher elevations and when annual precipitation exceeds 2500 mm, patches of the Valdivian rainforest appear, being the giant alerce or lahuan (*Fitzroya cuppressoides* (Molina) I.M. Johnst. (Mol.), southern larch in English) its most conspicuous tree species (Figure 12 b). Alerce is one of the oldest living tree species of the world (Lara and Villalba 1993), being the tallest of all native trees of South America.

a) b)

Figure 12. a) *Araucaria araucana* (Molina) K.Koch forests in the tree-line with *Nothofagus antarctica* (Forster) Oerst.at the front (photo by P. Pantaenius). b) Valdivian rainforest, showing young *Fitzroya cupressoides* (Molina) I.M. Johnst. (Mol.) trees (the tallest trees, intermixed with some *Nothofagus* spp.). Precipitation reaches 3000 mm/yr (photo by M.M. Godoy).

In the whole Andean forests, from lower elevations in the ecotone with the steppe and up to the timberline, wild- and man-caused fires have been present since immemorial times (Withlock et al., 2006). These fires are today more frequent and less severe at lower elevations, and less frequent and more severe at higher elevations (Tortorelli 1947; Veblen and Lorenz 1988; Veblen et al., 1999; Kitzberger 2003).

Fire Ecology and Management

While for the other regions of Patagonia fire has been relatively poorly studied, the Andean region presents a wide and very profuse literature on fire and fire effects. As already mentioned, wildfires in native Patagonian forests have always been present, as there are fire evidences even in pristine areas (Kitzberger 2003; Withlock et al., 2006).

Despite the fact that most modern fires have been caused by humans, previous research based on instrumental as well as pre-instrumental records has demonstrated strong controls of interannual climatic variability (e.g., ENSO related variability) on regional fire spread and occurrence in Andean Patagonia (Kitzberger and Veblen 1997; Kitzberger et al., 1997; Veblen et al., 1999). Tree ring records from northern Patagonia clearly document years of synchronous and widespread fires several times during both the late 19th century period (when European colonization began), and before permanent European settlement (Kitzberger et al., 1997; Veblen et al., 1999). According to Veblen et al., (1999), 1827 and 1897 were years of exceptional fires. Although those years coincided with severe droughts, the temporal and spatial pattern of all fires, including small fires, implied that Indians significantly influenced fire frequency, especially in the ecotone between the steppe and *Austrocedrus* woodlands (Veblen et al., 2008). More fires in Indian hunting grounds than in ecologically similar areas less frequented by these hunters were recordered in descriptions of 19th century explorers (Veblen et al., 1999; Veblen et al., 2008). According to reports of scientific explorers (Willis 1914; Rothkugel 1916; 1922), large fires occurred between 1890 and 1914. During this period, several drought periods also coincided with the greatest expansion of the European settlement of western Patagonia. This settlement was accompanied by an immense increase of introduced livestock (mainly sheep), that in the nearby Province of Santa Cruz went from less than 900 thousand in 1894 to more than 5 millon in the late 1910´s (Morrison 1917).

During the period that went from early settlement in 1850 and up to around 1930, the most common causes of ignition in a frequency order were: colonists that burned forest overgrowth or cleaned the ground for expanding pasturelands (Rothkugel 1916; Tortorelli 1947), negligences of people, intentional fires to move away cattle, and finally fire produced by lightning or incendiaries (Cox 1863; Musters 1871; Fonk 1896; Moreno 1897; Tortorelli 1947). Lightning is produced in the Andean Cordillera in North and Central Patagonia by dry storms, and they still are an important source of wildfires (Dentoni and Cerne 1999; Defossé et al., 1999; Dentoni et al., 2001; Defossé et al., 2003; del Valle et al., 2004; Defossé et al., 2006; González et al., 2006). Indians, instead, used to burn grasslands and adjacent forests to allow grasses grow, making them attractive places for game animals, such as rhea and guanaco. Sometimes they might have also used fire as a defense tool against enemies (Rothkugel 1916; 1922; Tortorelli 1947; Veblen and Lorenz 1988; Veblen et al., 1999).

At a regional scale, the establishment of National Parks in the 1930's, the changes in cultural patterns, and the fire-control policy, had strong effects on the fire history of Patagonian forests (Veblen et al., 1999). In dryer environments, the decrease in fire frequency might have been associated, in part, to the displacement and confinement of native Indians to reserves, restricting the previous use of fire for hunting and warfare (Veblen et al., 2003). This confinement was also accompanied with the increasing growth of sheep and goat herds in the reserves, which diminished the pressure to hunt wild game for feeding and then the need to burn for hunting purposes. Furthermore this reduction in fire occurrence in more mesic areas appears to be related to the high livestock pressure that, starting around 1900, may have also reduced the abundance of fine fuels and therefore fire activity (Willis 1914; Morrison 1917). A similar and synchronous pattern of fire reduction since the mid-20th century has been observed for *Austrocedrus* woodlands in northern Patagonia (Kitzberger and Veblen 1997). This reduction trend may help explain why after the implementation of the fire suppression policy, stands of *A. chilensis* started to expand toward the steppe (Veblen and Lorenz 1988).

Related to the wildfire effects on individual tree species, Kitzberger et al., (2005) found complex effects on postfire regeneration processes for the dominant *N. pumilio,* especially when stands of this species had been disturbed by logging, and/or its understory highly impacted by a long history of continuous grazing (Quinteros 2014). In fact, grazing and browsing by large (sheep, cattle and horses) and small introduced herbivores such as the European hare (*Lepus europaeus* Pallas) may change understory environmental and biotic conditions in the fire affected area (Urretavizcaya et al. 2012; Quinteros 2014). Among these changes, the increase in grasses and grasslike species (Quinteros 2014), the lack of available seed sources, and the aggressive resprouting from rhyzomes of the bamboo-like *Chusquea culeou* E. Desv. (González et al., 2006), have proven to have detrimental effects on *N. pumilio* postfire regeneration. These factors may have more negative effects if fire occurs on steep slopes exposed to strong winds, impeding or delaying the recolonization of *N. pumilio* on those sites (Veblen et al., 1996; Bertolin et al., 2013). Besides this, Veblen and Lorenz (1988) observed that closed and relatively undisturbed stands of *N. pumilio* grown at high altitudes in this forest develop trees relatively free of lower branches, presenting scarce understory vegetation. This condition makes difficult the initiation of surface fires and their propagation as crown fires. In contrast and at mid-altitudes, woodlands of the tree-like *N. antarctica* appears. This species develops multi-stem branches near the ground, and sprouts viguoursly from its crown following destructive fires. The structure of this tree plus the abundance of the bamboo-like *C. culeou* and other associatd shrubs, made the mid-slope communities more adapted to cope with frequent fires. Wildfires also seem to stimulate massive resprouting of *N. antarctica*. When regenerating fires do not frequently occur, this species can have high mortality rates because senescent trees are predisposed to accumulate a heavy load of epifits and mistletoes that finally kills the tree (Donoso et al., 2006). The different strategies with regard to fire presented by *N. pumilio* and *N. antarctica* stands may help explain why, in absence of other disturbances, fires are more frequent and less severe at lower altitudes, and less frequent and more severe at higher altitudes, especially in the central Andean Region of Patagonia (from the 40° to the 46° S).

In northern Patagonian forests, from the 36° to the 40° S, the dominant tree species is *Araucaria araucana*, being *N. antarctica* the co-dominant species (Figure 12 a). According to the recurrence of events reconstructed since 1700, fires have been the major disturbance in *Araucaria* forests (Mundo et al., 2013). As an example and in relatively recent times, Tortorelli (1947) described the quick spread of a fire in *Araucaria* forests during January 1943, in a 3500 ha allotment located in the north of Lanín National Park. These fires were also favored by the massive flowering and withering of the bamboo *C. culeou*. In these forests, *N. antarctica* also presents a fire dynamics associated to *Araucaria* (Burns 1993). The adaptation of *A. araucana* to fire includes thick bark, sprouting from epicormic buds, and branches with protected terminal buds. During the first 10 years after a fire, *A. araucana* seedlings or root suckers begin to establish beneath a resprouted *N. antarctica* tree, which acts as a facilitating species. Then *A. araucana* grows through the sparse *N. antarctica*, eventually overtopping and suppressing it in a few years. If this suppression persists uninterruptedly more than 150 years, succession may lead to pure stands of *A. araucana* and the disappearance of *N. antarctica*. Under natural conditions, frequent fires allow both species to persist in the system, acting as a medium of species coexistence (Burns 1993). The fire disturbance usually leads to the formation of clustered stands composed of 10 to 20 m tall *A. araucana* trees surrounded by 2 to 5 m tall *N. antarctica* thickets (see Figure 12 a).

Tortorelli (1947), and Veblen and Lorenz (1988) studied the age structures of mixed *N. dombeyi* and *A. chilensis* forests that are usually located at lower altitudes than *N. antarctica*, and established that both species form even-aged populations following stand-destroying fires. These stands cease to establish once a dense canopy has been formed. Villalba (1995) observed that changes in tree recruitment in ecotonal habitats were due in part to changes in the frequency of disturbances such as fire.

As mentioned in the former paragraphs, most studies already done in Andean Patagonia have mainly dealt with fire ecology at stand or individual levels. Recent studies, however, have focused on other ecological aspects such as the determination of species flammability (Blackhall et al., 2012; Bianchi 2014), the interactions of fire and grazing (Quinteros 2014), or the ways of restoring fire-disturbed forests (Urretavizcaya et al. 2012; Urretavizcaya and Defossé 2013; Pafundi et al., 2014). Other aspects of fire management, such as pre-suppression, suppression, and other combat activities, the development of fire rating danger systems for fire prediction (Dentoni et al., 2014), the analysis of large fires of extreme behavior (Dentoni et al., 1999, 2001) and social aspects of fire (Farías and Defossé 2011; 2012) are also studied in Andean Patagonia.

Among the social aspects of fire, it is important to mention the emergent problem that represents the incesant growth of urbanizations in fire-prone areas of Andean Patagonia used for both residence and recreational activities (Veblen et al., 2008). Population in these commonly called WUI, are growing at rates from two to three times faster in Andean Patagonia than in the rest of Patagonia or in the whole Argentina (Dirección Nacional de Estadísticas y Censos, Censos 2001 and 2010). The problem they face is similar to what is currently occurring in other areas of the world (Western North America, Australia, Chile or the Mediterranean countries). In some of these areas, the depopulation of people working in rural activities allowed subtantial increases of biomass once used to feed animals (Pausas and Fernández-Muñoz 2012). The resettlement of these areas are now mainly due to urban residents trying to find new lifestiles in those environments (Pausas and Keeley 2009; Farías and Defossé 2011).

The combination of high fuel availability and more ignition sources (arson or accidental) are exposing more people to wildfires, and the results have often been catastrophic (Pausas and Keeley 2009). Catastrophic fires have occurred in Andean WUI in northwestern Chubut in Patagonia during the hot and dry 2010-2011 and 2011-2012 fire seasons. Fortunately, fires occurred one week after the vacation period ended and thousand of tourists had left the region. Although several houses and buildings were completely destroyed, and 2700 and 3460 ha of WUI forests were consumed by both the 2011 and 2012 fires, respectively, no human casualties were reported (SAyDS 2012).

While fire management in Monte and Patagonian steppe depends on civil defense and local volunteer firefighters, fire management in Andean Patagonia relays on a well-organized system, including federal, provincial and local institutions. Since Argentina is a federal country, by law the primary responsibility for fire suppression in case of fires relays on provincial and local authorities. If their capabilities are overcome in case of big fires, they can ask for support to the National Fire Management Service (SNMF), who may provide the National Fire brigade (strategically located in central western Patagonia) or coordinate and dispatch aerial support.

Provincial forest fire brigades are spread out in the whole Andean region from northern Neuquén up to the southermost province of Tierra del Fuego. In the city of Esquel, the office of the SNMF brings technical advise and provide fire weather forecasts, not only for the Patagonian region but also nationwide. The University of Patagonia (in its Esquel and Puerto Madryn campuses) and the University of Comahue (located in Bariloche) have qualified scientists working in fire science in Patagonia. Both universities have strong ties which includes common projects and scientist exchages with other similar fire research institutions around the world.

AFFORESTATIONS IN THE PATAGONIAN FOREST – STEPPE ECOTONE

Importance

In the 1970´s, afforestations with North American conifers started to be promoted in Andean Patagonia, in response to governmental subsidies. The main objective was then to produce high density conifer stands for the celullose industry, to satisfy Agentina´s internal demands. Since these plantations mainly replaced native forests, many environmentalists, forest and land managers, politicians, and concerned citizens, started to oppose and disapprove them.

After a decade of meager results and pushed by the whole society, the idea of planting on Andean forests by replacing native forests was gradually abandoned. In the early 1990´s, however, several studies showed the potential of the adjacent ecotone with the Patagonian steppe for carrying out successful afforestation projects (Ferrer et al., 1990; Irisarri and Mendía 1991; Irisarri et al., 1995). Apart from a renewed legislation that promoted afforestations for timber production through subsidies and tax reductions, afforesting the steppe had other ecological and economical incentives. One of them was the high availability of low-cost suitable lands in vast overgrazed rangelands composed of sparse native shrubs and grasses. The other was that the soil upon which afforestations were proposed have a rich subsoil of volcanic ashes, to which roots of native plants generally did not reach (Colmet Daage et al., 1995). In these areas, isolated conifer afforestation trials showed that the root system of conifers reached the subsoil, allowing productivity and growth rates many times greater than in their sites of origin in northwestern North America (Urzúa 1991; Gonda and Lomagno 1995; Gonda 1998).

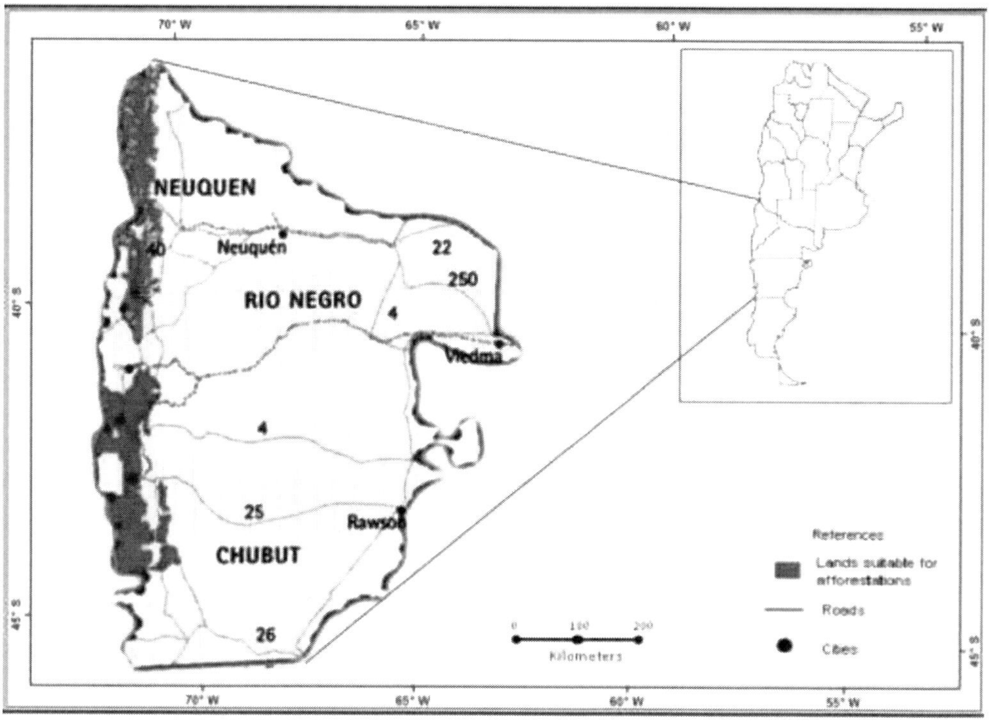

Figure 13. Area suitable for dryland afforestations (ca. 2 millon ha) in the northern Patagonian provinces of Neuquen, Rio Negro, and Chubut in Argentina (Drawn after Colmet Daage 1993, and SAGPyA 1999).

At present, the surveyed ecotonal area suitable for afforestations outside the native forests covers around 2 million ha (Ferrer et al., 1990; Irisarri and Mendía 1991; Irisarri et al., 1995; Loguercio et al., 2004). This area goes from the 36° to the 45° S, in a north to south narrow belt in which yearly precipitation varies from 400 to 800 mm (Figure 13) (Buduba 2006).

Environmental and economic constraints, however, may reduce commercially afforestable areas to about 800,000 ha (Loguercio and Dececchis 2006 a; b). Compared to other areas of the world, this steppe region in Patagonia presents other good competitive and comparative advantages to develop conifer afforestation projects. These advantages reside not only in the good growth rates already achieved, but also in the low amount of carbon found in native grass and shrub vegetation (the base line), which makes them attractive for developing carbon sequestration projects within the Clean Development Mechanism of Kyoto protocol (Loguercio et al., 2004).

Today, there are about 90 thousand ha already planted, the bulk of them still young (from 1-5 to about 20-25 yr old (Figure 14 a, b). The main conifer species used in these afforestations are ponderosa pine (*Pinus ponderosa* Douglas ex C. Lawson), which comprises about 80% of all afforestations, and then, in a decreasing order, Douglas fir (*Pseudotsuga menziesii* (Mirb.) Franco), lodgepole pine (*Pinus contorta* Dougl. var. latifolia Engelm), and radiata pine (*Pinus radiata* D. Don) (Loguercio et al., 2008). Some of these afforestations, however, are yearly affected by summer wildfires (Figure 15) (SAyDS 2009).

a) b)

Figure 14. a) Young ponderosa pine afforestation in the steppe (photo by G. E. Defossé), and b) the same afforestation 15 yr later. Precipitation at the site is only 400 mm/year. Note that this afforestation has recently been pruned and thinned, and residues taken away to reduce fire hazard (photo by M. M. Godoy).

a) b)

c)

Figure 15. a) Young ponderosa pine afforestation in Patagonia affected by a summer wildfire. The extreme severity of the fire is noticed by the exposure of stones at ground level. Precipitation at the site is 500 mm/year. Untreated (neither pruned nor thinned) 23 yr old Douglas fir (b) and a 13 yr old ponderosa pine stands (c), both burned by an extreme summer wildfire near the city of Esquel in Patagonia, Argentina (photos by M. M. Godoy).

Fire Ecology and Management

Afforestations in the forest steppe region of Patagonia could be considered a relatively new activity as compared to cattle or sheep raising, or timber production based on native forests. This relatively new activity was in general started by producers with scarce knowledge on afforestation tasks, and also by forest technicians who, formed in the European tradition of forestry, had low experience on the functioning and management of the new ecosystem they were creating. Although these technicians proposed in general sound silvicultural plans to conduct afforestations, not always these recommendations are followed by producers. One example that explains this behavior is the reluctance of some producers to conduct appropriate prunings and thinnings as afforestations age. Others, who do the appropriate pruning and thinning activities, do not handle these residues accordingly, leaving most of them in the stands´ understory. Apart from prunning and thinning residues (and as it happens in other temperate forest ecosystems with Mediterranean climate), Patagonian afforestations accumulate needles, herbs, and shrubs that grow in their understory, increasing fire hazard. As a consequence of these mismanagements, recent fire data showed that about 2 to 3% of all Patagonian afforested areas are burnt annually by wild or arson fires, and that this percentage may increase in the future as these afforestations age and climate changes (Defossé et al., 2011).

From the fire management perspective, it should be noted that almost all Patagonian afforestations are composed of ponderosa pine, Douglas fir, and lodgepole pine, all fire-prone species (Wright and Bailey 1982; Agee 1993; Fitzgerald 2005; Naficy et al., 2010; Fule et al., 2012). If we plan for healthier and productive stands up to their final cut, these species not only need the classic pruning and thinning treatments, but also proper residues treatments. Prescribed burnings may be then necesary to help reduce these residues, prodiving for productive stands and reducing at the same time the risk of unwanted wildfires (Weaver 1957; Peterson et al., 1994; Thies et al., 2005; Pritchard et al., 2010).

Taking into consideration those facts, a study was conducted in 1996 in an emblematic afforestation of Patagonia located in San Ramón Ranch. The idea was to get useful information for reducing fire hazard and proposing proper fire management practices that would serve later to be generalized to other Patagonian afforestations. By using the Fire BEHAVE prediction system (Burgan and Rothermel 1984; Andrews and Chase 1989) and a custom fuel model applied to real data, Rodriguez (1996) concluded that tree age and size, as well as fuel residues left on the ground (mainly 1 and 10 hr fuels that are the product of pruning), may have an important influence on ponderosa pine tree mortality should a wildfire occurs. He also pointed out that removing all fuels may also reduce the risk of high intensity fires and thus tree mortality, although this practice may be undesirable because nutrient cycles are altered, and can reduce tree growth rates (Rodriguez 1996). Three years later, in 1999, an extreme wildfire affected the afforestations that Rodriguez (1996) used as an example for his study. These afforestations comprised 654 ha, 80% of which corresponded to ponderosa pine, and the rest to Douglas fir (12%) and lodgepole pine (8%). Most of them (67%) were 10 to 25 yr-old stands which had received only one pruning and no thinning treatments since their establishment, and whose residues were left on the site. The rest were older (between 30 to 35 yr-old, 5%) and younger (less than 5 yr old, 28%) stands which had no treatments at all. The extreme meteorological conditions, combined to high understory fuels, caused the fire crowning, killing 94% of all trees (Dentoni et al., 1999).

This event strongly influenced not only forest producers, researchers, and technicians, but also the forest research and administration agencies at provincial and national levels. While some skeptical pleaded for discontinuing with the activity, others suggested increasing research on fuel treatments as the only way to reduce fire hazard and have productive afforestations. They proposed that the studies should be based on local examples, but taking into consideration the research already done for the same species in North America and other regions of the world. The idea was to resume and expand the work done by Rodriguez (1996), and the pioneer prescribed burn studies done by Kunst and Rodriguez (1997) and Kunst et al., (2003) in an experimental ponderosa pine afforestation site of Patagonia. These studies pointed out that stem injuries caused by prescribed burns may be very small, and that only basal scorch in some individuals was observed, in sites where fine fuels accumulated from previous prunings. No other negative effects in site quality were observed long after the prescribed burn was done.

Following with these studies in 2007, an Argentine team composed of fire scientists from CIEFAP in Esquel, Patagonia, and Santiago del Estero in north-central Argentina, joined the European-funded fire project *Fire Paradox*. One of the objectives of that project was "*the creation of the scientific and technological bases for new practices and policies under Integrated Wildland Fire Management in Europe and other target countries*". Within the project, the provision of science based technologies for using prescribed burning and other technologies to mitigate potential wildfire impacts was one of the main objectives. In 2008, Argentine (from CIEFAP and INTA Santiago del Estero), and Spanish (University of Lleida) fire scientists who participate in the project that was funded by the European Union, presented a series of prescribed burning courses and seminars in two places of Patagonia. The prescribed burn practices were done at different locations, comprising conifer afforestatins of different ages and under different silvicultural treatments. These prescribed burns achieved the objectives of reducing fuel loads from 40 to 65% in the understory of the different afforestations and at the same time decreased risk of unwanted fires. Sixty six Argentine and sixteen Spanish firefighters were trained in those courses (Figure 16 (Godoy et al., 2009).

a)

b)

Figure 16. a) Prescribed burn in a 52 year old ponderosa pine afforestation in Patagonia. The aveage fuel load was 26 ton/ha and was mainly composed of pine needles and other residues. b) Prescribed burn in a 35 year old ponderosa pine afforestation in which pruning and small thinning residues had been left on site. The average load of these residues was 62 ton/ha (photos by M. M. Godoy).

In a recent study, Godoy et al., (2013) analyzed the effects of a wildfire of extreme behavior on a 23 yr old Douglas fir afforestation previously pruned and thinned to different densities. These densities were 500, 700, and 900 tress/ha, plus the control (neither pruned nor thinned), which remainded at the initial density of 1,200 trees/ha. While all trees in the control stand died, stands thinned to densities from 500 to 700 trees/ha not only withstand the extreme wildfire, but also showed the highest postfire survival and growth. Besides thinning and removal of residues, these authors considered that it the near future, other silvicultural treatments (mastication or prescribed underburning) should be used to reduce wildfire risk in Patagonian afforestations. At present, CIEFAP scientists are planning and conducting several studies in these afforestations. The objectives are focused at finding the ways of reducing wildfire hazard and allowing for greatest growth through the application of appropriate silvicultural techniques (thinning densities, prescribed burning, mastication, residues removal, etc.). An associated objective is also to determine which of these techniques will produce the lower amounts of CO_2 emissions. The overall goal is to produce timber products and at the same time contribute to atmospheric carbon removal and help mitigate climate change (Laclau 2003; Defossé et al., 2011).

Fire Behavior Components of Patagonian Vegetation

Table 1 summarizes some fire behavoir variables in different fuels types found in Patagonia. The table is merely informative, since the data present single observations of specific wildfires. The lack of fire behavior predition models for the fuels present in the region is one of the most glaring gaps in our fire knowledge at present.

Conclusion

In Patagonia, three main vegetational provinces occur (Monte, Patagonian Steppe, and Andean Forest). All share similarities regarding climate (although dryest in the Monte and Patagonian steppe and more humid in the Andean Forest), and fire season. In this mediterranean-type of climate, the vegetation is prone to wildfires during the long summer drought. In ancient times, lightning from dry storms appeared as the main ignition source of these wildfires, helping shape the structure and dynamics of that pristine vegetation. Indians and settlers,from the 1800´s and to the early 1900´s, used fire to hunt, or to open areas for pastures. New disturbances appeared later (logging for either timber production or firewood, grazing for domestic animals, a fire exclusion policy, urban settlements in the WUI, and afforestations), exerting new pressures and producing subsequent changes on vegetation and patagonian landscapes. Even considering these changes, most individual species composing native Patagonian vegetation appear to have evolved by developing survival strategies to make them resilient to fire, if the frequency and severity of these fires are kept within their evolutionary range. As in other Mediterranean type of climate, recent unprecedented changes in land use may have altered part of the natural balance in terms of fire frequency, continuity, severity, and effects.

Table 1. Fire behavior variables for selected fuel types of Patagonia, Argentina

Fuel type	Fuel load (ton/ha)	Fire frequency (yr)	Fire type		Rate of spread (m/min)	References
Nothofaguspumilio (Poepp. & Endl.) Krasser forests (A)	44.1 [A1] 29.4 [A2]	~90	Backfire		0.2 to 2	Kitzberger et al., 2003 Sagarzazu and Defossé 2009
			Surface fire		2 to 6	
			Intermittent crown fire		12 to 19	
			Continuous crown fire		42 to 72	
Austrocedrus chilensis (D.Don) Florin et Boutelje forests (B)	67 ± 14.7	27-42	Surface fire		6 to 8	Kitzberger 1997, 2012 Veblen et al., 1992 Kitzberger 2003 Sagarzazu and Defossé 2009 Laclau 2003
			Intermittent crown fire		12	
			Continuous crown fire		>12 to 60	
Nothofagus antarctica (Forster) Oerst. woodlands (C)	9 to 27	~25	Backfire		1 to 1.5	Veblen et al., 1992 Sagarzazu and Defossé 2009 Bianchi et al., 2012
			Headfire	No continuous fuel	5 to 10	
				Continuous fuel	> 10 to 37	
Araucaria araucana (Molina) K.Koch forests (D)	-	5 - 20	-			Mundo et al., 2013
Forest – Steppe ecotone shrubland 1 to 4 m tall (E)	7.4 to 10.2		Surface fires		6 to 8	Bianchi et al. 2012 Sagarzazu and Defossé 2009
			Intermittent crown fire		12	
			Continuos crown fire		>8	
Patagonian steppe grassland with shrubs up to 2 m tall (F)	16.8± 2.6 (66% fine, 31% medium fuels)		Headfires		< 1 to 35, occasionally higher	Dentoni et al., 2001 Sagarzazu and Defossé 2009
Patagonian steppe grassland with shrubs up to 0.6 m tall (G)	2 – 4	7-15	Surface fires		10>60	Bianchi et al. 2012
Douglas-fir afforestations in the forest-steppe ecotone (H)	43.6 ± 5.4 (fine and medium fuels)	4-7	Continuous crown fire		2.5 to 30	Godoy et al. 2012 Sagarzazu and Defossé 2009

(A) Dense and humid forests, low wind effects inside the forest, deep organic layer. Fuel load does not consider tree trunks or living branches; (A1) irregular stands with considerable amounts of understory vegetation (i.e., the bamboo *Chusquea culeou*); (A2) regular stands with scarce understory vegetation. (B) Open *A. chilensis* forests exposed to winds, understory covered with shrubs and grasses, dead debris with a thin organic layer. (C) Dense and/or open woodlands, mainly composed of *N. antarctica* 3-7 m tall, highly flammable. (D) *Araucaria araucana* forests located in northern Patagonia in xeric sites. (E) Dense shrubs 1 to 4 m tall, surrounded by grasses. (F) Medium-height sparse shrubs (up to 2 m tall) surrounded by a matrix of grasses; (G) Half-shrubs and grasses up to 0.6 m tall; rate of spread highly depended on wind speed. (H) Afforestations. Similar to pine models pruned up to 2 m from the ground and thinned, with residues left on the ground. ROS varies depending on number of trees/ha and fuel type.

The interaction among these factors is complex, and could make the comprehension of further fire responses of individual plants to this changing environment very difficult. More studies will be necessary to understand how these unprecedented changes may affect survival and growth of Patagonian native plants affected by wildfires. With this in mind, it is also important to know how prescribed burns could be used as a management tool to restore damaged native ecosystems to meet ecological and societal needs. Related to Patagonian conifer afforestations, it seems that they may play an important role, not only in providing timber for Argentina but also as an important contributor to help mitigate worldwide GHG emissions. To achieve these goals, however, further research should be mainly focused at reducing fire hazard caused by mismanagement of residues and excessive understory vegetation growth.

References

Agee J.K (1993) Fire Ecology of Pacific Norhwest forests. (Island Press: Washington D.C.)

Ares J, Beeskow AM, Bertiller MB, Rostagno CM, Irisarri MD, Anchorena J, Defossé GE (1990) Structural and dynamic characteristics of overgrazed lands of Northern Patagonia, Argentina. In 'Managed Grasslands: Regional Studies' (Ed. A Breymeyer), pp. 149–175 (Elsevier Science Publishers: Amsterdam).

Barreda V, Anzótegui LM, Prieto A R, Aceñaloza P, Bianchi MM, Borromei AM, Brea M, Caccavari M, Cuadrado GA, Garralla S, Grill S, Guerstein GR, Lutz AI, Mancini MV, Mautino LR, Ottone EG, Quattocchio ME, Romero EJ, Zamaloa MC, y Zucol A (2007) Diversificación y cambios de las angiospermas durante el Neógeno en Argentina. Asociación Paleontológica Argentina. Publicación Especial 11 ISSN 0328-347X Ameghiniana 50º aniversario, pp. 173-191 (Buenos Aires, Argentina).

Bertolin ML, Urretavizcaya MF, Defossé GE (2013) Estructura y diversidad del bosque de lenga (*Nothofagus pumilio*) en tres sitios afectados por incendios en la región Central de la Patagonia Andina Argentina. In: Actas 2das Jornadas Forestales de Patagonia Sur y 2° Congreso Internacional Agroforestal Patagónico, p 49. (Calafate. Santa Cruz. Argentina).

Bianchi LO, Dentoni MC, Muñoz MM, Defossé GE (2012) Modelos de Combustibles en el Ecotono Bosque-Estepa de la Patagonia Andina: Caracterización de las asociaciones arbustivas de acuerdo a su respuesta al fuego. CIEFAP Publicación Técnica nº 37. (Buenos Aires, Argentina).

Blackhall M, Raffaele E, Veblen TT (2012) Is foliar flammability of woody species related to time since fire and herbivory in northwest Patagonia, Argentina? *Journal of Vegetation Science* 23, 931-941.

Bran DE; Cecchi GA, Gaitán JJ, Ayesa JA, López R (2007) Efecto de la severidad de quemado sobre la regeneración de la vegetación en el Monte Austral. *Ecología Austral* 17 (1), 123-131.

Buduba CG (2006) Modificaciones en el pH y contenido de materia orgánica en suelos del ecotono estepa /bosque andino patagónico por implantación de pino ponderosa. Doctoral Thesis, Escuela de Graduados Alberto Soriano de la Universidad de Buenos Aires. (Buenos Aires, Argentina).

Burns B (1993). Fire induced dynamics of *Araucaria araucana – Nothofagus antarctica* forest in the southern Andes. *Journal of Biogeography* 20, 669-685.

Byram GM (1959) Combustion of forest fuels. In 'Forest Fire: Control and Use' (Ed EK Davis) pp. 61-89 (Mc Graw-Hill: New York)

Cabrera AL (1976) Regiones Fitogcográficas Argentinas. Encic. Arg. Agric. Jard. Tomo II, (Editorial ACME: Buenos Aires, Argentina).

Cabrera A, Willinck A (1973) Biogeografía de América Latina. Monografía nº 13. (Organización de Estados Americanos: Washington).

Censo 2001. http://www.indec.gov.ar/censo2001s2/ampliada_index.asp?mode=01

Censo 2010. http//: www.censo2010.indec.gov.ar

Channing A, Zamuner AB, Zúñiga A (2007). A new Middle–Late Jurassic flora and hot spring chert deposit from the Deseado Massif, Santa Cruz province, Argentina. *Geological Magazine*, 144, 401-411.

Claraz J (1988) Viaje de exploración al Chubut 1865-1866. (Marymar Editora: Buenos Aires, Argentina).

Colmet Daage, F (1993) Principales áreas forestales de Argentina. Productividad elevada y sostenida: dos características favorables del Cono Sur. *Suelos Forestales* 10, 257-269.

Colmet Daage F, Lanciotti M, Marcolin A (1995) Importancia forestal de los suelos volcánicos de la Patagonia Norte y Central, Climo-topo-secuencias de suelos y vegetación. ORSTOM (Francia) – INTA (Argentina), Bariloche, Argentina. (Consejo Federal de Inversiones of Argentina: Buenos Aires, Argentina).

Cox G (1863) Viajes a las regiones septentrionales de Patagonia 1862-1863. *Anales de la Universidad de Chile* 23, 3-239 and 437-509.

D´Alia J (1988) Using Geographic information systems to examine fire ignition patterns and fire danger in arid and semiarid –western United States. MSc Thesis. University of Idaho, (Moscow, Idaho).

Damascos MA, Ghermandi L (2007) Morfología del arbusto Mulinum spinosum en sectores quemados de pastizales del NO de la Patagonia. *Ecología Austral* 17(1), 43-150.

Defossé GE, Rostagno C.M, del Valle HF, Dentoni MC (2003) El fuego en la porción austral de la Región del Monte. In 'Fuego en los Ecosistemas Argentinos' (Eds Kunst C, Bravo S, Panigatti JL), pp. 167-179 (INTA Santiago del Estero: Santiago del Estero, Argentina).

Defossé G.E, Loguercio G., Oddi FJ, Molina JC, Kraus PD (2011) Potential CO2 emissions mitigation through forest prescribed burning: A case study in Patagonia, Argentina. *Forest Ecology and Management* 261, 2243–2254.

Defossé GE, Sagarzazu MS, Godoy MM (2006) Incendios por rayos en los Bosques Andino Patagónicos. In: Actas de la Segunda Reunión Patagónica y Tercera Nacional sobre Ecología y Manejo del Fuego. pp. 211-216 (Esquel, Argentina).

Defossé GE; del Valle H, Bertiller M (1999) Respuesta de la vegetación al fuego en el Noreste del Chubut. Informe Final PIA – CONICET N 638/97 (Centro Nacional Patagónico Puerto Madryn, Argentina).

del Valle HF, Novara M, Rostagno CM, Defossé GE, Coronato F (2004) Cartografía de áreas quemadas con sensores remotos ópticos y de radar en ecosistemas secos del noreste de Patagonia Central. In 'Teledetección Aplicada a la Problemática Ambiental Argentina' (Eds SM Navone, HG Rosatto, F Vilella.) pp. 1-16 (CIATE Ed. Fac. Agronomía UBA: Buenos Aires, Argentina).

Dentoni MC, Cerne B (1999) La atmósfera y los incendios. Secretaría de Recursos Naturales y Desarrollo Sustentable. PAGES (Gráfica San Lorenzo: Buenos Aires. Argentina).

Dentoni MC, Defossé GE, Rodríguez NF, Muñoz MM, Colomb H (1999) Estudio de Grandes Incendios: El caso de la Ea. San Ramón en Bariloche, Río Negro – Patagonia Argentina PNMF – CIEFAP – GTZ - SAyDS (Buenos Aires, Argentina). http://www.ambiente.gov.ar/archivos/web/PNMF/file/Bibliografia/San%20Ram%C3%B3n.pdf.

Dentoni MC, Defossé GE, del Valle HF, Labraga JC (2001) Atmospheric and fuel conditions related to the Puerto Madryn fire of 21 January, 1994. *Journal of Meteorological Applications* 8(3), 361-370.

Dimitri MJ (1972) La Región de los bosques andino-patagónicos. Sinopsis general. Colección Científica del INTA. (INTA Editorial: Buenos Aires, Argentina).

Donoso ZC, Steinke L, Premoli A (2006) *Nothofagus antarctica* (G. Forster) Oerst. In 'Las especies arbóreas de los bosques templados de Chile y Argentina. Autoecología'. (Ed C Donoso Zegers), pp. 401-410 (Marisa Cuneo Ediciones: Valdivia, Chile)

Farías, CA, Defossé GE (2011). Social representations of fire risk in a wildland urban interface Proceedings 5th International Wildfire Conference, pp. 140 (Sun City, Sud Africa).

Farías CA, Defossé GE (2012) Aportes de la comunicación estratégica a la gestión del riesgo de incendios de interfase. Actas de la Reunión Eco-Sociedad 2012: Bosque, Ruralidad y Urbanismo. pp. 173 (Esquel, Argentina).

Ferrer J, Irisarri J, Mendía J (1990) Estudio regional de suelos de la provincia del Neuquén, Vol. 5, pp. 131 (Secretaría del Estado del COPADE, provincia de Neuquén – Consejo Federal de Inversiones de Argentina)

Fischer MA, Di Bella CM, Jobbágy EG (2012) Fire paterns in central semiarid Argentina. *Journal of Arid Environments* 78, 161-168.

Fitzgerald SA (2005) Fire ecology of ponderosa pine and the rebuilding of fire-resilient ponderosa pine ecosystems. In 'Proceedings of the Symposium on Ponderosa Pine: Issues, Trends, and Management' 18-21 October 2004, Klamath Falls, OR (Eds MW Ritchie, DA Maguire, A Youngblood), USDA Forest Service, Pacific Southwest Research Station. General Technical Report PSW-GTR-198, pp 197-225 (Albani, CA).

Fonk F (1896) Libro de los Diarios de Fray Francisco Menéndez. (Ed CF Niemeyer: Valparaíso, Chile).

Frost C (1998) Presettlement fire frequency regimes of the United States: a first approximation. In 'Shifting the paradigm from suppression to prescription fire in ecosystem management', Proceedings Tall Timbers Fire Ecology Conference (Eds T Prudent, L Brennan), 20, 70-81. (Tallahasee, Florida).

Fulé PZ, Crouseb J E, Roccaforteb JP, Kaliesb EL (2012) Do thinning and/or burning treatments in western USA ponderosa or Jeffrey pine-dominated forests help restore natural fire behavior? *Forest Ecology and Management* 269, 68-81.

Ghermandi L, de Torres Curth MI, Franzese J, Gonzalez S (2010) Non-linear ecological processes, fires, environmental heterogeneity and shrub invasion in northwestern Patagonia. *Ecological Modelling* 221, 113–121.

Ghermandi L (2006) Los incendios en los pastizales del noroeste de la Patagonia. Investigaciones y aplicaciones. Actas de la 2da Reunión Patagónica y 3ra Reunión Nacional Sobre Ecología y Manejo del Fuego. pp. 54-64. (Esquel, Argentina).

Ghermandi L, Gonzalez S (2009) Diversity and functional groups dynamics affected by drought and fire in Patagonia grasslands. *Ecoscience* 16, 408–417.

Ghermandi L, Guthmann N, Bran D (2004) Early postfire succession in northwestern Patagonia grasslands. *Journal of Vegetation Science* 15, 67–76.

Ghermandi L, Franzese J, Parodi J (2006) Arbustos en los pastizales del Noroeste de la Patagonia: ¿Persistencia o avance postfuego? Actas de la 2da Reunión Patagónica y 3ra Reunión Nacional sobre Ecología y Manejo del Fuego. pp. 157-160. (Esquel, Argentina).

Gittins C (2001) Respuesta post-fuego de dos coirones patagónicos: *Stipa speciosa* var major y *Festuca pallescens*. Tesis de licenciatura, Universidad Nacional del Comahue, San Carlos de Bariloche, Argentina.

Gobbi ME (1994) Regeneración de la vegetaciónen incendios recientes de bosques de ciprés de la cordillera (*Austrocedrus chilensis*) en el área del Parque Nacional Nahuel Huapi. *Medio Ambiente* 12(1), 9-15.

Godoy MM, Defossé GE, Kunst C, Bravo S, Leiva M, Ledesma R (2009) Prediction of pine plantations and/or native Patagonian and Chaco trees resistance to fire. Deliverable D3.2-6 of the Integrated project "Fire Paradox", Project no. FP6-018505, European Commission. 38 pp.

Godoy MM, Defossé GE, Thren M (2007) Especies forestales promisorias para la diversificación de las forestaciones en la Patagonia Argentina. *Bosque,* 28(1), 25-32.

Godoy MM, Defossé GE, Bianchi LO, Davel MM, Withington TE (2013) Fire-caused tree mortality in thinned Douglas-fir stands in Patagonia, Argentina. *International Journal of Wildland Fire* 22, 810-814.

Gonda H (1998) Height-diameter and volume equations, growth intercept and needle length site quality indicators, and yield equations for young ponderosa pine plantations in Neuquén, Patagonia, Argentina. Doctoral Dissertation. College of Forestry, Forest Resources Department. Oregon State University.

Gonda H, Lomagno J (1995) El raleo de plantaciones jóvenes de pino ponderosa puede producir un importante volúmen de materia prima comercial. IV Jornadas Forestales Patagónicas, Vol. I. pp. 299–305. (Asentamiento Universitario, San Martín de los Andes, Neuquén, Argentina).

González ME, Donoso Z C, Ovalle P, Martínez Pastur G (2006) *Nothofagus pumilio* (Poepp. et Endl) Krasser. In 'Las especies arbóreas de los bosques templados de Chile y Argentina. Autoecología' (Ed C Donoso Zegers), pp 486-500 (Marisa Cuneo Ediciones: Valdivia, Chile)

González S (2011). Estrategias de regeneración post-fuego en pastizales del Noroeste Patagónico: un enfoque experimental. Doctoral Thesis. Univesridad Nacional del Comahue, San Carlos de Bariloche, Argentina.

Graham RT, McCaffrey S, Jain TB (2004) Science basis for changing forest structure to modify wildfire behavior and severity. USDA Forest Service, Rocky Mountain Research Station, General Technical Report RMRS-GTR-120 (Fort Collins, CO).

Hardtke L A, del Valle HF, Sione W (2011) Spatial distribution of wildfire risk in the Monte biome ((Patagonia, Argentina). *Journal of Maps* 7(1), 588-599.

Iglesias A, Artabe E, Morel EM (2011) The evolution of Patagonian climate and vegetation from the Mesozoic to the present. *Biological Journal of the Linnean Society* 103, 409–422.

Irisarri J, Mendia JM, Roca C, Buduba C, Valenzuela F, Epele F, Fraseto F,Ostertag G, Bobadilla S, Amdenmatten E (1995). Zonificación de tierras para la forestación en la provincia del Chubut, Dirección General de Bosques y Parques de la Provincia del Chubut (Esquel, Chubut, Argentina).

Irisarri JA, Mendía JM (1991) Reconocimiento de suelos y evaluación de la aptitud forestal de la región precordillerana de la Pcia. de Río Negro. (CFI: Buenos Aires, Argentina).

Kitzberger T (2003) Regímenes de fuego en el gradiente bosque-estepa del noroeste de la Patagonia: variación espacial y tendencias temporales. In 'Fuego en los Ecosistemas Argentinos' (Eds CR Kunst, S Bravo, JL Panigatti), pp. 79-92. (INTA Santiago del Estero: Santiago del Estero, Argentina).

Kitzberger T, Raffaele E, Heinemann K, Mazzarino MJ (2005) Effects of fire severity in a north Patagonian subalpine forest. *Journal of Vegetation Science* 16, 5-12.

Kitzberger T, Veblen TT (1997) Influences of human and ENSO on fire history of *Austrocedrus chilensis* woodlands in northern Patagonia, Argentina. *Ecoscience* 4, 508-520.

Kitzberger T, Veblen TT, Villalba R (1997) Climatic influences on fire regimes along a rainforest-to-xeric woodland gradient in northern Patagonia, Argentina. *Journal of Biogeography* 23,35-47.

Kröpfl AI; Deregibus VA, Cecchi GA (2007) Disturbios en una estepa arbustiva del Monte: cambios en la vegetación. *Ecología Austral* 17(2), 257-268.

Kunst C, Rodríguez NF (1997). Fuego prescripto en plantaciones de pino. Proyecto Forestal de Desarrollo SAGPyA. (SAGPyA: Buenos Aires, Argentina).

Kunst C, Rodríguez NF, Maldonado JA (2003) Fuego prescripto en plantaciones de pino. In 'Fuego en los ecosistemas argentinos' (Eds CR Kunst, S Bravo, JL Panigatti), pp 225-237 (INTA Santiago del Estero: Santiago del Estero, Argentina).

Kunst C, Rodríguez N, Maldonado J (2002) Uso del fuego prescripto en plantaciones comerciales de pino. Informe preliminar PIA 18/97. (INTA EEA: Santiago del Estero, Argentina).

Labraga JC (1998) Escenario de cambio climático para la Argentina. *Ciencia Hoy* 8, 44 http://www.cienciahoy.org.ar/ch/hoy44/clima3.htm

Laclau P (2003) Biomass and carbon sequestration of ponderosa pine plantations and native cypress forests in northwestern Patagonia. *Forest Ecology and Management* 180, 317-333.

Lara A, Villalba R (1993) A 3620-year temperature record from *Fitzroya cupressoides* tree rings in southern South America. *Science* 260, 1104-1106.

León RJC, Bran D, Collantes M, Paruelo JM, Soriano A. (1998). Grandes unidades de vegetación de la Patagonia extra-andina. *Ecología Austral* 8, 125-144.

Loguercio GA, Dececchis, F (2006a). Forestaciones en la Patagonia Andina: potencial y desarrollo alcanzado. Parte 1. *Patagonia Forestal* 1, 4–6

Loguercio GA, Dececchis F (2006b) Forestaciones en la Patagonia Andina: potencial y desarrollo alcanzado. Parte 2. *Patagonia Forestal* 1, 7–8.

Loguercio G, Lencinas J, Antequera S (2004) Bases Estratégicas para Proyectos de Forestación en la Patagonia como un Mecanismo para un Desarrollo Limpio. Estudio de Caso en la Provincia del Chubut. Proyecto Forestal de Desarrollo. Secretaría de Agricultura, Ganadería, Pesca y Alimentación (SAGPyA-CIEFAP: Esquel, Argentina).

Loguercio GA, Jovanovsky A, Molina JC, Pantaenius P (2008) Residuos de biomasa de forestaciones y aserraderos de la región Andina de las Provincias de Chubut y Neuquén. Evaluación preliminar de oferta Centro de Investigación y Extensión Forestal Andino Patagónico (CIEFAP) and Japan International Cooperation Agency JICA, Publicación técnica 34. (CIEFAP: Esquel, Argentina).

Moreno FP (1897) Reconocimiento de la Región Andina de la República Argentina. Apuntes preliminares sobre una excursión a los territorios del Neuquén, Rio Negro, Chubut y Santa Cruz. *Revista del Museo de La Plata* 8, 1-10.

Morrison JJ (1917) La ganadería en la región de las mesetas australes del territorio de Santa Cruz. BSc Thesis. Facultad de Agronomía y Veterinaria, Universidad de Buenos Aires. Buenos Aires, Argentina.

Mundo IA, Kitzberger T, Roig Juñent FA, Villalba R., Barrera M. D. (2013) Fire history in the *Araucaria araucana* forests of Argentina: Human and climate influences. *International Journal of Wildland Fire* 22, 194–206.

Musters G (1871) At home with Patagonians: a year´s wandering over untrodden ground from the Strait of Magellan to the Rio Negro. (John Murray: London, England).

Naficy C, Sala A, Keeling EG, Graham J, DeLuca TH (2010) Interactive effects of historical logging and fire exclusion on ponderosa pine forest structure in the northern Rockies. *Ecological Applications* 20, 1851–1864.

Pafundi L, Urretavizcaya MF, Defossé GE (2014). Early response of Austrocedrus chilensis seedlings to the associated shrubs removal in NW Patagonia, Argentina: an approach to the ecological restoration of temperate forests. *Environmental Management* (*in press*).

Pausas JG, Fernández Muñoz S (2012) Fire regime changes in the Western Mediterranean Basin: from fuel-limited to drought-driven fire regime. *Climatic Change* 110(1-2), 215-226.

Pausas JG, Keeley JE (2009) A burning story. The role of fire in the history of life. *Bioscience* 59(7), 593-601.

Peterson DL, Sakectt, SS, Robinson LJ, Haase SM (1994) The effects of repeated prescribed burning on *Pinus ponderosa* growth. *International Journal of Wildland Fire* 4, 239-247.

Prichard SJ, Peterson DL, Jacobson K (2010) Fuel treatments reduce the severity of wildfire effects in dry mixed conifer forest, Washington, USA. *Canadian Journal of Forest Research* 40(8), 1615-1626.

Quinteros CP (2014) Grandes herbívoros en bosques de lenga (*Nothofagus pumilio*): uso espacio-temporal de los recursos y sus efectos sobre la regeneración y el sotobosque. Doctoral Thesis, Universidad Nacional del Comahue, Bariloche, Argentina.

Rodríguez NF (1996) Risk of mortality from wildfire in ponderosa pine *(Pinus ponderosa)* plantations in the Andean Region of Patagonia, Argentina. MSC Thesis. College of Graduate Studies, University of Idaho. (Moscow, Idaho).

Rostagno CM, Defossé GE, del Valle HF (2006) Postfire vegetation dynamics in three rangelands of northeastern Patagonia, Argentina. *Rangeland Ecology and Management* 59, 163-170.

Rothkugel M (1916) Los Bosques Patagónicos. Oficina de Bosques y Yerbales. (Talleres Gráficos del Ministerio de Agricultura de la Nación: Buenos Aires, Argentina).

Rothkugel M (1922) Los incendios en los Andes Patagónicos. Dirección General de Agricultura y Defensa Boletín nº 3 (Ministerio de Agricultura: Buenos Aires, Argentina)

SAGPyA. (1999) Argentina, oportunidades de inversión en bosques cultivados. Secretaría de Agricultura, Ganadería, Pesca y Alimentación (SAGPyA), (Buenos Aires, Argentina).

SAyDS Secretaría de Ambiente y Desarrollo Sustentable (2009) Plan nacional de manejo del fuego. In: Estadísticas., http://www.ambiente.gov.ar

Soriano A (1956) Los distritos florísticos de la Provincia Patagónica. *Revista de Investigaciones Agrícolas* 10(4), 324-346.

Thies WG, Westlin DJ, Loewen M (2005) Season of prescribed burn in ponderosa pine forests in eastern Oregon: impact on pine mortality. *International Journal of Willand Fire* 14, 223-231.

Tortorelli LA (1947) Los incendios de bosques en la Argentina. (Ministerio de Agricultura: Buenos Aires, Argentina).

Urretavizcaya MF, Defossé GE, Gonda H (2012) Effect of Sowing Season, Plant Cover, and Climatic Variability on Seedling Emergence and Survival in burned *Austrocedurs chilensis* forests. *Restoration Ecology* 20(1), 131–140.

Urretavizcaya MF, Defossé GE (2013) Effects of nurse shrubs and tree shelters on the survival and growth of two *Austrocedrus chilensis* seedling types in a forest restoration trial in semiarid Patagonia, Argentina. *Annals of Forest Science* 70, 21–30.

Urzúa JD (1991) Desarrollo de las plantaciones forestales del Neuquén y su aprovechamiento industrial. CIEFAP. Publicación Técnica 3 (Esquel, Argentina).

Veblen TT, Lorenz DC (1988) Recent vegetation changes along the forest/steppe ecotone of northern Patagonia. *Annals of the Association of American Geographers* 78(1), 93-100.

Veblen TT; Kitzberger T, Burns BR, Rebertus AJ (1996) Perturbaciones y dinámica de regeneración de bosques andinos del sur de Chile y Argentina. In 'Ecología de los bosques nativos de Chile' (Eds JJ Armesto JJ, C Villagrán, MK Arroyo), pp 169-198 (Editorial Universitaria: Santiago, Chile).

Veblen TT, Kitzberger T, Lara A (1992) Disturbance and forest dynamics along a transect from Andean rain forest to Patagonian shrubland. *Journal of Vegetation Science* 3, 507-520.

Veblen TT, Kitzberger T, Raffaele E, Lorenz D (2003) Fire history and vegetation changes in northern Patagonia, Argentina. In 'Fire and Climatic Change in temperate ecosystems of the western Americas, Ecological Studies' (Eds TT Veblen TT, W Baker, G Montenegro, TW Swetnam). pp. 265-295 (Springer Verlag: New York).

Veblen TT, Kitzberger T, Raffaele E, Mermoz M, González ME., Sibold JS, Holz A (2008) The historical range of variability of fires in the Andean-Patagonian *Nothofagus* forest region. *International Journal of Willand Fire* 17, 724-741.

Veblen TT; Kitzberger T, Villalba R, Donnegan J (1999) Fire history in Northern Patagonia: the role of humans and climatic variation. *Ecological Monographs* 69(1), 47-67.

Villagra PE, Defossé GE, del Valle HF, Tabeni S, Rostagno CM, Abraham E (2009) Land use and disturbance effects on the dynamics of natural ecosystems of the Monte Desert. Implications for their conservation, restoration and management. *Journal of Arid Environments* 73, 238-243.

Villalba R (1995) Climatic influences on forest dynamics along the forest-steppe ecotone in Northern Patagonia. PhD Thesis. University of Colorado, Department of Geography. Boulder, Colorado.

Walter H, Lieth H (1967) Klimadiagramm Weltatlas. (VEB Gustav Fisher: Jena, Germany).

Weaver H (1957) Effects of prescribed burning in ponderosa pine. *Journal of Forestry* 55(2), 133-138.

Willard EE (1973) Effect of wildfires of woody species in the Monte region of Argentina. *Journal of Range Management* 26, 97-100.

Willis B (1914) El Norte de la Patagonia. Comisión de Estudios Hidrológicos Tomo 1. (Ministerio de Obras Públicas: Buenos Aires, Argentina).

Withlock C, Bianchi MM, Bartlein PJ, Markgraf V, Marlon J, Walsh M, McCoy N (2006) Postglacial vegetation,climate, and fire history along the east side of the Andes (lat.41-42,5° S), Argentina. *Quaternary Research* 66, 187–201

Withlock C, Moreno PI, Bartlein P (2007) Climatic contrls of Holocen fire patterns in southern South America. *Quaternary Research* 68, 28-36.

Wright HA, Bailey A (1982) Fire Ecology. United States and Southern Canada. (John Wiley and Sons: New York).

In: Current International Perspectives
Editors: Brigitte Leblon and Martin E. Alexander
ISBN: 978-1-63463-682-7

Chapter 8

FIRE HISTORY, FIRE ECOLOGY AND MANAGEMENT IN THE ARGENTINE CHACO

Carlos Kunst[1,*], Roxana Ledesma[1], Sandra Bravo[2], Ada Albanesi[3], Analía Anriquez[3], Juan Silberman[3], Guillermo E. Defossé[4] and Joaquin Casillo[5]

[1]Instituto Nacional de Tecnología Agropecuaria (INTA), Santiago del Estero Agricultural Experimental Station, Santiago del Estero, Argentina
[2]Universidad Nacional de Santiago del Estero, Faculty of Forest Sciences, Botany Chair, Santiago del Estero, Argentina
[3]Universidad Nacional de Santiago del Estero, Faculty of Agronomy, Soil Microbiology Chair, Santiago del Estero, Argentina
[4]Centro de Investigación y Extensión Forestal Andino Patagónico (CIEFAP-CONICET) and Forest Department, Universidad de la Patagonia San Juan Bosco, Esquel, Chubut, Argentina
[5]Private consultant, Buenos Aires, Argentina

ABSTRACT

In this chapter, we review some facts related to the ecology, history, and management of fire in the Chaco region of Argentina, South America. The Chaco is a fire prone ecosystem due to (a) the interplay of anticyclones and a low pressure center that results in cold and dry winters as well as rainy and hot summers, and (b) vegetation prone types made of hardwood forests, savannas and grasslands, which make available both fine and heavy fuels in the same space and time. As reported by historians, travelers, and ethnographers, native people used fire extensively. The extent in which these fires shaped the vegetation is unknown, but empirical evidences, such as historical pictures and accounts, indicate that fire kept the balance between the woody and grassy plant formations of the Chaco. They also suggested that return intervals of fires were long. The native tree and shrub species of the Chaco, belonging to the *Aspidosperma*, *Schinopsis*, *Prosopis*, *Acacia,* and *Schinus* genera are fire resistant. They have thick bark and/or

* Corresponding Author address: Jujuy 850, Santiago del Estero, 4200, Argentina, Email: kunst,carlos@inta.gob.ar.

resprout from the crown, if fire kills all aboveground structure. The fire return interval of the Chaco savannas is estimated being 3-4 years, similar to those of tropical and subtropical world herbaceous formations. Average fire frequency in the savanna helps maintaining its diversity and does not negatively affect soil properties. Mean flame lengths of headfires in savannas and grasslands are about 3-4 m and are considered of high intensity. In some cases, volatile fuels may act as fuel ladders, contributing to the occurrence of crown fires. The increase of the woody component of the vegetation and the loss of suitability of the current vegetation for livestock operations have brought the widespread use of mechanical clearing treatments and seeding of exotic grasses, such as *Panicum* and *Cenchrus* species. Such changes may affect the fire regime of the Chaco, generating complex fire management situations.

Keywords: Disturbances, fire regime, Chaco, fire ecology, fire management, fire behaviour

INTRODUCTION

"Burning grasslands or the bush is a common hunting method throughout the Chaco. Even if the fire does not raise large game, it always puts to flight hundreds of small rodents at which the hunters hurl short clubs with bulging heads. The charred carcasses of animals overtaken by the fire are gathered up and eaten on the spot. Later the Indians return to the fired area to stalk the countless deer lured by the salty ashes or the thick and tender new grass. The winter hunting drives of the Bermejo River Indians also require the collaboration of many people. Two parties of about 100 men set fire to the bush along parallel lines; the animals caught between two walls of fire seek to escape at the ends, where they are met by the hunters, who kill them with spears, clubs, or arrows."

Métraux (1946)

"The smoke was the burning grassland, which the Indians set afire almost always in the place they stop...as a signal for any other tribe...and also to find, as they return to the same site, fresh and abundant grass."

Jacques (1858)

'...fire is one of the most fearsome Indian weapons. They take great care in keeping some specific sites unburned, in order to use fire for defense, or to triumph in the offensive. Although they burn in certain times for hunting... to clear their forests, or to prepare agricultural land, it is also their custom to announce themselves with a big cloud of smoke"

F. Castro-Boero, Argentine Army Captain (1873) cited by Morello (1970)

The Chaco is a vast region extending throughout western Paraguay, south-eastern Bolivia, southern Brazil, and central and northern Argentina, making it the second largest biome of South America, after Amazonia (Figure 1). The Chaco region contains the most extensive seasonally dry forest of South America (Cabrera and Willink 1976; Hannah et al., 1994; Pennington et al., 2000; Bucher and Saravia Toledo 2001).

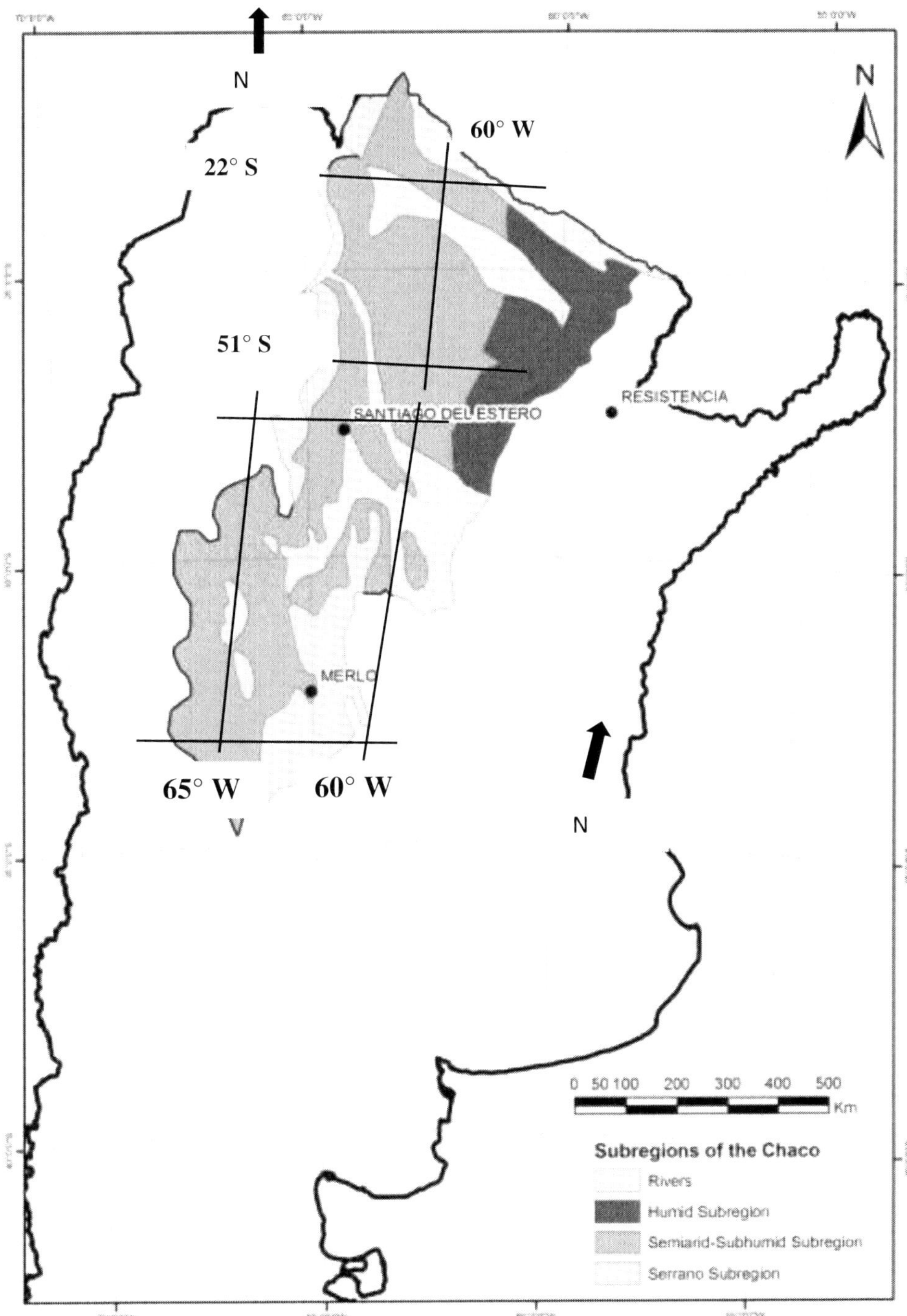

Figure 1. Chaco sub-regions: humid, semiarid-sub-humid, and 'Serrano' (modified from Cabrera 1976) and location of Resistencia, Santiago del Estero, and Merlo. 'Rivers' indicates soil and vegetation types recently influenced by geomorphic processes produced by rivers.

In Argentina, Chaco is bordered to the north by the Pilcomayo River, to the southwest and east by the Monte and Espinal regions, and to the northwest by the Yungas, which are arid, humid, and tropical biomes, respectively (Figure 1). The region comprises an alluvial plain, but there are low to mid altitude 'sierras' (1000-3000 m above sea level) in the west and south, forming a specific sub-region called the 'Chaco Serrano'.

Fire has been recognized as an ecological factor in the Chaco since the 50' (Papadakis 1951; Morello and Saravia Toledo 1959). However, fire ecology studies in Chaco began just recently, including research on species chemical composition, flammability, fire origin, fire effects on growth of some species, and burning season (Tortorelli 1947; Burkart 1969; Bordón 1993; Giménez and Moglia 1993a).

In this chapter, we review the fire history, the fire ecology and fire management in Chaco. Fire research related to pasture and livestock management will be also analyzed and discussed. Such research is off importance as there was recently an increasing fire frequency in the sierras and ecotonal grasslands located in the Chaco 'Serrano'.

Ecosystem Features and Vegetation Types of the Chaco

Climate of the Chaco ranges from humid to arid, being strongly seasonal. Winters are dry and cold with mean air temperature of 14 ºC in July, while summers are hot and rainy with a mean air temperature of 28 ºC in January. The most distinctive feature of the climate is the rainfall pattern, varying within and between years. Since it rains only during the summer (Figure 2), the climate of Chaco is described as 'monzonic'. Rainfall in Chaco is controlled both by a low air pressure center located in northwest Argentina (Ferreira et al., 2004) and by the relative positions of anticyclones located in the southern Atlantic and Pacific oceans, respectively. The low pressure system is generated by the heating of the Earth surface by sun radiation, being more stable during summer than in winter (Ferreira et al., 2004). In the same time, both anticyclones are located in their most southern position. As a result, humid air masses generated by the southern Pacific and Atlantic anticyclones are attracted in the summer to the Chaco, causing rainfall. In contrast, during winter, the low air pressure center is less stable and the anticyclones are located toward the north, near the continental mass, precluding the entering of humid air masses and generating dry and cold winters in the region.

Toward the east, soils of the Argentine Chaco are formed by sediments originated in the Tertiary and Quaternary periods, which are partially mixed with volcanic ashes or other materials of aeolic origin, and classified as entisols and aridisols. In the western part of Chaco, there are more developed soils (mollisols, argiustols, and haplustols) because of high water availability. In dry environments such as in the southeast, haplic xerosols dominate (Morello et al., 2012).

The fire-prone environment of the Chaco is caused by several concurrent factors: (a) the seasonal pattern of rainfall and air temperature, (b) the annual amount of rainfall, and (c) soil fertility. All these factors generate weather conditions suitable for burning for several months and various vegetation types that produce significant available amounts of fine and coarse fuels throughout the region.

At a regional level, Chaco has been traditionally divided in three phyto-geographical sub-regions: humid in the east, sub-humid-semiarid in the center, and arid in the southwest. On a more detailed level, the vegetation of the region as a whole is a mosaic of forests, woodlands, savannas, and shrublands (Morello and Saravia Toledo 1959; Bucher 1982).

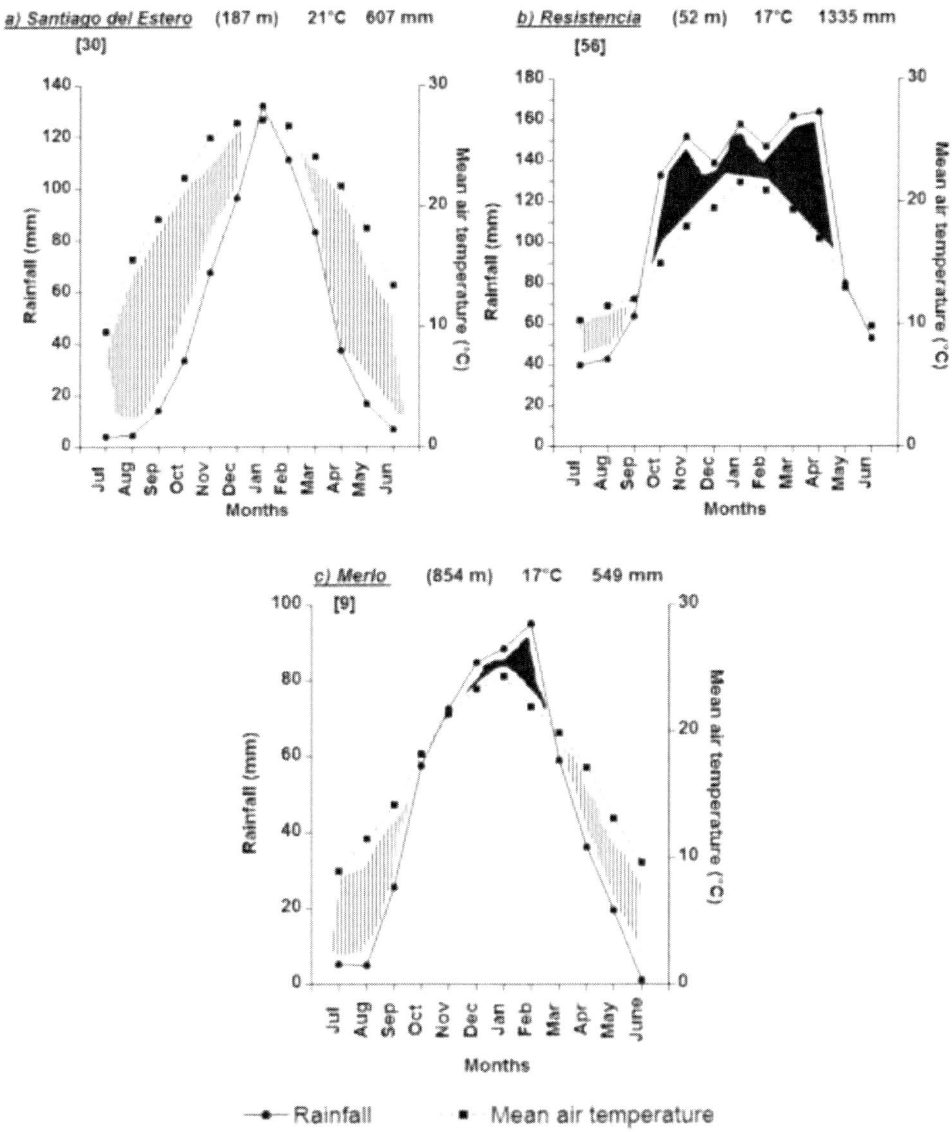

Figure 2. Climodiagrams of three towns ((A) Resistencia, (B) Santiago del Estero, and (C) Merlo) located in the Chaco region.

At a scale of less than 1:20.000, soils and vegetation in the region are distributed along a topographic gradient (known in soil science as a 'catena') from the upland, with soils of coarse texture; to the lowland, with soils of finer texture and greater development than the former. Hardwood forests are usually located in the uplands, while savannas are located in the lowlands (Morello and Saravia Toledo 1959; Kunst et al., 2006) (Figure 3). 'Parques' (parks) are a vegetation type consisting of grasslands with scattered tree and shrub groves that are located in the midlands (Adámoli et al., 1972; Morello and Adámoli 1974) (Figure 3). In the humid sub-region, lowlands may have intermittent/permanent swamps, creeks or lagoons in the lowest level. This array of vegetation and soils types could be described by the following 'ecological sites': upland, midland, and lowland, respectively. They are characteristic for the whole Chaco region (Kunst et al., 2006). Ecological sites have different soil types and drainage capacity related to the geomorphological processes associated with river levees and flats (Bucher 1982).

Figure 3. An oblique aerial photography of the Chaco showing ecological sites and associated vegetation types at a cartographic scale of approximately 1:10,000 and corresponding ground views.

Hardwood forests of Chaco are dominated by *Aspidosperma quebracho blanco* Schlecht[1] and *Schinopsis lorentzii* Griseb. in the sub-humid-semiarid sub-region, while *S. balansae* is dominant in the humid sub-region. *A. quebracho blanco* is an evergreen hardwood species that may reach up to 20 m height and 80 cm diameter at breast height (DBH) (Tortorelli 2009). *S. lorentzii* and *S. balansae* are hardwood deciduous species that shed their leaves in early-spring (October), reaching a height of 25 m and a DBH of 1.50 m. *Prosopis nigra* (Griseb.) Hieron., *P. alba* Griseb., *P. kuntzei* Harms., and *Zyziphus mistol* Griseb. emerge in the midland and upland sites. A conspicuous feature of Chaco tree species is their high wood density, ranging from 0.85 to 1.16 kg. dm^{-3} (Tortorelli 2009). The shrub species *Acacia aroma* Hook et Arn. and *Schinus spp.* are common in lowland sites, whereas *A. gilliesii* Steud., and *Celtis ehrenbergiana* (Klotzsch) Liebm. are common in upland sites. Morello and Saravia Toledo (1959) described the 'quemados' ('burns'), which are grasslands developing in forest sites after a fire event, as having a different botanical composition compared to savannas. Grass species in the 'burns' of the upland and midland sites belong to the *Trichloris*, *Setaria*, *Digitaria*, *Chloris*, and *Gouinia* genera. All aboveground structures of the

[1] Plant nomenclature follows the guidelines of Instituto de Botánica Darwinion. 2011. Catálogo de Plantas Vasculares de la República Argentina. [www.darwin.edu.ar/Proyectos/FloraArgentina]

native woody vegetation are eliminated and replaced immediately after fire by a homogeneous herbaceous stratum. In the savannas of semiarid-sub-humid sub-region, the dominant grass species is *Elionorus muticus* (Spreng.) Kuntze. It is a bunchgrass of 0.5 m tall containing citral in the foliage, making it highly flammable (Burkart 1969): Other species of the savanna are *Pappophorum pappipherum* Lam. Kuntze, *Heteropogon contortus* (L.) P. Beauv. Ex. Roem.and Schult., *Schizachyrium tenerum* Nees, *Botriochloa spp*, *Aristida mendocina* Phil., and several herbaceous legumes species.

Fire Regime of the Chaco

A fire regime is a description of the role that fire plays in an ecosystem. It may include a description of the fire frequency, fire intensity, the nature of the fire disturbance, and effects of fire on vegetation (Agee 1993).

Fire History

Fire was, and still is, a frequent event in Chaco (Lorea and Kunst 2011). Native peoples used it extensively for war, hunting, and agricultural purposes, as recorded in travelers and soldiers' reports (Métraux 1946; Morello and Saravia Toledo 1959). Other reports suggested that the vegetation types dominated by 'fire prone' grasses were more frequent in old times, since travelers such as Lizárraga (1530) recorded abundance of ostridges (*Rhea americana* L.) and guanacos (*Lama guanicoe* M.), both wildlife species associated to grassland habitats.

Hieronymus (1874), a naturalist contracted by the Argentine government, was one of the first to mention the use of fire for livestock raising and pasture management. He described not only the use of fire, but also large savannas and grasslands along the route from Santiago del Estero to Tucuman (Figure 4), where today only exist agricultural fields, secondary forests, and low height shrub vegetation types. Fire was used to create green forage during early spring. As a naturalist, he condemned the use of fire for several reasons, the main being the loss of valuable grass species due to a negative interaction with grazing and a loss of soil nutrients. Such negative view was also shared by Tortorelli (1947) who condemned the use of fire in livestock and agricultural operations, because of the potential damage to native forests. However, Papadakis (1951), an ecologist working for the Argentine Ministry of Agriculture, recommended the use of fire in Chaco savannas for brush control. Later, Morello and Saravia Toledo (1959), Morello and Adamoli (1974), and Morello et al., (2012) recognized fire as an ecological factor helping in maintaining the balance between tree/shrub and grass dominated vegetation types. They also recommended research in the use of fire as a tool for management of native vegetation.

Fire Frequency

Fire is a disturbance that eliminates biomass. As a disturbance, it should be characterized by its intensity, severity, frequency, return times, and other attributes.

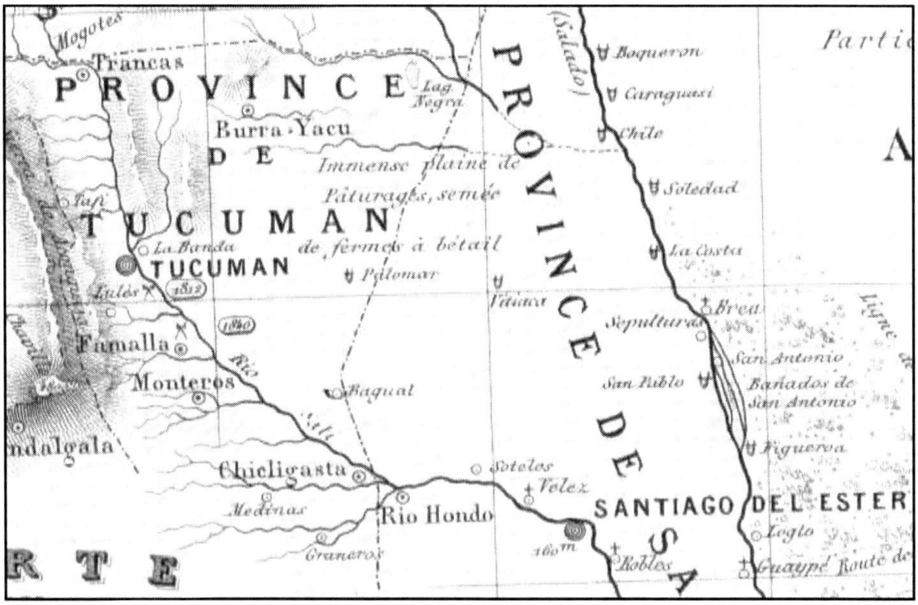

Figure 4. A map of the semiarid-subhumid Chaco around 1870 with some vegetation, types, according to Moussy (1873). The map showed that between Santiago del Estero and Tucumán (center) there was a huge plain, mainly composed of pastures and cattle ranches. Today, there are brush thickets and secondary forests.

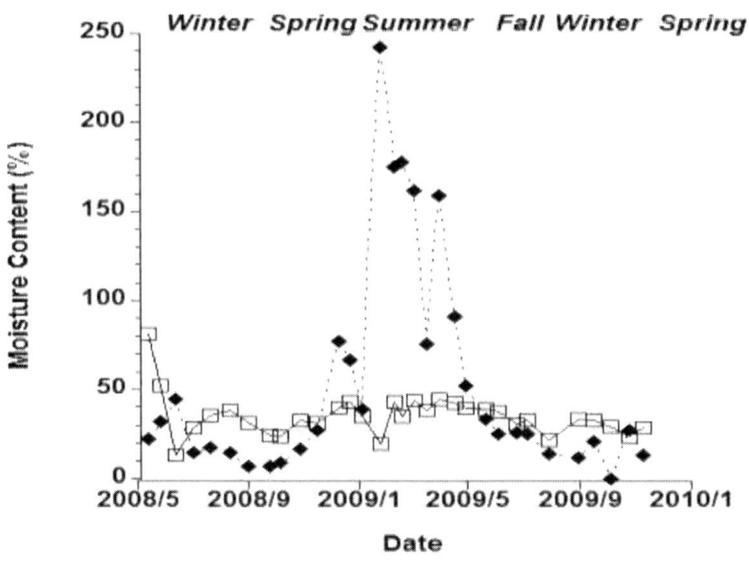

Figure 5. Seasonal dynamics of moisture content in two representative species of Chaco: twigs and leaves of *Schinus spp*, and leaves and stems of *Pappophorum pappipherum* L. Kuntze. Midland ecological site '*La Maria' Experimental Ranch. INTA EEA Santiago del Estero* (adapted from Kunst et al. 2014a and b).

The fire season in Chaco starts with the first frosts, usually in June, and may last until the first rains, that may fall from September to December due the annual climatic variability. Fuel desiccation results from a combination of decreasing soil moisture and freezing air temperature (Figure 5). Fine fuels accumulate in savannas and grasslands during summer and fall, reaching appropriate burning conditions during late winter and early spring. Drying effects of northern winds during spring also enhance fuel desiccation, increasing ignition probability. Atmospheric conditions during late winter and early spring include steady winds from the north/north-east (having mean velocity of 20-40 km*h^{-1}), air temperatures up to 35-40 °C, and relative air humidity as low as 20% at noon and early afternoon (Bordón 1993, Bravo et al., 2001a) These conditions are used for setting the fires, generally for shrub clearing and control, and in order to promote new grass growth. Fires lit in the savanna borders spread quickly into this vegetation type, killing all aboveground shrub biomass within the savanna and in the midland parks. Fire rarely propagates into the hardwood forests, although few crown fires have been reported in the literature (Morello and Adamoli 1974), suggesting that savannas and forests have different fire regimes. Wood and leaves of native Chaco species are rich in terpenes and other volatile compounds. Presence of these compounds may be interpreted as an herbivore repellent and/or as a drought adaptation. However, they deeply increase vegetation flammability, and thus influence fire ecology.

The two main vegetation types of Chaco, savanna and hardwood forests seem to have different fire regimes as deduced by their dominant growth forms. There is no information about the fire regime of the forests. Empirical evidences, such as oral communication from rural inhabitants and studies about fire resistance in tree species of the *Prosopis* genus reported fire scars up to 7 m in height in crown and branches, suggesting that crown fires can occur under extreme weather conditions (Bravo et al., 2001b; Bravo et al., 2008). Grass species of the forests quickly recover after fires, and in the savannas, fire is associated to an increase in herbaceous diversity in the first 3- 6 years after a fire event (Kunst et al., 2003).

In savannas, Bravo et al., (2001a) used dendrochronological techniques to find that the mean fire return interval (~fire frequency) of an *E. muticus* savanna is 3-3.5 years, which is similar to other world savannas and grasslands. This return time is associated to rainfall pattern, but there is a time delay of around 2-3 years between above average rainfall and fire events (Bravo et al., 2010).

Fire Behavior

Fine fuel loads (standing above ground biomass) of parklands and savannas in good range condition may vary between 2000-6000 kg DM*ha^{-1} (Figure 6). This fuel buildup could be achieved in a relatively short time (one wet season), if rainfall is average or above mean values and the pasture is deferred from grazing.

Several factors may affect fire behavior in Chaco:

- Weather, as described in earlier paragraphs.
- Abundance of grass-dominated plant formations. This 'open' plant formations are the parklands and savannas, located in the midland and lowland ecological sites, respectively.

- The foliage of tree and shrub species are rich in aromatic compounds such as essential oils and resines making them quite volatile fuels (Nuñez et al., 2006, Moriconi et al., 2009). This may therefore contribute to fire intensity and work as fuel ladders. Individuals of the *Acacia, Prosopis, Schinus, Lippia, Aloyssia, Geoffroea* and *Larrea* genera are interspersed in savannas and midlands, providing a quite flammable fuel despite relatively low fuel loads compared with the grass component.
- Relative topographic position of savannas, surrounded by the hardwood forests. This landscape feature deeply affects fire behavior, especially in the savannas, where winds blow without almost any obstacles and even accelerates near the ground, as forests act as funnels.

a) Grassland in a midland ecological site

b) Pastures created by severe roller-chopping

Figure 6. Rangelands and pastures of the Chaco region corresponding to: (a) grassland in a midland ecological site and (b) Pastures created by severe roller-chopping and seeding of *Panicum*.

While *E. muticus* and *P. pappipherum*, the two dominant species in the savannas and parklands, are bunchgrasses, *Trichloris pluriflora* E. Fourn, *Digitaria californica* (Benth.) Henrard, and other grass species present in forests and 'burns' are erect, and stems predominate over leaves. Flame lengths and fire scars in woody species observed in experimental and actual wildfires vary between 1- 7 m depending on the available fuel amount, vegetation type and ignition technique. Fires could be classified as being of high intensity (Figures 7 and 8). Therefore, fire control should be indirect and blacklines wider than 30 m should be used (Wright and Bailey 1982). Air temperature profiles in fires occurring in fine fuels of midland grasslands and lowland savannas have been reported in the literature for Chaco, and are quite similar to other ecosystems reported around the world (Casillo et al., 2011; Ledesma et al., 2011).

Currently, fine fuel loads in Chaco forests are usually low, due to the lack of sunlight reaching the soil surface and livestock overgrazing. However, when grazing is impeded, a ground cover of moss develops, propagating surface fires that may climb to the trees through their bark and start a crown fire (Kunst, personal observation).

Figure 7. Burning blacklines in a savanna and in a grassland of the Chaco region.

a) Grassland in a midland ecological site

b) Savanna in a lowland site

Figure 8. Headfires in the Chaco: (a) grassland in a midland ecological site and (b) savanna in a lowland site.

Fire Effects on Chaco Soils

Fire is both a limiting and a regulating factor for the growth of plant communities, animals, and soil biota, particularly for the species that are 'fire species', i.e., those whose selective pressure and source of variability is precisely fire. Fire acts like a biological decomposer of accumulated and dry fallen leaves, over which the microbial population cannot work or would work very slowly. In the Chaco region, *E. muticus* responds to this pattern, being the dominant grass in the savannas.

Several factors of the ecosystem, such as the amount of fuel and soil-related factors, like water-air relationship, soil texture, structure, nutrient content, and organic matter quality, recovery rates, etc., make more difficult the analysis of cause-effect relationships in Chaco than in other ecosystems. These factors modify the heat transmission into the soil, thus it is not easy to relate fire intensity and severity with the reached temperature.

High severity fires that consume all fuels produce negative effects on the soil. In Chaco, the formation of 'hot spots' is scarce in savannas and grasslands because of the total lack and/or paucity of heavy fuels. However, hot spots are more frequent in forests and woodlands because of high density of trunks, branches, and dry trees. The same happens in locations where severe roller-chopping of the current vegetation has been applied.

Ecological site influences fire effects. In the lowland, soil organic matter decreases through the action of prescribed fire, if grass species composed of fragile materials, such as annuals, predominate. The size distribution of soil aggregates also varies because the decrease in soil carbon associated to the macro and micro-aggregates (Montenegro 2002). However, in a lowland site subjected to a low intensity roller chopping treatment, the content of total organic carbon was not modified when prescribed fire was applied after two seasons following the roller-chopping (Albanesi et al., 2001; Anriquez et al., 2005). Soil total nitrogen content (N_t) decreased immediately after the fire, because a large quantity of the dead biomass is volatilized generally as N_2 in the savanna and/or bushland fires. Additionally, the increase of the post-fire soil temperature accelerates the microbial nitrification. Nevertheless, the total N is returned in similar quantities as unburned controls two seasons after the fire. As a consequence of nitrogen loss caused by fire, the C/N ratio in the lowlands rises immediately after burning (Albanesi et al., 2001; Anriquez et al., 2005).

Prescribed fire does not affect the other soil variables, such as moisture, bulk density, soil respiration, particulate carbon, and C associated to silt and clay particles (Anríquez et al., 2005). Soil total cation content nor interchangeable cation content are also not influenced by fire. Incomplete combustion does not modify ionic relationships, not altering soil pH (Abril and Gonzalez 2000). Gonzalez et al., (2002) have reported that savannas burned at their 'average' fire return interval presented the same nitrogen and carbon status as savannas burned at 10-11 year intervals. However, savannas burned every year had their carbon and nitrogen pools depleted, with a decrease in the total and organic soil carbon and nitrogen content, in the carbon and nitrogen content associated to soil macro and micro aggregates, and in the microbial biomass nitrogen. These results demonstrated that burn frequency above 'average' return intervals reduces organic matter as well as soil biological activity (González et al., 2002)

In semiarid and arid Chaco sub-regions, the pulse of mineralization of the organic matter accumulation is slow because of the low productivity of the plant community. Any disturbance tending to diminish the humus content is dangerous for soil stability. This is the risk of using prescribed fire to reduce woody residues after roller chopping. Prescribed fire must be managed in very controlled conditions and should act over the biomass and aerial dead mass, but not over organic horizons nor organic-mineral horizons, in order to enrich the soil with organic matter and to make the ashes being a source of fertility but not of degradation (Ibáñez et al., 1983),

The NO_3 contents in the soil increases significantly after burning cover of three representative shrub species of the Chaco region (*Schinus molle* L., *A. gilliesi*, and *C. eherenbergiana*) (Albanesi et al., 2013). After a fire, N availability can increase due to the high rates of mineralization (Dooley et al., 2012). This increase has been documented in several ecosystems (Shenoy et al., 2013). In turn, fire severity depends on the species. Weber et al., (2014) reported N post fire's level changes depending on plant populations. Soil temperature and fuel load are two factors that control the N mineralization behaviour. It has been reported that in low fine fuel loads ($\leq$ 5000 kg DM. ha^{-1}), the soil reaches low temperatures and the liberation of N-NO_3 is high. This behaviour is contrary to Weber et al., (2014) who reported that, depending on vegetation, the level of NO_3 rises 30 times, on average, as a function of the intensity of fire.

On degraded forests of the Chaco Region, it is very unlikely that return N exists in low intensity burns, but a decrease in N during the following growing season has been observed (Abril and González, 2000; González et al., 2002).

Dead macro-organisms are source of mineralization from the most unstable organic forms, supplying available nutrients to the plant growing cycle. In the short term, dead micro-organisms are returned to the cycle in a short time after the burns at the expense of the fractions of more rapid mineralization (Abril and Gonzalez 2000) This changes the rates and forms of more rapid mineralization, generating a different quality of organic matter and rates of return. In general, rates of mineralization and the unstable forms of more rapid mineralization rise in the short term and then decrease (Fernández et al.,1999)

After the burning, micro-organisms fixing nitrogen in the soil can also be favoured, most of all those which constitute cryptogamic crusts (Loftin and White 1994; Evans and Johansen 1999), like those observed in the Chaco Region (Albanesi, unpublished data).

In the Chaco Region, as in other regions of the world, severe fires deeply affect the soil. Its structure and hydrophysical properties deteriorate and the status of nutrients is affected in such a way that the growth of post fire vegetation is harmed, making possible its degradation and/or erosion. In less severe fires, the supply of ashes or uncompletely carbonized matter improves the aforementioned properties, generating the inverse effect on post fire vegetation.

Fire Effects on Vegetation

The main effect of fire on local vegetation is to keep the balance between the woody and grassy components, as in other similar world ecosystems (Morello and Saravia Toledo 1959). Casillo et al., (2011) studied the effects of fire in two ‘bottlenecks’ of pasture management: the germination and recruitment of grasses and shrubs and its interaction with water availability (rainfall above historical average). Their conclusion was that fire and water availability stimulate grasses rather than shrub and broadleaf plants, therefore perpetuating the ‘grass state’ over time.

Studies about fire tolerance of the native hardwood tree and shrub species composing the medium and high strata of the forests and savannas suggest high fire resistance and fire tolerance for both species types (Bravo et al., 2001b; 2006; 2008; 2010a). They are related to traits such as bark thickness, bud features, resprouting pattern, and phenology (Bravo et al., 2001b; 2006; 2008; 2010a). The dominant species of the upper stratum of Chaco forests, *A. quebracho-blanco* and *S. lorentzii* are remarkably resistant to fire at the mature state and reestablish damaged aerial structures by profuse resprouting (basal and epicormic). Both species have thick bark, a trait allowing them to effectively protect the cambium tissue of the bole, branches, and buds. This great ability to resprout after fires produces some architectural changes, such as an increase of vegetation cover and multi-stemmed forms, mainly on sapling and pole-sized trees (Bravo et al., 2014).

Fire usually eliminates aerial structures of shrub species, such as *A. aroma*, *A. gilliessi*, *Capparis atamisquea* Kuntze, *Celtis spp.*, and *Schinus spp.*, because of their lower plant height and thinner bark thickness than tree species. Shrub species recover mainly from basal resprouts, and vegetation cover could even increase after a burn, depending on fire severity, aerial bud bank, and reserve levels (Kunst et al., 2000; 2008). Secondary tree species of Chaco forests, such as *P. nigra*, *P. alba*, *Z. mistol*, *Geoffroea decorticans* (Gillies ex Hook. &Arn.) Burkart, and *Cercidium praecox* (Ruiz & Pav. Ex. Hook) Harms. have different fire resistance according to different combinations of bark thickness and density, which allow keeping alive their main bole and buds. Although the main bark feature related to fire resistance is thickness, a high bark density may contribute to delay the time to reach ignition (Bravo et al., 2014). Bark protection to buds could be insufficient in high severity fires. This has been observed in *P. nigra* and *Z. mistol*. Other thin bark species, like *G. decorticans* and *C. praecox,* continuously exfoliate bark tissues, thus becoming very susceptible even in low intensity fires. High fire intensity and high fire frequency could overwhelm the bark protection, causing loss of epicormic buds of stem and branches. Even in these cases, the species show high ability to differentiate adventitious buds from the crown zone, producing vigorous basal resprouts (Kunst et al., 2000; 2008).

Stress caused by fire usually adds to the climatic dryness and coldness, which are typical of fall and winter in Chaco region, and this synergism may influence also fire tolerance of native vegetation. Mortality after medium to high intensity prescribed fires reached up to 10% of individuals (Bravo et al., 2014) and this magnitude could diminish if fire occurs during late fire season (end of spring and early summer), when better environmental conditions promote the fast recovery of vegetation. The effect of fires on the sexual propagation and dispersion of native woody species is an interesting future research because it will allow understanding its relationship with vegetation dynamics in spatial and temporal scales.

PRESCRIBED FIRE IN THE CHACO

Many farmers and cattlemen use fire as a complementary practice in pasture management and many of them do it through 'prescribed fire'. Fire is usually applied every two years, a frequency that could be considered as high, since it is above the world range of frequency proposed for savannas and grasslands. An 'appropriate' fire frequency would be 4 to 8 years for the arid sub-region, and 3 years, for the humid-semiarid sub-region (Gonzalez et al.,

2002). Farmers and cattlemen are aware of this excessive frequency and ask for more training in prescribed fire, and range and pasture management (Feuillade et al., 2011a; 2011b).

From a cattlemen's point of view, the main objectives for burning pastures in the region are twofold: to eliminate poor quality old growth and biomass, so 'creating forage', and to control 'undesirable' vegetation that hinders and compete with 'desirable' vegetation. Fire is also used to eliminate woody residues left by reclamation treatments for developing agricultural systems, although the use of fire for this purpose is banned by provincial laws.

Kunst et al., (2000) and Casillo et al., (2006a) have studied the effect of fire on shrub populations, especially in *A. aroma* and *C. ehrenbergiana,* considering the timing of fire as a function of the phenological status of the species. Fires at the early and mid-winter season kill the above ground structure of shrubs, but shrubs easily regenerate because they are at a dormant stage (Figure 9). So, fires only control the shrub population, but do not eradicate it. Fires at the early spring season are more effective, since target plants receive the treatment when all reserves are up in their aerial structure. So, fires should be planned late in the fire season in order to achieve maximum killings.

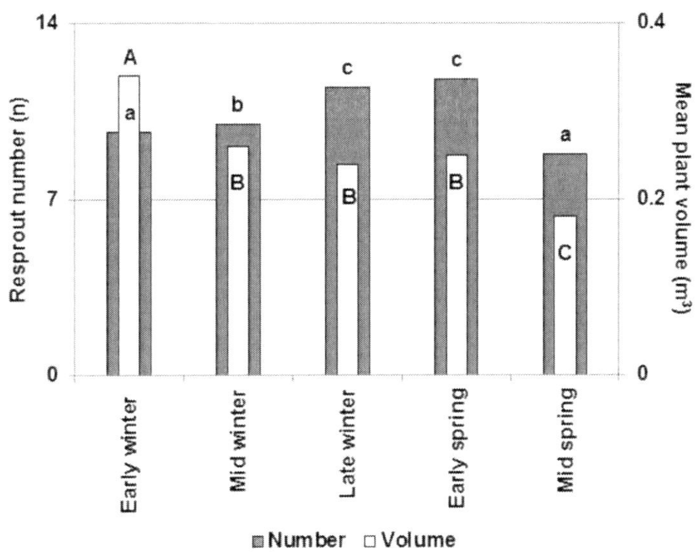

Figure 9. Effect of fire timing on mean plant volume and mean resprout number per plant of *Acacia aroma* Hook et Arn. Bars with different letters are significantly different at $\alpha = 0.05$, Duncan test. (Lowland ecological site; 'La Maria' Experimental Ranch. INTA EEA Santiago del Estero, from Kunst et al., 2000). The mean values are computed over a period of 3 years after the burn.

Kunst et al., (2000) reported that fire controls *A. aroma* for up to 3-4 years after the burn. On the other hand, burned savannas increases their diversity up to 3-4 years after fire, because of the appearance of broadleaf species (especially herbaceous legumes) and of short grass species with high nutritive value (Kunst et al., 2003). Cornacchione et al., (2001) reported that the nutritive quality of roller chopped and burned pastures was higher than unburned pastures (18 % versus 8 % mean crude protein). The improvement occurred early in the spring season, which clearly benefits livestock since pastures in late winter and early spring lack of forage quality.

Research conducted by Navall et al., (2005) and Bravo et al., (2009) reported that fire do not eradicate the population of tree species, only 'controls' them by killing their aboveground structure. Plants easily regenerate from buds located in the crown branches that were not affected by fire.

Nowadays, the Chaco vegetation is dominated by a homogeneous, almost impenetrable woody strata dominated by shrub species. The shrub growth form is considered as an evolutionary answer to the complex interaction of factors generating stress in the region, such as herbivory, seasonal drought, and fire (Morello et al., 2012). These changes in landscape and vegetation physiognomy have been caused by livestock overgrazing, indiscriminate logging, a decrease of fire frequency, and fencing (Adamoli et al., 1972; Kunst et al., 2006). Lack of fire caused an increase in the dominance of woody species changing the fuel complex. The increase of heavy fuels over fine fuels has created complex management situations. The spatial distribution of these fuels has been studied (Kunst et al., 2006a; 2006b, Rueda et al., 2011).

The conversion of a woodland, shrub thicket or a dense secondary forest in a vegetation type that is better fitted for livestock operations is done in the Chaco by 'roller chopping' (Kunst et al., 2013). This mechanical treatment crushes the woody undergrowth and shrub thickets, creating a park-like structure and enhancing accessibility of both livestock and personnel. Trees with DBH larger than 10-15 cm and tall shrubs are left standing, so a roller chopped paddock has reduced sunlight availability by comparison to treeless pastures. Standing forage and stocking rates are increased by the sowing of grass simultaneously with the roller chopping treatment (Kunst et al., 2013). Species of the *Panicum* and *Cenchrus* genera are commonly used. Introduction of these species may change the fire regime of Chaco, since they tend to replace native species and usually have more standing biomass (Kunst et al., 2014). Woody residues also create a complex problem, since fire is usually used to dispose them. Woody residues may represent a fuel load several times higher than fine fuel loads (Kunst et al., 2006).

A mechanical treatment followed by fire usually keeps the pastures 'clean' of 'undesirable' vegetation (Kunst et al., 2013). However, this result conflicts with a lack of clear definition of 'undesirable' species, and the potential change of the fire regime of the Chaco. These facts bring the discussion to an important question: how to get rid of the woody residues? There are highly valuable native tree species in the Chaco, such as *S. lorentzii*, *A. quebracho blanco*, etc. that are source of wood, fire wood, and other goods. The burning of large piles of wood also creates concerns about fire effects on soils and the untreated nearby vegetation. Studies on the use of fire in mechanically treated areas with respect to coarse fuel loads and dynamics of their moisture content recommended that mechanical disturbances should be of medium to low intensity (Kunst et al., 2006a; 2006b). Such a disturbance generates low coarse fuel loads that are quickly degraded naturally, therefore fire may not be mandatorily used as a complement.

FIRE IN THE CHACO SERRANO

The 'Chaco Serrano' covers a long and narrow belt along the western side of the Chaco, from the 23° to the 33° South Parallels in Central Argentina (Figure 1). In the Serrano district, the vegetation occupies altitudinal zones leading to a succession of semi-deciduous woodlands, scrub and tussock grassland with the altitude (Zak et al., 2004). Vegetation of the southern part of this district has been described by Zak and Cabido (2002) and Zak et al., (2004).

The 'Serrano' sub-region in general, and its southernmost part in particular, comprises original lowland, mountain dry forests, and woodlands. Since the beginning of the twentieth century to the 1960s and early years of the 2000s, the area was subjected to diverse land-use changes, similar as in the whole Chaco region, which include land clearing for agriculture, increasing areas for pasture, and development of incipient afforestation projects based on *Pinus* species (Zak et al., 2004). Natural fire regimes that once modelled the Chaco Serrano landscapes for centuries were then disrupted. The native vegetation became fragmented and went through different processes of secondary succession (Zak et al., 2004).

During the last decade of the nineteen century, the depopulation of rural areas and consequent migration to cities was another factor that affected the way the native vegetation of this district was managed (Díaz et al., 1987; Zak et al., 2004). The phenomenon of repopulation of rural areas around small towns, defined in English fire science as "*Wildland Urban Interface areas*" (WUI) are now occurring worldwide, also impacting the Chaco Serrano and changing its former landscapes. One of the first impacts is produced by the expansion of town boundaries through different forms of urbanizations of the WUI. Many WUI residents may use their dwellings either for permanent housing or recreational activities during weekends, trying to find new lifestyles in those environments (Pausas and Keeley 2009; Farías and Defossé 2011). It is interesting to analyze the rate of population growth in one representative village located in the Chaco Serrano, district and compare the results to those at either provincial or national levels. The village of Merlo (San Luis, Argentina) had 11,159 permanent residents in 2001, and 17,049 in 2010, an increase of 52.7% of its population over the 9 year period. During the same period, the Province of San Luis, where Merlo is located, grew 18%, while nationwide in Argentina, the population grew only by 10.6% (Dirección Nacional de Estadísticas y Censos, Censuses 2001 and 2010). The high percentage of population growth rates in Merlo is almost 5 times higher than the national growth rate. This was also the case in most of the small towns located at both sides of the sierra range that crossed the southernmost part of the Chaco Serrano district.

Related to fire and fire management, the problem faced by these high growing WUI areas is similar to what is currently occurring in other WUI areas of the world (Western North America, Australia, Chile, or the Mediterranean basin countries). Rural population leaves these areas and many pastoralism activities ceased. As it happened in other areas of the world (Pausas and Fernández-Muñoz 2012), lack of grazing allows for substantial increases of biomass once used to feed animals, and the combination of high fuel availability and more ignition sources (arson or accidental) are exposing more people to wildfires in these environments (Pausas and Keeley 2009). In fact, the Chaco Serrano is prone to wildfire during the long winter dry season. The lack of an integrated fire management policy has led to devastating fires during the winter fire season of 2013 in some parts of the Serrano of the province of Córdoba and severely affected some of its WUI towns. The law in this province forbids the use of fire for any rural activity and its consequences are proven to be catastrophic. The excessive accumulation of dead biomass makes that, given an ignition source, thousands of hectares burn in hours, affecting homes, buildings, and natural and cultivated vegetation. Changes in the law for allowing prescribed fires and other silvicultural management options for reducing biomass growth, without allocating more material and human resources for fighting fires, seems to be the way to reduce future risks and negative effects of wildfires in the Chaco Serrano of Argentina.

Conclusion and Management Implications

Chaco is a fire prone region due to weather and vegetation features. Native peoples used fire extensively for several purposes. The extent in which these fires shaped the Chaco vegetation is unknown, although the main effect of fire was to maintain the balance between the grass and woody plant formations, i.e., the savannas and forests. These two physiognomies are located in different ecological sites along a soil catena, from the lowland to the upland, respectively. It is likely that the decrease in fire frequency caused by overgrazing of savannas and grasslands may have caused the increase of the woody component of the vegetation, causing at the same time a change in the fire ecology of the Chaco as a whole.

The fire ecology and fire regime of the savannas have been more studied than those of the forests, suggesting a shorter fire return in the former than in the latter, i.e., 3 to 4 years depending on the rainfall pattern. Species diversity of the savannas is directly related to fire frequency. If the fire return interval is similar or less than the average return interval, there is no damage to the soil properties. In savannas and midland grasslands, shrub species with volatile compounds in twigs and leaves may contribute significantly to fire behavior. Head fires in grasslands and savannas move fast, and their average flame lengths are 3-4 m. They are considered very intense. Fires in the savanna only control the woody component, rarely killing it.

The lack of suitability for livestock operations (low accessibility of livestock and personnel and low stocking rates) brought the widespread use of mechanical treatments such as roller-chopping and seeding of tropical grass species, changing the fire regime of the Chaco. Changes are caused by the woody residues of high density trees, and high fine fuel loads generated by the seeding of *Panicum* and *Cenchrus*. Both factors may significantly affect woody species distribution and soil sustainability.

All these above results should be interpreted in the context of a ranch and livestock management. To use fire properly to control bush, enough fine fuel must be present, which means that pastures should be deferred, a matter of concern for cattlemen. A ranch is a complex system where factors such as commercial concerns, herd management, sanitary plans, etc., interact with climatic and other ecosystem features. So the use of fire may be or may be not appropriated in all circumstances. As usual, the diagnostic of a professional is essential for success.

Acknowledgments

Research and extension personnel of the Santiago del Estero Experimental Station, the Microbiology of the Faculty of Agronomy, the Denchronology , Botany and Bionenergy Chairs of the Faculty of Forestry Sciences and the Fire Paradox project are the source of most of these findings and observations, and should be accordingly thanked: Rodolfo Renolfi, Arnaldo Fumagalli, Mónica Cornacchione, Marcelo Navall, Niní Carrizo, Pablo Tomsic, Eliseo Monti, Miguel Perotti, José L. Gelid, Alejandro Radrizzani, María C. Sánchez, Fabio Moscovich, Julio D. Herrera, Marta Leiva, Luciano Lorea, Marta Rueda, Celia Gonzalez, Silvia Velez, Ana Giménez, Graciela Moglia, José Godoy, Francisco Lopez, Victor Navarrete and Jose Lopez. The owners and administrators of the following ranches should be also properly thanked: 'El Mangrullo', 'Los Itines', 'Santa Teresita', 'Toro Negro', 'Buena Agua', Maulhardt SA, 'La Canoa', 'Cambá-Cuá', and 'Grupo CREA Semiárido Norte, Santiago del Estero'.

REFERENCES

Abril A, Gonzalez C (2000) Dinámica de la fertilidad y de las poblaciones microbianas en suelos afectados por incendios en las Sierras de Córdoba. *Agriscientia* 16, 63-68.

Adámoli J, Neumann R, de Colina A, Morello J (1972). El Chaco aluvional salteño. *Revista. de Investigaciones Agropecuarias* 3 (9), 165-237.

Agee J (1993) Fire Ecology of Pacific Norhwest forests. (Island Press, Washington DC)

Albanesi A, Anriquez A (2003) El fuego y el suelo. In 'Fuego en los ecosistemas argentinos' (Eds C Kunst, S Bravo, J Panigatti), pp. 47-60 (Ediciones INTA, Buenos Aires, Argentina).

Albanesi A, Anriquez A, Polo Sánchez A (2001) Efectos de la agricultura convencional en algunas formas del N en una toposecuencia de la Región Chaqueña, Argentina. *Agriscientia* 18, 3-11.

Anriquez A, Albanesi A, Kunst C, Ledesma R, Lopez C, Rodriguez Torresi A, Godoy J (2005) Rolado de fachinales y calidad de suelos en el el Chaco occidental, Argentina. *Ciencia del suelo* 23, 145-157.

Anriquez A, Silberman J. Kunst C, Albanesi A, Ledesma R, Bravo S (2011) Postfire effects on soil nitrate under tree and shrub cover in a chronosequence. In 'Diverse Rangelands for a Sustainable Society', Proceedings of the International Rangeland Congress, Rosario, Argentina. (Eds. S Feldman, G Oliva, M Sacido), p. 428.

Bordón A (1993). Notas sobre incendios en la región chaqueña con base en una encuesta para información sintética sobre incendios de campos. In 'Memoria Seminario-Taller Ecología y manejo de fuego en ecosistemas naturales y modificados' (Eds C Kunst, A Sipowicz, N. Maceira, S. Bravo) pp. 69-74 (Ediciones INTA, Santiago del Estero, Argentina).

Bravo S (2008). Manejo de rolados, caracteres estructurales de leñosas que influencian la producción de rebrotes frente tratamientos mecánicos. In 'RBI, Rolado Selectivo de Baja Intensidad' (Eds C Kunst, R Ledesma, M Navall) pp. 126-131 (Ediciones INTA, Santiago del Estero, Argentina).

Bravo S (2010). Anatomical changes induced by fire-damaged cambium in two native woody species of argentine Chaco region. *IAWA Journal* 31, 283-292.

Bravo S, Giménez A, Moglia G (2001b). Efectos del fuego en la madera de *Prosopis alba* Griseb. y *P. nigra* (Griseb.) Hieron. *Mimosaceae. Bosque* 22, 51-63.

Bravo S, Giménez A, Moglia G (2006) Caracterización anatómica del leño y evolución del crecimiento en ejemplares de *Acacia aroma* y *Acacia furcatispina* en la Región Chaqueña. *Bosque* 27, 146-154.

Bravo S, Kunst C, Grau R (2008) Suitability of the native woody species of the Chaco region, Argentina, for use in dendroecological studies of fire regimes. *Dendrochronologia* 26, 43-52.

Bravo S, Kunst C, Giménez A, Moglia G (2001a) Fire regime of *Elionorus muticus* Spreng. Kuntze savanna, western Chaco region, Argentina. *International Journal of Wildland Fire* 10, 65-72.

Bravo S, Kunst C, Basualdo M, Leiva M, Ledesma R (2011) Assessment of the aerial bud bank of native tree and shrub species of the Chaco region. In 'Diverse Rangelands for a Sustainable Society', Proceedings of the International Rangeland Congress, Rosario, Argentina. (Eds. S Feldman, G Oliva, M Sacido) p. 428.

Bravo S, Kunst C, Leiva M, Ledesma R (2014) Response of hardwood tree regeneration to surface fires, western Chaco region, Argentina. *Forest Ecology and Management* 326, 36-45

Bravo S, Kunst C, Leiva M, Ledesma R (2012) Espesor de corteza y tolerancia al fuego en especies nativas del Chaco en estadio juvenil. In 'Proceedings 2do Congreso Nacional de Sistemas Silvopastoriles' (Eds J. Arroquy et al.,) p. 915 (Ediciones INTA, Santiago del Estero, Argentina).

Bravo S, Kunst C, Grau R, Aráoz E (2010). Fire-rainfall relationships in Argentine Chaco savannas. *Journal of Arid Environments* 74, 1319-1323.

Bucher E (1982) Chaco and Caatinga. South American arid savannas, woodlands and thickets. In 'Ecology of tropical savannas' (Eds. B Huntley, B Walker), *Ecological Studies* 42, pp. 48-79 (Springer Verlag, Berlín).

Bucher E, Saravia Toledo C (2001) Restauración y manejo sustentable del Gran Chaco. In 'Fundamentos de conservación biológica' (Eds. R Primack, R Roız, P Feinsinger, R Dirzo, F Massardo) pp. 579–580. (Perspectivas Latinoamericanas, Fondo de Cultura Económica, México).

Burkart A (1969) Flora ilustrada de Entre Ríos (Argentina), Parte II: Gramíneas, (Ediciones INTA, Colección Científica, Tomo VI, Parte II, Buenos Aires, Argentina) 500 pages.

Cabrera A, Willink A (1973) Biogeografía de América Latina. *Monografía* nº13, (Organización de Estados Americanos, Washington DC).

Casillo J, Kunst C, Ledesma R, Godoy J (2006a) Control de *Celtis spp* (tala) mediante fuego prescripto. *Actas Ecofuego*, 185-188.

Casillo J., Kunst C., Ledesma R., Godoy J (2006b) Dinámica temporal de la temperatura en la superficie y a 1 cm de profundidad del suelo en quemas en sabanas del Chaco semiárido. *Actas Ecofuego*, 126-128.

Casillo J, Kunst C, Semmartin M (2011) Effects of fire and water availability on the emergence and recruitment of grasses, forbs and woody species in a semiarid Chaco savanna. *Austral Ecology* 37, 452-459.

Cornacchione M, Kunst C, Argañaraz M (2001) Prácticas de control del fachinal, II. Efectos sobre la calidad del forraje disponible. In 'Memorias 1er Congreso de Pastizales Naturales', Asociación Argentina de Manejo Pastizales Naturales, pp 84-85 (San Cristóbal, Santa Fé, Argentina).

Díaz S, Bonnin M, Laguens A, Prieto M (1987) Estrategias de explotación de los recursos naturales y procesos de cambio de la vegetación en la cuenca del Rıo Copacabana (Dpto. Ischilín, Pcia. de Córdoba). I. Mediados del siglo XVI. Mediados del siglo XIX. Nueva Serie 45, p. 320-330 (Publicaciones del Instituto de Antropología, Córdoba, Argentina)

Dirección Nacional de Estadísticas y Censos de Argentina (2010) Censo Nacional de Población, Hogares y Vivienda. Censo del Bicentenario Serie B.2, Tomo 1. 500 pages.

Dooley, S, Treseder K (2012) The effect of fire on microbial biomass, a meta-analysis of field studies. *Biogeochemistry* 109, 49–61.

Evans R, Johansen J (1999) Microbiotic crusts and ecosystem processes. *Critical Reviews in Plant Sciences* 18, 183-225.

Farías A, Defossé G (2011) Social representations of fire risk in a wildland urban interface. In Proceedings 5th International Wildfire Conference, Sun City, South Africa, pp. 330-332.

Fernández I, Cabaneiro A, Carballas T (1999) Carbon mineralization dynamics in soils after wildfires in two Galician forests. *Soil Biology and Biochemistry*. 31, 1853-1966.

Ferreira L, Seluchi M, Saulo C (2004) Climatología sinóptica de la baja del Noroeste Argentino, campos medios de enero y julio. Publicación del Depto. de Cs. de la Atmósfera y los Océanos – FCEyN – UBA, Buenos Aires, Argentina, 50 pages.

Feuillade D, Ledesma R, Kunst C (2011a) Perceptions of integrated management of vegetation fires in Argentina. In 'Diverse Rangelands for a Sustainable Society', Proceedings of the International Rangeland Congress, Rosario, Argentina. (Eds. S Feldman, G Oliva, M Sacido) p. 773.

Feuillade D., Ledesma R., Kunst, C. (2011b) Technology and pedagogy, methodology of technology dissemination in fire ecology. In 'Diverse Rangelands for a Sustainable Society', Proceedings of the International Rangeland Congress, Rosario, Argentina. (Eds. S Feldman, G Oliva, M Sacido) p 774.

Gonzalez C, Studdert G, Kunst C, Albanesi A (2002) Comportamiento de algunas propiedades del suelo en una sabana del Chaco semiárido occidental bajo distintas frecuencias de fuego. *Ciencia del Suelo* 19, 92-100.

Hannah L, Lohse D, Hutchinson C, Carr J, Lankerani A (1994) A preliminary inventory of human disturbance of world ecosystems. *Ambio* 23, 246–250.

Hieronymus J (1874) Observaciones sobre la vegetación de la Provincia de Tucumán. *Boletín de la Academia Nacional de Ciencias Exactas de Córdoba* 1, 185-234.

Ibañez J, Lobo M, Almendros G, Polo A (1983) Impacto del fuego sobre algunas ecosistemas edáficos de clima mediterráneo continental en la zona centro de España. *Boletín Estación Central de Ecología* 12, 24-30.

Jacques A (1858) Una excursión al Salado y al Chaco. (Emecé Editores, Buenos Aires, Argentina).

Kunst C, Ledesma R, Bravo S, Albanesi A, Anriquez A, van Meer H, Godoy J (2012) Disrupting woody steady states in the Chaco region (Argentina): responses to combined disturbance treatments. *Ecological Engineering* 42, 42-53.

Kunst C, Bravo S, Moscovich J, Herrera J, Godoy J, Vélez S (2003) Fecha de aplicación de fuego y diversidad de herbáceas en una sabana de *Elionorus muticus* (Spreng) O. Kuntze (aibe). *Revista Chilena de Historia Natural* 76, 105-115.

Kunst C, Ledesma R, Casillo J, Godoy J. (2006a) Rolados y residuos leñosos, I. Estimación de la carga de combustibles. Actas Ecofuego, 197-198.

Kunst C, Ledesma R, Casillo J, Godoy J (2006b) Rolados y residuos leñosos, II Dinámica del contenido de humedad de combustibles de origen leñoso. Actas Ecofuego, 217-219.

Kunst C, Ledesma R, Cornacchione M, Castañares M, van Meer H, Godoy J (2014) Yield and growth features of *Panicum maximum* (Jacq.) var Trichoglume cv Petrie (Green Panic) under woody cover, Chaco region, Argentina. *Agroforestry Systems* 88, 157-171.

Kunst C, Ledesma R, Navall M (2008) RBI. Rolado selectivo de baja intensidad. Informe Técnico 57. (Ediciones INTA, Santiago del Estero, Argentina)

Kunst C, Bravo S, Moscovich F, Herrera J, Godoy J, Vélez S (2000) Control de tusca (Acacia aroma Gill ap. H. et A.) mediante fuego prescripto. *Revista Argentina de Producción Animal.* 20, 199-213.

Kunst C, Ledesma R, Bravo S, Defossé G, Godoy' J, Navarrete V (2012) Comportamiento del fuego en un pastizal del sitio ecológico media loma, región chaqueña occidental. *Revista de Investigaciones Agropecuarias* 38, 70-77.

Kunst, C., Ledesma, R., Bravo, S., Defossé, G., Godoy' J., Navarrete V. 2014. Dinámica de la humedad de los combustibles y su relación con la ecología y manejo de fuego, región chaqueña occidental (Argentina) II: follaje y residuos de árboles y arbustos. *Revista de Investigaciones Agropecuarias* 40, 165-181.

Ledesma R, Kunst C, Bravo S, Leiva M, Lorea L, Godoy J, Navarrete V (2011) Fire intensity and air temperature profiles in experimental burns in Chaco shrublands. In 'Diverse Rangelands for a Sustainable Society', Proceedings of the International Rangeland Congress, Rosario, Argentina. (Eds. S Feldman, G Oliva, M Sacido) p. 456.

Lizárraga, R. de (circa 1600). Descripción colonial. Buenos Aires. 1916. http,//www.lluisvives.com/servlet/SirveObras/acadLetArg/68016141545467830532279/p0000001.htm#I_0_consulted 26/03/2014.

Loftin S, White L (1994) Potential nitrogen contribution of soil cryptogams to post-disturbance forest ecosystems in Bandelier National Monument, NM. In 'Fire Effects in Southwestern Forests', Proceedings of the second La Mesa fire symposium, pp. 140-148.

Lorea L, Kunst C (2011) Historical aspects about fire, herbivory and vegetation types of the Argentine Chaco. In 'Diverse Rangelands for a Sustainable Society', Proceedings of the International Rangeland Congress, Rosario, Argentina. (Eds. S Feldman, G Oliva, M Sacido) p 442.

Métraux A (1946) Ethnology of the Chaco. In 'Handbook of South American Indians Vol. I The Marginal Tribes' (Ed. J Steward), *Bulletin of the Bureau of American Ethnology of the Smithsonian Institution* 143, 197-379. http://biodiversitylibrary.org/page/26734435 consulted 27/03/2014.

Molina M, Llinares J (1998) Effects of fire intensity on the soil properties related to structure, organic matter, aggregate stability and water retention capacity. In 'Fire, Landscape and Dynamics on the Mediterranean Area' (Eds L Trabaud, R Prodon), France, pp. 67-72.

Montenegro O (2002) Calidad del Suelo, Efecto del rolado y fuego en un ecosistema degradado (fachinal) de la Región Chaqueña. Trabajo final de Intensificación de grado. FAyA-UNSE, Santiago del Estero. 42 pages.

Morello J (1970). Modelo de relaciones entre pastizales y leñosas colonizadoras en el Chaco argentino. Revista *IDIA* 276, 31-52.

Morello J, Adámoli J (1974) Las grandes unidades de vegetación y ambiente del Chaco argentino. 2da Parte. INTA Serie fitogeográfica 13, 200 pages.

Morello J, Saravia Toledo C (1959) El bosque chaqueño I y II. *Revista Agronómica del Noroeste Argentino* 3, 5-81/209-258.

Morello J, Matteucci S, Rodriguez A, Silva M (2012) Ecorregiones y complejos ecosistémicos argentinos. (Orientación Gráfica Editora. Buenos Aires, Argentina)

Moriconi J, Fernández E, Martínez E, Giulietti A (2009) Rendimiento de materia seca y composición del aceite esencial de *Lippia turbinata* Griseb. "poleo" (Verbenaceae), en plantas de San Luis (Argentina). *Horticultura Argentina* 28, 34-38.

Moussy M de (1873) Description geographique et statistique de la Confédération Argentine. 2ème edition, Atlas. (Librairie de Fermin Didot Frères, Paris, France).

Navall M, Kunst C, Ledesma R, Casillo J, Godoy J (2005) Efecto de la quema prescrita sobre la regeneración natural de especies forestales en un bosque rolados del Chaco semiárido. Proceedings 3º Congreso Nacional sobre manejo de Pastizales Naturales. Facultad de Ciencias Agropecuarias de Oro Verde - UNER y Asociación Argentina para el manejo de Pastizales Naturales, Entre Ríos, Argentina. 60 pages.

Nuñez M, Sosa A, Aguado M, Raisman J (2006) Caracterización de extractos fluidos de Lippia turbinata Griseb. Resumen E-052, Comunicaciones científicas y tecnológicas, Universidad Nacional del Nordeste. www.unne.edu.ar/Web/cyt/cyt2006/08-Exactas/2006-E-052.pdf. Consulted 16/08/2014

Papadakis J (1951) Posibilidades agrícolas de La Rioja, Catamarca, Tucumán, Salta, Jujuy, Corrientes, Santiago del Estero, sudoeste de Buenos Aires, sudeste de La Pampa y noreste de Río Negro. Ministerio de Agricultura y Ganadería, Buenos Aires, Argentina. 350 pages.

Pausas J, Fernández Muñoz S (2012). Fire regime changes in the Western Mediterranean Basin, from fuel-limited to drought-driven fire regime. *Climatic Change* 110, 215-226.

Pausas J, Keeley J (2009). A burning story. The role of fire in the history of life. *Bioscience* 59, 593-601.

Pennington R, Prado D, Pendry C, (2000) Neotropical seasonally dry forests and Quaternary vegetation changes. *Journal of Biogeography* 27, 261–273.

Rueda M, Kunst C, Velez S, Ledesma R (2011) Range fuels, high calorific value of two shrub species of Chaco Region (Argentina), In 'Diverse Rangelands for a Sustainable Society', Proceedings of the International Rangeland Congress, Rosario, Argentina. (Eds. S Feldman, G Oliva, M Sacido) p 443.

Shenoy A, Kielland K, Johnstone J (2013) Effects of fire severity on plant nutrient uptake reinforce alternate pathways of succession in boreal forests. *Plant Ecology* 214, 587-596

Tortorelli L (1947) Los incendios de bosques en la Argentina. Ministerio de Agricultura, Dirección Forestal, Buenos Aires, Argentina, 400 pages.

Tortorelli L (2009) Maderas y Bosques Argentinos. 2da. (Orientación Gráfica Editora, Buenos Aires, Argentina).

Weber C, Lockhart J, Charaska E, Aho K, Lohse K (2014) Bacterial composition of soils in ponderosa pine and mixed conifer forests exposed to different wildfire burn severity. *Soil Biology and Biochemistry* 69, 242-250.

Wright H, Bailey A. (1982). Fire Ecology: United States and Southern Canada. (John Wiley and Sons, New York).

Zak M, Cabido M (2002) Spatial patterns of the Chaco vegetation of central Argentina, integration of remote sensing and phytosociology. *Applied Vegetation Science* 5, 213–226.

Zak M, Cabido M, Hodgson J (2004) Do subtropical forests in the Gran Chaco, Argentina have a future? *Biological Conservation* 120, 589-598.

In: Current International Perspectives
Editors: Brigitte Leblon and Martin E. Alexander
ISBN: 978-1-63463-682-7

Chapter 9

ASSESSING FIRE-RELATED CASUALTIES AND SOME CRITICAL FIRE FIGHTING PROCEDURES IN SOUTH AFRICA OVER TWO DECADES

Cornelis de Ronde *
Silva Forest Services CC, Sedgefield, South Africa

ABSTRACT

In Sub-Saharan African regions, South Africa is the most advanced in fire fighting resources, specialised equipment, advanced fire management and manpower training levels, in the field of Veld and Forest fires. However, some serious shortcomings have been identified in this country, which still need to be addressed, such as some specialised training at most levels. Wildfire fatalities peaked during the years 1994 and 2001, but have since then been reduced significantly. The single most outstanding problem regarding wildland fire fighting procedures, remains the use of "fire to fight fire" by means of the application of counter firing. When the success rate of these fire applications was recently assessed, it was found that – until today – only approximately10% of counter fires applied during wildfires can be rated as being "successful", while in the case of the remaining 90% wildfires, wildfire damage and areas burned-over increased substantially as a result of counter fires applied, and even doubled or tripled during some wildfire events. Although training in fire fighting procedures in general improved significantly at all levels, there is still a lack of practical knowledge of vegetation and fuel dynamics assessment procedures and related wildfire behaviour prediction, when wildfires events occur in the main vegetation biomes in South Africa. This resulted in some disastrous decisions made by fire bosses, ICS commanders and fire managers. It is recommended that specialised, advanced, training courses are developed and presented in Southern African (SADC) countries to reduce this unnecessary increase in wildfire damage as a result of counter fire application, as a matter of urgency, also using lessons learned during the FP6 research projects in Europe.

---

* Corresponding Author address: Silva Forest Services CC, P.O. Box 835, Sedgefield 6573, South Africa. Email: nderonde@brd.dorea.co.za.

Keywords: Wildfire fatalities, Southern Africa, fire behaviour predictions, fuel dynamics, fire dynamics, fire behaviour modelling, counter fires, advanced fire-related training, wildfire assessment

INTRODUCTION

Wildfire Fatalities in South Africa

Some of the worst wildfire fatalities were experienced in this country during 1994 (in Industrial Plantations) and during 2001 (in Savanna Grassland). Both events occurred in the Mpumalanga Province of South Africa, during the dangerous, dry, fire season, within the period September to November. In both cases the lack of understanding and knowledge of fuel dynamics and related wildfire behaviour assessment, played an important role in fatal decisions made, although desperation and fear by the commanders played a significant role, as this can be expected from any human being under such stressed circumstances. In the case of the 1994 "Sabie fire", ten fire fighters lost their lives, while eight fire fighters were seriously injured. In the 2001 "Kruger fire", 24 fatalities were experienced, of which 23 were civilians and one was a park ranger fighting the fire. In the case of the rangers, a few of them as well as a few civilians, were also seriously injured (de Ronde, 2002a; de Ronde 2002b).

Although the 1994 fatalities experienced during the Sabie fire were mainly caused by the lack of proper fuel and fire dynamics assessment, some other "wrong" decisions made by the fire managers, which also contributed to this tragic disaster. The 2001 Kruger fire fatalities were also mainly caused directly (or indirectly) by wrong fuel and fire-related assessment on site, but also by some other contributing factors, such as the civilians having been working (grass-cutting) nearby, during an extreme fire danger situation, inside a very dangerous – fully cured - Montane Grassland fuel. This should never have been allowed.

Assessing Counter Fire Applications to Fight Wildfires

During the past two decades I have investigated more than one hundred wildfires in all eight South African provinces, in all major biomes, macro-climatic regions and man-made fuels in disturbed ecosystems. Although I identified a few experienced fire managers during the wildfires I investigated over the past two decades, which could apply counter fires successfully, most other counter fire ignitions "went wrong" (approximately 90%). It has to be remembered that in a number of wildfires no counter fires were applied at all for various reasons, while in others as many as ten to twelve were applied, during a single wildfire. Many times such counter fire attempts were hidden from investigators and fire reporters, but – in the process of the fire investigation and wildfire reconstruction - most if not all were "found out" with sufficient evidence to prove such actions. Hiding such evidence should not have been necessary in any of these cases, as the intention of the fire managers would almost certainly have been honest attempt to stop or restrict these wildfires, and vital decisions had to be made under extremely stressful circumstances.

I found that in many wildfires I investigated such counter fire attempts did not reach acceptable goals, but can still be regarded as "honest attempts" to restrict wildfire spread. Some attempts "gone wrong" were even applied by experienced, senior, fire managers. In other cases inexperienced persons (such as farmers, labourers or other non-professional fire fighters) attempted the ignition of such fire lines.

I also identified that each counter fire "gone wrong" had its own unique set of problems to consider on site, but that in most cases the lack of knowledge about fuel and related fire dynamics, played a highly significant role (de Ronde, 2002a and various wildfire investigation reports written by the writer during the past 20 years).

I developed and tested a range of custom fuel models for Southern African conditions over the past twenty years and used these in both experimental prescribed burning as well as wildfire situations. In the case of the latter, I developed and tested more than 150 fuel models for specific vegetation types run over by the wildfire investigated, using real time fire behaviour observations and other evidence (such as photographs, aerial photography and satellite images and other data). Many times such models had to be adjusted and "tuned-up" to improve accuracy after testing it against real time evidence. I also found that in the case of each wildfire, very few existing fuel models could be used "as is" and most had to be adjusted to some degree to meet existing evidence of fuel status and related fire behaviour. The fuel model database used thus grew over time to more than 150 models, and newly-developed fuel models are still being added to this database.

BEHAVE and later BehavePlus (Andrews, 1986; Andrews and Bevins, 2000; Andrews and Chase, 1986; Burgan and Rothermel, 1984; Rothermel, 1972) formulas and methodology were used for the development and testing of the fuel models used for (in particular) wildfire investigations, with an improved degree of accuracy with time, achieved over the past two decades. This fuel model development and testing experience – with its advantages and disadvantages – will be discussed later in this paper.

BACKGROUND INFORMATION

South African Biomes and Man-Made Fuel Situations

The main biomes in South Africa can be found within specific rainfall regions, namely (i) in the Winter Rainfall region of the Western Cape Province, (ii) the Constant Rainfall Region of the Western and Eastern Cape Provinces and (iii) the Summer Rainfall Region covering the remainder of Southern Africa (Kruger, 2004). Some of these biomes are seldom experiencing wildfires (e.g., the Karoo biome), but most others are exposed to wildfires to some degree (Bond et al., 2004). The characteristics of these biomes are summarised in Table 1.

In addition, some important vegetation/fuel situations have been created by man for e.g., agricultural and industrial purposes, and these can be divided in two main groups, namely (i) agricultural crops (e.g., maize fields, wheat lands and sugar cane) and (ii) even-aged industrial timber plantations (e.g., *Acacia*, *Eucalyptus* and *Pinus* stands). The latter can be linked to stand age, species, site quality and silvicultural treatments (e.g., pruning, thinning and clear felling), for fuel model development purposes. These so-called "disturbed ecosystems" have been established in South Africa in Fynbos shrublands, Montane Grassland and Savanna Grassland vegetation (Bond et al., 2004).

Table 1. Summary of the main South African biomes regularly exposed to serious wildfires (Bond et al., 2004; Kruger, 2004 and Low and Rebelo, 1998)

Sub-biome	Climatic region	Yearly rainfall	Fuel types
Fynbos Biome			
Fynbos	South-Western Cape and Southern Cape	More than 600mm/year, Winter and Constant Rainfall regions	Shrubland
Renosterveld	Little Karoo and Western Karoo	Less than 600mm/year, Constant and Summer Rainfall regions	Low profile shrubland
Grassland Biome			
Highveld Grassland	Moist Highveld Grassland, Eastern Grassland, Dry Highveld Grassland and South-Eastern Coastal Grassland	600 – 900 mm/year, Summer Rainfall region	Dynamic grass, curing during winter season
Mountain Grassland	Eastern Mountain Grassland and Alpine Heathland	More than 900 mm/year, Summer Rainfall region	As above
Savanna Biome			
Moist Savanna	Lowveld Mountain Bushveld, Eastern Coastal Bushveld and Kwazulu-Natal Central Bushveld,	700 – 1000 mm/year, Summer Rainfall region	Dynamic grassland curing during the winter season and a degree of tree and shrub component
Dry Savanna	Kalahari Bushveld, Kalahari Hardveld Bushveld, Central Bushveld, Northern Arid Bushveld, South-Eastern Thornveld and Lowveld Bushveld	250 – 650 mm.year, Summer Rainfall region	As above

Agricultural fuels can only produce a fire hazard situation during relatively-short periods of time just before and after harvesting. In contrast, however, it should be remembered that even-aged industrial timber plantations are fast growing and that fire hazard is here also influenced by silvicultural treatments (e.g., weed control, pruning and thinning) and in particular after clear felling and timber exploitation. During the latter process, an extremely hazardous situation can be created within a few days (own observations). During the early years of a timber plantation stand, the natural biome vegetation where the trees were planted can still create hazardous fuels, and also if the natural vegetation is mixed with timber slash from earlier rotations. Because the fire hazard inside timber plantations can range from "extreme" to "almost nil", fuel modelling by age, species and site quality may be required to arrive at a representative fuel model set for a specific region (own observations).

Progress with Fire Fighting Training in South Africa: A Short Review over the Past Two Decades

While two decades ago fire fighting training ranged from basic fire fighting training to almost no training at all, this has now changed in most regions - where a regularly-high fire hazard is experienced – to improved training at all levels. A comprehensive training programme has now been developed by organizations such as "Working on Fire", local and regional fire brigades, from basic fire fighting training to specialised ICS courses. These training programs have now created a much better standard-base for fire fighting procedures being applied throughout the country. However, this does not apply to the dry savanna biomes/regions because of lack of understanding of fuel and related fire dynamics, causing underrated assessments of fire dangers in these areas. As a result, some of the worst wildfires were during the past three to five years experienced during the dry winter season in the dynamic Montane grassland of the Western Free State and Dry Savanna of the Northern Cape Provinces (own unpublished results of wildfire investigations).

In regions such as the Northern Cape Province, the lack of understanding why there are such serious wildfires occurring in this otherwise dry Savanna region, has resulted in an almost total lack of training being available for fire fighting in Veld and Forest fires, at all levels. This problem requires urgent attention in South Africa. In the rest of Southern Africa with the exception of South Africa, Swaziland and Lesotho, fire fighting training levels for fire fighting in Veld and Forests also require priority attention, particularly in SADC countries (own observations).

Fire Management and Incidence Command Systems

With regard to fire management in general, there has been a steady improvement in formal training, mainly through the Nelson Mandela Metropole University (NMMU Saasveld campus in George, South Africa) in the form of certificate, diploma and degree courses in fire-related subjects, such as wildfire management. However, specialist training in wildfire investigation (particularly wildfire reconstruction) and fuel and fire dynamics, remain a serious shortcoming, which needs to be addressed to provide the experts needed in wildfire investigations, required for the increased number of civil High Court cases now experienced as a result of wildfires and wildfire damage (own observations).

Regarding Incidence Command Systems (ICS), most fire fighting commanders are today receive specialist courses in the subject, presented by highly-qualified and experienced local trainers supplemented by some trainers from the USA, and few wildfires are today fought without trained commanders in control. However, even ICS-trained commanders have not been trained properly in fuel and fire dynamics assessment, or in the application of counter fires, with the result that a high percentage of these leaders are still not capable to assess wildfires in progress properly. As a result, wrong decisions are still being made particularly with regard to counter fire application, leading to unnecessary increases in wildfire damage and even exposure to danger for fire fighters and/or the public, The latter can fortunately still mostly be avoided as a result of the general improvement of fire commanders' experience.

Methods

Wildfire Fatalities

Both examples of wildfires used where wildfire fatalities were experienced – the 1994 Sabie and 2001Kruger fires – were selected because these are on record as being the worst experienced in South African history. However, fatalities were also experienced during other wildfire events in South Africa, during these two decades. Unfortunately we do not have any statistical evidence in this regard, and no analyses could thus be conducted for a more detailed picture.

I did attend the High Court proceedings in connection with the 1994 Sabie fire Legal civil case in the Northern Gauteng High Court in Pretoria. Unfortunately, I did not conduct a full field investigation of this wildfire, but only some selected tasks. However, I did report on the circumstances surrounding this wildfire event. I can thus report on the basic problems relating to this tragedy, about the cause(s) for these fatalities.

Regarding the 2001 Kruger wildfire fatalities, there was a full investigation into the circumstances surrounding the cause(s) for these mortalities, including investigating the origin of this fire. I was called in to assist here at the eleventh hour (a week before the Legal hearing was concluded), thus leaving little time for full report-writing, preparation and consultation. A field investigation was also conducted at the site of the tragedy, by the writer and other experts. We managed to collect sufficient evidence of the cause and origin of this wildfire, and then reconstructed further wildfire spread events from this site with the assistance of the FARSITE 3-D fire spread simulation program (de Ronde, 2002a; Finney, 1996) and with the assistance of A. Emery (a GIS expert), who drew up the GIS and FARSITE input and other data required for these computer runs.

It should also be noted that the fatalities during both wildfires examined, discussed in this paper, were not taken lightly, and an immediate follow-up in the form of various forms of training were formally or informally presented to fire fighters, fire bosses and IC commanders after these events. This resulted in a significant reduction of fatalities in South Africa. However, I have observed that in many SADC countries south of the Sahara, the situation is still extremely disturbing, with most training in Veld and Forest fire-related subjects almost non-existing at all levels. It should also be noted that fatalities are seldom reported in many of those countries outside South Africa. Thus basically a situation still exists as was experienced in South Africa two to three decades ago (own experience and personal assessment). As a result, fatalities north of South Africa could not be properly assessed until today.

Assessing Counter Fires

I investigated more than one hundred serious wildfires over a period of about 20 years, and I used these experiences (of which in about 40% of cases, settlements were reached before developing into Legal Court proceedings) to draw some conclusions of the success of such fire applications or not, and the consequences thereof, particularly in relation to added wildfire damage. The reasons for "counter fires gone wrong" were also evaluated, and the results of these studies will be highlighted with specific examples. Counter fires were applied as many as twelve times per wildfire event, but also in some cases not at all, or only once or twice.

Many counter fires applied were not officially reported, particularly where these caused added damage, or sometimes even developed into life-threatening situations. Once I became aware of this "hush-hush" situation in some cases with regard to counter fires applied, I developed some methodology to reconstruct wildfire events with the assistance of (i) field evidence collected after the fire (e.g., scorch patterns on e.g., trees, shrubs or grasses), (iii) the use of satellite images or other satellite-based data, (iv) weather data, (v) photographic evidence from e.g., spotters or helicopters, (vi) aerial photography and (vii) 2-D and 3-D fire behaviour and fire spread simulation. This could then supplement factual evidence where this existed. Some of such information was presented during Legal Court proceedings, but this was many times avoided on purpose, because of the sensitively of such information. However – as an independent wildfire investigator – I made sure that the truth was always presented in some form inside or outside Legal proceedings, regardless the consequences.

RESULTS

Wildfire Fatalities

1994 Sabie Fire

During the 1994 Sabie fire, ten fire fighters died and another eight were injured. It was then extremely dry, with a typical north-westerly wind blowing most days, becoming at its strongest during the afternoon, when this fire was ignited, most probably by honey hunters. The fire started in *Eucalyptus* slash created as a result of clear felling a mature – even-aged - *Eucalyptus* stand, in the lee (wind) side of a mature stand of the species, with the wind blowing over the tree crown canopy, leaving the micro-site basically untouched by the wind at first, until such time that the spreading fire front became directly exposed to the wind.

The first fire fighting team reached the fire scene within minutes after the fire was ignited. However, they failed to bring the fire under control because *Eucalyptus* slash typically produces flare-ups reaching 10m in flame height or more (own observations of fire behaviour in this fuel type). When the water bombers with a spotter also arrived minutes later, they were helpless in containing this fire, basically because such slash fuels just "stand-up" after dropping a water load, because of the long fire "residence time". This is another unique characteristic of these fuels, as was also observed by spotter and bomber pilots in the past. A typical example of such "long residence" fuel burning in a timber depot is illustrated in Figure 1 (not taken from the 1994 Sabie fire).

The spotter pilot in the air at the 1994 Sabie fire site took some pictures at a fairly early stage of fire spread development, which illustrated the extreme fire behaviour experienced, even before it was exposed to strong winds later. Note that the fire was at this stage still burning in the lee-side of the mature *Eucalyptus* trees nearby (Figure 2). Seconds later the fire intensity and rate of spread had already increased dramatically, also developing multiple spotting (Figure 3).

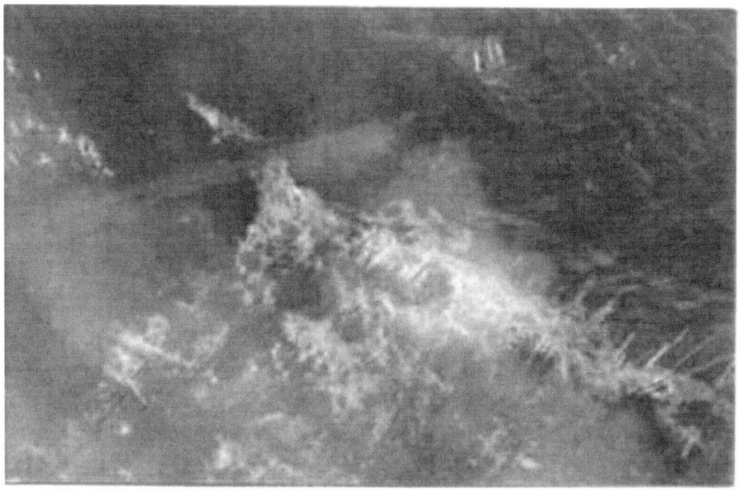

Figure 1. Stacks in a *Eucalyptus* timber depot burning inside an industrial plantation, Mpumalanga Province, South Africa, as seen from the air by a spotter pilot, who took this picture at the time, also during the early 1990's (photograph courtesy of FFA).

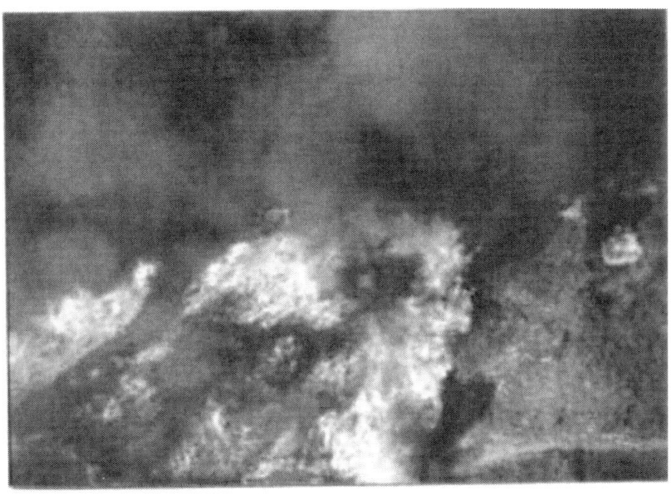

Figure 2. Fire burning in *Eucalyptus* slash slightly uphill and already spotting. Note mature even-aged *Eucalyptus* stand the stems of which are just visible on the left of the picture. Note depth of fire front and obvious uncontrollable state of the fire at this stage (Courtesy of FFA, with photograph taken by the spotter pilot, Johan Heine).

At this critical time (or somewhat later, as no time was recorded) the fire escaped from this site with the fire now developing into a typical "fire storm", with multiple spotting occurrence. Then the fire fighting team fighting the fire on site, was ordered to proceed with their light vehicles and a water tanker to a position in front of the head fire direction of maximum fire spread (thus ahead of the head fire front) to attempt to fight the fire from there. They thus left the scene of fire origin along a plantation road to reach where they were mopping up the rear section of the fire, to the site in question at the fire's head front, of the (by now) fast-spreading and spotting fire. To make things worse, there was no escape route from there, so the team was basically trapped the moment they arrived there. The only picture the spotter pilot managed to take from the fire at about this time is provided in Figure 4. The poor quality of the picture is understandable, particularly in the light of the turbulence the spotter pilot must have experienced above this fire.

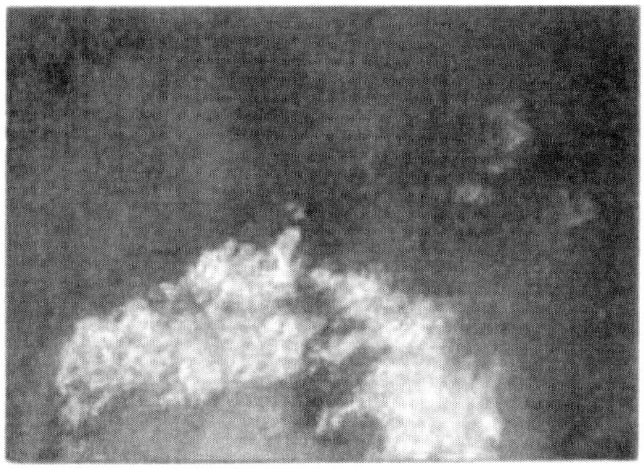

Figure 3. Photograph taken by Johan Heine seconds after he took the previous picture, Figure 2 (Courtesy FFA).

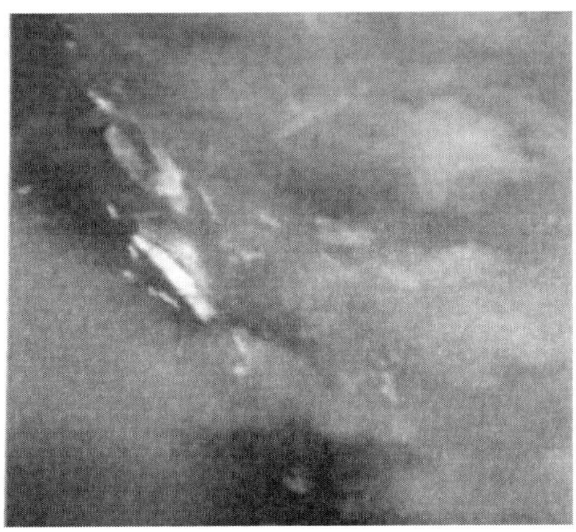

Figure 4. This photograph was taken at the approximate time that the fire fighters on the scene were trapped, some not coming out of the area alive. Note multiple spotting occurring (Courtesy FFA, photograph taken by spotter pilot Johan Heine). The poor quality of the picture can be appreciated as it was taken very close to the fire, with serious turbulence above the fire being experienced.

When the team realized that they were trapped by this fast-approaching fire - by this time spotting in abundance (Figure 4) - most of the team members trapped were jumping at the back of an open light vehicle with restricted carrying capacity, and drove as fast as they could on the rough dirt road, right through the wall of fire, to reach safety.

Most fire fighters at the back of this vehicle were seriously injured by the scorching heat of the fire, and had to be evacuated to the nearby hospital by helicopter. Those unable to leave the fire scene, were overtaken by the fire, and died.

A full investigation was conducted after the fire, but no one was found to be accountable for the fatalities experienced.

2001 Kruger Fire

I determined that this fire started at a picnic site inside the nature reserve, at about 13h00 that day the fire started, next to one of the main (tarred) roads between two rest camps. This fire ignition was most probably caused when tourists attempted a "braaivleis" (barbecue). This was contrary to the allegations that the fire was caused by a research team conducting an experimental fire within a research experimental area not far away from the fire, proving that this was definitely not the case. The fire spread very fast from its fire origin site (picnic site), inside typical Savanna-Grassland (fully cured) fuel. Weather conditions when this fire escaped were typical of a high fire danger, with (i) an air temperature of 38^{0}C, (ii) relative humidity of 13% and (iii) and an average wind speed of > 20km/h with wind gusts reaching 40 km/h at times. The moisture content of the cured grass biomass was also only approximately 3%. The rate of fire spread estimated from the area burned over at the time, provided a calculated area of approximately 70 000 ha, burned over within about ten hours. By the time the fire fighters (park wardens) arrived at the scene, it was clear that this fire was beyond control. Tourists had to be evacuated from some rest camps in the area, as the fire spread uncontrolled right through the afternoon and into the night of day one. The fire was only brought under control the next day (reports and statements from the park wardens in charge of fire fighting in the park).

We managed to simulate fire spread 3-D, by using FARSITE (Finney, 1996), starting with developing a 2-D custom fuel model set of the vegetation/fuels of the area that burned at the time, based on known grassland ages and vegetation types before the fire. The fire spread simulated, was made to start from the picnic site as was determined by the investigation, and the simulated 3-D fire spread was run as long as the end of the afternoon of day one (sun set), then through the night and also through day two. One tourist being evacuated from one of camps observed the fire spreading at about 16h50 that late afternoon of day one, not far away from where he was being evacuated. This provided a useful "check" of fire spread with FARSITE. The FARSITE fire spread progress was found to be deviating about 10 minutes from the time the tourist reported as having observed this fire. Thus the real-time fire progress was very close to the simulated spread results, considering that by that time the fire had already spread for approximately four hours, over an approximate area of 28 000 ha.

During night/next morning the day after the fire started, three counter fires were attempted by the park wardens, and the spread of these fires could also be simulated simultaneously with the main fire, with known times of ignition and position of these counter fire lines, in the landscape. The outcome of the simulation conducted clearly indicated that by the time the first two fires were ignited during the night/early morning during day two, these added fire fronts were way off the flanks of the fire where such counter fires should be attempted, because of significant changes in wind direction as well as cooler conditions experienced during that part of the night/day. These two fires subsequently never played a role in the spread of the main fire, because of wrong assessment of the weather conditions at the time.

The third counter fire attempt was according to report conducted later during the morning of the second day, also by park wardens, all along a tarred road kilometres in front of the fire head. Again, we obtained the time of ignition of the fire line as well as its site of fire ignition and its route.

Ignition of the 3rd counter fire line was attempted while (i) the main fire front was still too far away from this head fire line, and (ii) some critical changes took place in wind direction and windspeed/gusts experienced. A third factor affecting this created counter fire line, was the complete lack of knowledge the wardens had of fuel and fire dynamics of the fuel they ignited. This fuel base in fact consisted of an extremely dangerous fully cured grassland, of about 2.0m in height, in contrast to the average height of the savanna grassland nearby, which was only about 0.5m (own estimations and measurements on site, when I investigated these fire events).

To illustrate how this roadside grass fuel differed from the adjoining savanna grassland fuels, I have simulated both by means of 2-D fire behaviour simulation, using two specially-developed fuel models for the purpose with BEHAVE and BehavePlus (Burgan and Rothermel, 1984; Andrews, 1986; Andrews and Bevins, 2000; Andrews and Chase, 1986). The results from this simulation – with typical windspeed input for this day – are provided in Tables 2 and 3.

Although the Savanna grassland fuel model was a typical example of the fuel models used as input for the FARSITE runs, the roadside grassland was not used for this purpose because the area covered (only a strip not more than 5–7m wide) was too small for meaningful using this fuel model. However, as this critical fuel type was one of the main causes for this counter fire becoming uncontrolled, it was included in the fuel model set used, as set out in Table 3.

Table 2. Kruger 2001 fire fuel status where the 2001 fire occurred: Fuel model biomass input as measured, calculated and estimated where applicable (custom fuel model database for South African fuels and vegetation, as updated from time to time (de Ronde, own custom fuel model database). Grassland was also assumed to be fully cured at the time

Description of fuel	KNP roadside dynamic grassland	KNP Savanna dynamic grassland
1 h fuel load (Tons/ha)	5.0	2.5
10 h fuel load (Tons/ha)	0	0
100 h fuel load (Tons/ha)	0	0
Live herb fuel load (Tons/ha)	0	0
Live woody fuel load (Tons/ha)	0	0
1 Hr surface area-volume-ratio (SAV) (m^2/m^3)	3500	4500
Live Herb SAV(m^2/m^3)	N/A	N/A
Live Woody SAV (m^2/m^3)	N/A	N/A
Fuel depth (m)	2.0	1.0
Fuel moisture extinction (%)	13	13
Dead fuel heat content (KJ/kg)	18900	18900
Live fuel heat content(KJ/kg)	18900	18900

Table 3. Illustrated fire behaviour of the Kruger 2001 wildfire, with an average 30 km/hour windspeed input, as example of the typical wind data experienced during this day (de Ronde, 2001; unpublished custom fuel model database for South African fuels and vegetation, as updated from time to time).

Vegetation type	Rate of fire spread (m/min)	Heat per Unit Area (KJ/m^3)	Fire Line Intensity (KW/m^2)	Flame Length (m)	Spotting distance (km)
Roadside grassland.	159.2	361.1	9581	5.3	0.9
Average Savanna grassland.	76.2	2007	2550	2.9	0.6

Take note of the significant differences between the two fuel types, (in particular) in the dry grass biomass and fuel depth (Tables 2 and 3). This contrasting difference is also illustrated in Figures 5 and 6. The higher fuel load in the roadside grassland is a feature along most major roads in South Africa, and is caused by the higher run-off of rainwater, causing greater moisture availability for the grasses (and other vegetation) growing there next to roads (own observations).

Figure 5. Illustration of higher grassland biomass development along a dirt road in the Kwazulu-Natal Province, at this stage not yet fully cured at the start of the dry winter season (Courtesy Clive Henderson).

Figure 6. Illustration of grassland height along a Kwazulu-Natal Province road, here about 2.0m high. The grassland outside this zone was only about 0.4m in height (not visible in the picture, Courtesy Clive Henderson).

This counter fire (3rd fire line applied during the 2001 Kruger fire) was ignited by park wardens along a section of the tarred road, along the roadside grassland. At this critical time, the wind direction changed to blowing towards the tarred road (and the fire fighters igniting the fire line there) and soon "leaned" across this road, starting to spread further from there. The wardens panicked and ran away in the front of this fire, and then climbed into some Savanna trees growing inside the grassland. The head fire overtook them while they were hiding for the fire in the trees and they were seriously injured by the heat of the fire. One of them subsequently died, and the other two were seriously hurt in the process.

This "counter fire" then spread fast in the direction of a temporary grass cutters camp nearby, where the people (cutting grass for thatched-roofs) had not yet been evacuated, because the main fire was at the time still a number of kilometres away. They were soon surprised by the counter fire, and also panicked, running away before the fast spreading head fire, which then caused the balance of the fatalities and injuries experienced. Ironically the main fire never reached this camp area and all fatalities were thus caused by the 3rd counter fire applied. The wardens were not held responsible for this unfortunate tragedy, in the following Legal investigations.

The History of Counter Fires Applied

The following table provides a selected list of counter fires applied in South Africa and main reasons why they failed (Table 4).

Table 4. List of selected examples of counter fires applied in different fuel/vegetation types in South Africa to attempt to control wildfires or restrict its fire spread, which I found in subsequent investigations had failed for particular reasons

Vegetation or fuel type	Why the counter fire was incorrectly applied	What should have been done by the fire fighters instead	Impact of incorrect counter fire application
Dry savanna with significant bush component.	Counter fire applied directly in the path of the head fire from run-away prescribed burn, with strong NW winds.	Apply counter fire only during light, stable, wind conditions, parallel to fire flanks, first anchoring the fire line to the fire flank.	600 ha farmland destroyed.
Pinus clear felling slash after exploitation.	Counter fire applied with the wind to "close-up" unburned area at back of wildfire, escaping parallel to right flank of fire.	No such fire should have been applied at all, because it was unnecessary.	Fire escaped from nature reserve, damaging > 1000ha of adjoining farmland.
Montane grassland, fully cured.	When fire escaped from single ignition point, the single point was extended to a 500m fireline at the back of the wildfire area.	Counter fires should have been applied parallel to the flanks, after anchoring to the fire perimeter.	Damage to industrial plantation was approximately doubled (approximately 500 ha).
Dry savanna, fully cured	Applied in front of the head fire, which was still kilometres away.	Too dangerous to apply any counter fire, with very strong NW winds.	More than 20 human fatalities.
Sei-moist Montane grassland, fully cured.	Applied approx. Two kilometres in front of a slow-moving, light intensity fire, which was blocked by a tarred road.	Nothing. No additional fire line was necessary.	The counter fire spread on its own, the original fire never reaching it. Added damage to farms: 800ha.

In all examples provided in Table 4, the situation on the ground was incorrectly assessed, particularly because the fire boss/fire manager did not have adequate knowledge of expected fuel and fire dynamics and subsequent fire behaviour that could be expected. The impact of constructing the counter lines incorrectly and/or at the wrong site in the landscape, also contributed to the added damage experienced. A general lack of experience and training can thus be regarded as the main cause of these disasters.

Discussion and Recommendations

During both the 1994 Sabie fire and the 2001 Kruger fire, it was clear that lack of knowledge and experience (particularly in training in fuel and fire dynamics at supervisory level) was the main cause for the mistakes made during fire fighting action.

More fatalities (over and above those experienced during 1994 and 2001) were also recorded during wildfires in South Africa during the past two decades, bringing e.g., the total fatality figure during 2001 to thirty-one.

After the 2001 Kruger wildfire, intensive training was provided in various fire fighting and fire management subjects, resulting in a significant improvement of fire-related action as well as the development of an integrated fire management plan for the involved nature reserve. This included a built-in prescribed burning programme, based on ecological principles well-researched in the park over a few decades in a range of various field experiments, conducted by the park. Over the years there has been a significant improvement of fire-related training levels for all fire fighters and fire managers, involved in Veld and Forest fire fighting throughout South Africa.

One remaining problem I have noticed in South Africa until today is that still many counter fires have "gone wrong" in some way, even where applied by experienced fire fighters, fire bosses and IC commanders, regardless all the fire-related training provided at a regular basis. From the various wildfire investigations I conducted over the past two decades, it became clear to me that this was not only caused by lack of experience in fire fighting procedures with the emphasis on counter fire application.

Although fire fighting training in general in South Africa has reached satisfactory levels and fire fighting equipment has been improved significantly over the years in most regions, it is clear that there is still a lack of knowledge in fuel and related fire dynamics for fire managers, sometimes causing wildfire areas burned over to increase and even double during some wildfire events. Although wildfire fatalities have been reduced significantly since 2001, there is still an unnecessary exposure to increased fire damage during Veld and Forest Fire events and focussed training of all fire managers in this respect is urgently recommended. It is recommended that some basic options of the BehavePlus programs are used, with the assistance of the existing fuel model base, for wildfire prediction exercises, to illustrate basic fuels and fire behaviour.

To overcome the problem with fuel models not providing a realistic output in terms of some wildfire behaviour parameters, I have always tested such models against some real data, photographic examples within known variable performance or historical information. Because of the significant variation and contrasts found in the field, even on relatively small sites, such models normally had to be developed and tested with such adjusted fuel situations, to arrive at the "best possible representative fit". From experience, I have learned that this approach works best in African conditions where site quality, altitude, aspect, climate and man-made changes to vegetation and fuels are very contrasting, and specific (existing) fuel models seldom present the correct picture.

The policies regarding the use of counter fires in countries other than South Africa is not at all "clear cut", and wide range of procedures is used by fire commanders, ranging from "regular fire-use" to "total exclusion, backed by Law". I could not identify any formal training in this or related fields having been provided for fire commanders anywhere, with the exception of a project presented and executed as part of the EU Fire Paradox FP6 program, in the application of (what they call) "Suppression Fire", which is in fact just another term for "Counter Fire" (Silva et al., 2010). Regarding the Fire Paradox FP6 program I must emphasise that "the use of fire to fight fire" was one of the cornerstones of these R&D programme's objectives, with a range of e.g., prescribed burning projects researched (own experience as one of the participating project leaders).

It is only fitting to state here that (i) I have also taken note of the wildfires where counter fires were applied successfully in South Africa, causing a significant reduction in potential damage, (ii) that counter firing ("Suppression of Fire" in EU terms) has tremendous potential, also in South Africa, but (iii) that the most important problem for such fire application to be applied at this stage in South Africa, is the lack of specialized training in counter fire application, including the knowledge and experience of assessment of fuel/vegetation and fire. I therefore would like to recommend that such training should be provided in South Africa as a matter of urgency. In the mean time counter firing should only be applied by the few fire commanders who have a proven record of experience in successful application of the technique in the field. Under no circumstances should the technique of counter firing be applied by untrained fire commanders in this field, nor by their crews.

Acknowledgments

This paper is contributed to all the fire fighters and fire managers, who bravely fought wildfires over the years, sometimes against all odds. Without them, I could not have conducted my research studies - including prescribed burning experiments - as well as (during the past two decades) the numerous wildfires I investigated throughout South Africa. To all relatives and family of fire fighters which died or were injured while attending to fire fighting actions I can only say that they are not forgotten, and that these tragedies greatly contributed to a safer fire fighting environment today. Various persons also contributed to my knowledge base about wildfires in South Africa, its history and statistical data. They are too many to mention or reference in this paper. Their assistance is highly appreciated.

References

Andrews PL (1986) Fire Behavior Prediction and Fuel MODELLING System – BURN Subsystem, Part 1. USDA Forest Service, Intermountain Research Station, General Technical Report INT-194 (Ogden, UT).

Andrews PL, Bevins CD (2000) BehavePlus Fire Modeling System Users' Guide Beta Text Version October 2000. *USDA Forest Service Rocky Mountain Research Station*, General Technical Report. (Misoula, USA).

Andrews PL, Chase CH (1986) BEHAVE: Fire Behavior Prediction and Fuel Modeling System – BURN Subsystem – Part 2. USDA Forest Service, *Intermountain Research Station, General Technical Report* INT-260 (Ogden, UT).

Bond WJ, Geldenhuys CJ, Everson TM, Everson CS, Calvin MF (2004) Fire ecology characteristics of some important biomes of Sub-Sahara Africa. In '*Wildland Fire Management Handbook for Sub-Sahara Africa*' (Eds J.G. Goldammer and C. de Ronde), pp. 11-59.

Burgan RE, Rothermel RC (1984) BEHAVE: Fire Behavior Prediction and Fuel Modeling System – FUEL Subsystem. USDA *Forest Service Intermountain Research Station, General Technical Report* INT-167 (Ogden, UT).

De Ronde C (2002a) Investigating the origin of a wildfire that occurred in the Kruger National Park (Pretoriuskop area) on 4 September 2001. Unpublished report.

De Ronde C (2002b) Wildland fire-related fatalities in South Africa: A 1994 case study and looking back at the year 2001. Proceedings of the 4th International Conference on Forest Fire Research – Wildland Fire Safety Summit Section, Luso, Portugal, Paper D2, 1-7.

Finney MA (1996) FARSITE Fire Area Simulator Version 2.0. Users' Guide and Technical Documentation. USDA Forest Service, *Rocky Mountain Research Station*, General Technical Report. (Misoula, USA)

Kruger AC (2004) Climate in South Africa: Climatic Regions (South Africa Weather Service Publication WS45, Pretoria, South Africa).

Low AB, Rebelo G (1998) Vegetation of South Africa, Lesotho and Swaziland. (South Africa Department of Environment Affairs and Tourism, Pretoria, South Africa).

Rothermel RC (1972) A mathematical model for predicting fire spread in wildland fuels. USDA Forest Service Intermountain Research Station, General Technical Report INT-115 (Ogden, UT).

Silva JS, Rego F, Fernandes P, Rigolot E (2010) Towards Integrated Fire Management – Outcomes of the European Project Fire Paradox. European Forest Institute Research Report 23, 177-218 (European Forest Institute Research, Joenssuu, Finland).

INDEX

C

D

E

F

G

H

M

N

Q

R

S

T

U

V

W

Y

Z